The Intelligent Universe

Scrivener Publishing
100 Cummings Center, Suite 541J
Beverly, MA 01915-6106

Publishers at Scrivener
Martin Scrivener (martin@scrivenerpublishing.com)
Phillip Carmical (pcarmical@scrivenerpublishing.com)

The Intelligent Universe

AI's Role in Astronomy

Edited by

Yogesh Chandra
Dept. of Physics, Government P.G. College Bazpur,
Kumaun University, Uttarakhand, India

Manjuleshwar Panda
Independent Astronomy Researcher, New Delhi, India

and

Mahesh Chandra Mathpal
Govt. I. C. Lohali, Uttarakhand, India

Scrivener
Publishing

WILEY

This edition first published 2025 by John Wiley & Sons, Inc., 111 River Street, Hoboken, NJ 07030, USA and Scrivener Publishing LLC, 100 Cummings Center, Suite 541J, Beverly, MA 01915, USA

For more information about Scrivener publications please visit www.scrivenerpublishing.com.

Wiley Global Headquarters
111 River Street, Hoboken, NJ 07030, USA

For details of our global editorial offices, customer services, and more information about Wiley products visit us at www.wiley.com.

The manufacturer's authorized representative according to the EU General Product Safety Regulation is Wiley-VCH GmbH, Boschstr. 12, 69469 Weinheim, Germany, e-mail: Product_Safety@wiley.com.

Library of Congress Cataloging-in-Publication Data

ISBN 978-1-394-35548-8

Cover image: Generated with AI using Adobe Firefly
Cover design by Russell Richardson

Set in size of 11pt and Minion Pro by Manila Typesetting Company, Makati, Philippines

Printed in the USA

10 9 8 7 6 5 4 3 2 1

This book is dedicated to truth seekers—those who have devoted their lives to the pursuit of knowledge, the advancement of science, the unwavering promotion of scientific temperament, and the continuous growth of scientific literacy in every aspect of life. To the fearless minds who have dared to question, relentlessly explored, and brought light to the depths of the unknown—those who have never wavered in their pursuit of truth, refusing to let ignorance persist where the light of understanding must shine. We honor the visionaries whose steadfast commitment has not only pushed the frontiers of human knowledge but has also ignited generations with the courage to think critically, question established paradigms, and embrace scientific thinking as the guiding force of progress. Their passion for scientific inquiry has shaped the course of history, and their legacy continues to guide humanity toward a future driven by discovery, wisdom, and progress. With profound respect and deepest gratitude, we salute their lifelong allegiance to the unyielding pursuit of truth.

Contents

Foreword

In the vast expanse of the cosmos, the fusion of artificial intelligence (AI) and astronomy heralds a transformative era. **"The Intelligent Universe: AI's Role in Astronomy"** is a testament to this convergence, offering a comprehensive exploration of how AI is reshaping our understanding of the universe. This volume brings together contributions from esteemed scholars across national and international institutions, delving into topics ranging from data mining and machine learning in astrophysics to the search for extraterrestrial intelligence.

As we stand at the intersection of technological innovation and astronomical discovery, this work serves as a beacon for future research. The chapters within not only highlight current advancements but also inspire new avenues of inquiry, emphasizing the pivotal role AI plays in unraveling the mysteries of the cosmos. It is with great enthusiasm that I commend this volume to all who seek to understand the intelligent universe we inhabit.

This book is a valuable guide to the future of AI-driven astronomical exploration.

Prof. Waleed Hamdy Elsanhoury

Professor of Astrophysics
National Research Institute of Astronomy and Geophysics (NRIAG)
Astronomy Department – Stellar Astronomy Lab
Helwan, Cairo, Egypt

Preface

This book is a journey into the exciting and rapidly evolving intersection of artificial intelligence (AI) and astronomy. It offers a comprehensive exploration of how AI technologies are reshaping the way we observe, understand, and interpret the universe. From automating complex data analysis to enhancing our capabilities in space exploration, AI is becoming a crucial ally in our quest to unlock the mysteries of the cosmos. This book aims to guide readers through these groundbreaking developments, highlighting AI's innovative contributions to astronomy while addressing the profound questions it raises for the future of scientific discovery.

The choice of this topic was inspired by the extraordinary potential AI holds for transforming astronomy. As astronomical observations generate increasingly vast amounts of data, traditional analysis methods struggle to keep pace. AI, with its ability to process and learn from massive datasets, offers new solutions to this challenge. This convergence reflects our innate desire to explore the unknown, bolstered by advances in machine learning, neural networks, and computational power. AI's ability to predict, simulate, and uncover new celestial phenomena is expanding the frontiers of space research.

AI has had a profound impact on astronomy. Today, it drives discoveries in fields ranging from exoplanet detection to galaxy formation. AI enables astronomers to automate the analysis of enormous datasets from telescopes, satellite missions, and observatories. Pattern-recognition algorithms help identify objects that might otherwise go unnoticed. AI also enhances our ability to model complex cosmic events, enabling more accurate predictions of phenomena like supernovae and asteroid trajectories. In short, AI is not merely augmenting astronomy—it is fundamentally transforming how we conduct research and conceptualize the universe. This integration is accelerating, making AI ever more relevant as space exploration expands.

This book is intended for a wide audience: professionals and students in astronomy and artificial intelligence, as well as enthusiasts eager to grasp the profound changes underway in space science. Whether you're an aspiring astronomer, a researcher in AI, or simply curious about the convergence of these fields, this book offers accessible, in-depth content to help you navigate the complexities of AI's role in astronomy.

Chapter 1 explores the transformative role of AI in astronomy, highlighting key advances in data analysis, galaxy classification, transient event detection, and image enhancement. It examines AI's role in simulating cosmic phenomena and optimizing telescope operations—crucial as observatory data volumes continue to grow.

Chapter 2 highlights how data mining and machine learning (ML) are revolutionizing astrophysics by enabling more efficient analysis of complex datasets. Techniques like deep learning, supervised learning, and anomaly detection are driving discoveries in exoplanet research, gravitational waves, and galaxy classification. The chapter also explores ML's integration with theoretical models and the challenges posed by big data.

Chapter 3 focuses on AI's pivotal role in the discovery and characterization of exoplanets. With over 5,700 exoplanets confirmed, AI has become essential for detecting and analyzing these distant worlds. The chapter explores how machine learning improves detection methods and spectral data analysis to assess habitability, especially as next-gen telescopes produce vast datasets.

Chapter 4 delves into dark matter's role in shaping the cosmos. It discusses theoretical models like Λ-CDM and MOND, observational methods such as CMB studies, and how AI is aiding the analysis of dark matter data. The chapter bridges cosmology, quantum physics, and AI-driven research into one of the universe's greatest mysteries.

Chapter 5 examines gravitational waves and their role in understanding cosmic events. From LIGO's first detection in 2015 to future observatories like LISA, the chapter explores how AI enhances detection, noise reduction, and real-time signal analysis, enabling breakthroughs in multi-messenger astronomy and fundamental physics.

Chapter 6 focuses on AI in radio astronomy. It covers signal detection, data processing, and image reconstruction, particularly in mitigating interference and classifying fast radio bursts. With the Square Kilometre Array on the horizon, AI is essential for managing vast datasets and identifying rare phenomena.

Chapter 7 discusses the convergence of image processing, computer vision, and AI in astronomy. It highlights techniques like noise reduction and object classification, and how deep learning and reinforcement learning address challenges like noise and scale. Real-world case studies demonstrate these technologies' research impact.

Chapter 8 explores astroinformatics—the intersection of astronomy, data science, and computation. It highlights data storage, pattern recognition, and visualization techniques, while discussing ethical concerns around open science and data sharing. AI and cloud computing are shown to be vital in managing the petabyte-scale data from modern observatories.

Chapter 9 outlines the impact of autonomous telescopes and observatories, powered by robotics, AI, and ML. These systems enable real-time data processing and transient detection with minimal human input. The chapter examines benefits like global telescope networks and accessibility, alongside challenges in reliability and cybersecurity.

Chapter 10 shows how AI is transforming space missions. It explores AI's role in real-time decision-making, autonomous navigation, and hazard avoidance in missions from Mars rovers to distant planetary probes. Case studies highlight how AI is expanding humanity's reach beyond Earth.

Chapter 11 presents astrobiology and SETI as interrelated fields studying life beyond Earth. It explores the search for biosignatures and technosignatures, discussing planetary habitability and signals of intelligent life. These pursuits reflect our enduring curiosity and raise profound scientific and philosophical questions.

Chapter 12 surveys how AI, ML, and Deep Learning are reshaping astronomy through data-centric approaches. It discusses AI's role in detecting exoplanets and transients, explores the potential of quantum computing, and addresses ethical considerations in algorithm design and interpretability.

This book bridges theory and practice in both AI and astronomy, presenting complex ideas in an engaging, accessible manner. It not only explores the technologies but also reflects on the ethical and philosophical implications of AI's role in discovery. What are the responsibilities that come with AI-driven breakthroughs? How do we ensure ethical use of these powerful tools? These questions are woven throughout, encouraging deeper reflection on AI's scientific and societal impact.

At its heart, this work is driven by a belief in the transformative power of AI and astronomy to shape our future. As we reach further into space, our questions multiply—many of which only AI can help us answer. This book seeks to inspire curiosity, promote innovation, and challenge readers to imagine the possibilities that lie ahead.

We extend our deepest gratitude to the researchers, developers, and astronomers whose pioneering work made this book possible. We thank our mentors, colleagues, families, and friends for their invaluable support.

In conclusion, we hope this book serves as both a resource and a springboard for future exploration. May it spark new ideas, inspire collaboration, and illuminate the ways AI can help us better understand our universe. Let us journey together into the intelligent cosmos.

Dr. Yogesh Chandra
Dept. of Physics, Government P.G. College Bazpur, Kumaun University, Uttarakhand, India
Mr. Manjuleshwar Panda
Independent Astronomy Researcher, New Delhi, India
Dr. Mahesh Chandra Mathpal
Govt. I. C. Lohali, Uttarakhand, India
August 2025

Acknowledgement

The successful completion of this book would have been impossible without the collaborative efforts, expertise, and contributions of researchers, scientists, and professionals from around the world.

First and foremost, we extend our deepest gratitude to all the contributing authors. Their insightful research and commitment have shaped this book into a valuable resource for the scientific community, highlighting the transformative role of Artificial Intelligence in Astronomy and Space Sciences.

We are immensely grateful to Martin Scrivener and the team at Scrivener Publishing for believing in our vision and providing the platform to create this book. The guidance and encouragement from the editorial team have been instrumental in refining and shaping the manuscript.

A special acknowledgment goes to our reviewers and advisory board members, whose meticulous feedback and constructive suggestions have greatly enriched the quality and depth of this work. Their expertise has been instrumental in enhancing its clarity, coherence, and overall impact.

We also appreciate the invaluable support from the broader scientific community, whose pioneering research and discussions continue to inspire advancements at the intersection of AI and astronomy.

Lastly, we extend our heartfelt thanks to our families, friends, and colleagues for their patience, encouragement, and relentless support throughout the journey of compiling this book. Their belief in our vision has been a constant source of motivation.

With this book, we hope to inspire future research and innovation at the intersection of artificial intelligence and astronomy, paving the way for discoveries that deepen our understanding of the universe.

Editors

Dr. Yogesh Chandra
Mr. Manjuleshwar Panda
Dr. Mahesh Chandra Mathpal

Part I

FOUNDATIONS AND CORE APPLICATIONS OF AI IN ASTRONOMY

1

Introduction to AI in Astronomy

Rahul Barnwal[1]*, Aman Kumar[2], Kala S.[1] and Sree Ranjani Rajendran[3]

[1]*Department of Electronics and Communication Engineering, Indian Institute of Information Technology Kottayam, Kerala, India*
[2]*Department of Physics, Savitribai Phule Pune University, Pune, India*
[3]*Department of Electrical Engineering and Computer Science, Florida Atlantic University, Boca Raton, Florida*

Abstract

Astronomy is undergoing a transformative phase driven by the explosion of data from modern telescopes, satellites, and space missions. Due to the vast amount and complexity of these data, conventional analytical methods are no longer adequate, and artificial intelligence (AI) has emerged as a revolutionary tool. AI, through machine learning (ML) and deep learning (DL), allows astronomers to process vast amounts of data, detect intricate patterns, and make predictions that would be impossible through manual methods alone. AI is driving substantial advancements in several crucial domains, including galaxy classification, where machine learning algorithms can analyze millions of galaxies, identifying their shapes, structures, and characteristics far more efficiently than human classification. Similarly, AI is advancing the discovery of transient astronomical events like gamma-ray bursts and fast radio bursts, providing astronomers with the ability to detect and observe these short-lived phenomena in real time. In exoplanet research, AI techniques have revolutionized the way scientists sift through light curve data from missions like Kepler, leading to the discovery of many new exoplanets that would have otherwise gone unnoticed. A specialized type of deep learning, known as convolutional neural networks (CNNs), is also proving indispensable in the enhancement of astronomical images, enabling researchers to reduce noise, improve resolution, and extract meaningful features from the images of galaxies and other celestial objects. Beyond observational data, AI is playing a critical role in simulating large-scale cosmic phenomena, such as galaxy formation, dark matter distribution, and the behavior of black holes, which helps astronomers refine

**Corresponding author*: rahulbarnwal20bec3@iiitkottayam.ac.in

Yogesh Chandra, Manjuleshwar Panda and Mahesh Chandra Mathpal (eds.) The Intelligent Universe: AI's Role in Astronomy, (3–46) © 2025 Scrivener Publishing LLC

their models of the universe. Reinforcement learning is another innovative application, allowing AI systems to autonomously control and optimize telescope operations, improving the efficiency of astronomical observations. This integration of AI with autonomous telescopes enables real-time decision-making, ensuring that critical observations are not missed, particularly when dealing with transient or rapidly evolving events. Historically, AI's journey in astronomy began with simple automation tasks, but it has evolved into an indispensable part of cutting-edge research, driving discoveries at an unprecedented rate. As next-generation observatories, such as the James Webb Space Telescope, become operational, AI's role will become even more critical in processing the exponentially increasing volumes of data. The synergy between AI and human expertise promises to push the boundaries of what we know about the universe, leading to deeper insights and helping astronomers address some of the most profound questions about the cosmos. As AI continues to gain significance, it is clear that it is not merely a tool but a fundamental collaborator in unraveling the mysteries of the universe. This chapter explores the diverse applications of AI in astronomy and highlights the ways in which it will shape the future of cosmic discovery.

***Keywords*:** Artificial intelligence, astronomy, machine learning, deep learning, data analysis, galaxy classification, cosmic phenomena, telescope optimization, celestial discovery

1.1 Introduction

Imagine standing beneath a vast star-studded sky on a clear night, gazing at the twinkling dots scattered across the heavens. For millennia, these celestial wonders have captivated poets, dreamers, and scientists alike. What secrets do they hold? How did they form? How do they evolve, and what can they tell us about the origins of our universe? These profound questions have fueled human curiosity since the dawn of civilization. Astronomy [1], at its core, is a journey of exploration—a quest to unravel the mysteries of existence, from the birth of stars to the enigmatic nature of black holes and the exploration of planets outside our solar system. The tools of astronomy have evolved dramatically over the centuries. Early civilizations charted the heavens with the naked eye, creating elaborate maps of the night sky. The invention of the telescope in the 17th century revolutionized our understanding of space, allowing astronomers to peer deeper into the cosmos. In the modern era, spacecraft, space telescopes, and radio observatories have expanded our vision beyond what was once unimaginable. Today, we can detect gravitational waves rippling through space-time, witness distant galaxies colliding, and analyze the chemical signatures of exoplanet atmospheres—all from millions or even billions of light-years away. However, with these advances comes an overwhelming flood of data. The universe is vast, and the amount of information collected by modern telescopes is staggering. Every second, space observatories capture terabytes of data, imaging galaxies at unimaginable distances, detecting faint signals from pulsars, and even searching for potential signs of extraterrestrial life. The sheer scale of this information presents an enormous challenge: how do we analyze it efficiently? Traditional methods, relying on manual examination and conventional algorithms, simply cannot keep up. If sifting through astronomical data was once like piecing together a thousand-piece puzzle, today's challenge is akin to assembling a trillion-piece mosaic within a day. Enter artificial intelligence (AI), the modern-day navigator in our cosmic voyage [2]. AI has revolutionized nearly every scientific field, and astronomy is no exception. Utilizing ML and DL approaches, scientists can identify patterns, forecast outcomes, and tackle challenges that were previously considered unsolvable. Consider the case of exoplanets—planets beyond our solar system. In the past, astronomers had to painstakingly sift through telescope data to identify tiny dips in a star's luminosity, indicating a planet passing in front of it. This process could take years. Today, AI accomplishes this in mere hours, identifying thousands of new exoplanets with remarkable accuracy. Similarly, AI-driven algorithms help classify galaxies, detect transient cosmic events like supernovae, and even interpret gravitational wave signals from black hole mergers. What once required months of human effort is now completed in fractions of the time. The true beauty of AI in astronomy lies not just in its speed but in its ability to learn and adapt. Neural networks, inspired by the human brain, recognize patterns with astonishing precision. Just as we can instantly distinguish a constellation from a cluster of stars, AI can differentiate between different types of galaxies, identify celestial anomalies, and generate highly accurate simulations of the early universe. These advancements are more than just conveniences; they are fundamental to pushing the boundaries of our knowledge. Beyond exoplanet discovery and galaxy classification, AI is making its mark in other critical areas of astrophysics. For instance, AI-powered algorithms are helping scientists map dark matter distributions by analyzing the gravitational lensing effects visible in astronomical images. Dark matter, an unseen and enigmatic substance, is thought to constitute a substantial part of the universe, yet it cannot be directly observed. AI's ability to detect subtle distortions in light has provided deeper insights into this elusive component of the cosmos. Additionally, AI is revolutionizing the study of black holes by sifting through radio telescope data to construct detailed images, such as the now-iconic first-ever photograph of a black hole's event horizon, captured by the Event Horizon Telescope. Another groundbreaking use of AI in astronomy is the automated capturing of fast radio bursts (FRBs) [3], enigmatic and powerful bursts of radio waves originating from deep space. Their transient nature makes them difficult to capture using conventional methods, but AI has enabled real-time identification of these signals, resulting in new discoveries and a deeper insight into their potential origins. Looking ahead, AI's role in astronomy will only continue to grow. With the upcoming launch of next-generation telescopes like the Nancy Grace Roman Space Telescope, the volume of data will expand exponentially. AI-driven methodologies will be indispensable in handling this deluge, allowing researchers to automate discovery processes, identify new cosmic phenomena, and even refine our understanding of fundamental physics. But how does this magic work? What exactly is AI, and what tools does it offer? How do concepts like neural networks, CNNs, and DL algorithms connect to the world of astronomy? In the chapters ahead, we will delve into the fascinating realm of artificial intelligence, beginning with its core principles and exploring its deep impact on astronomical research. We will break down complex ideas, provide real-world examples, and showcase how AI is revolutionizing our understanding of the cosmos. The universe has always been a frontier of discovery, and now, with AI as our guide, we stand at the brink of an unprecedented era of exploration. The mysteries of the stars are waiting—let us embark on this extraordinary journey together.

1.2 Understanding AI: Key Concepts and Techniques

1.2.1 What is AI?

AI [4] is the science and engineering of developing systems that can carry out tasks typically requiring humanly thinking. These tasks cover an enormous range, such as pattern recognition, problem-solving, and much more. AI is not synonymous with systems that are fully independent or conscious; rather, it refers to the ability of machines to emulate aspects of human cognition through programmed logic, data-driven learning, or a combination of both. A key component of AI is ML, which enables systems to improvise their performance by learning from data without explicit programming. Within ML, deep learning (DL) represents an even more specialized subset that leverages artificial neural networks to process and extract patterns from large datasets. This hierarchical structure—where DL

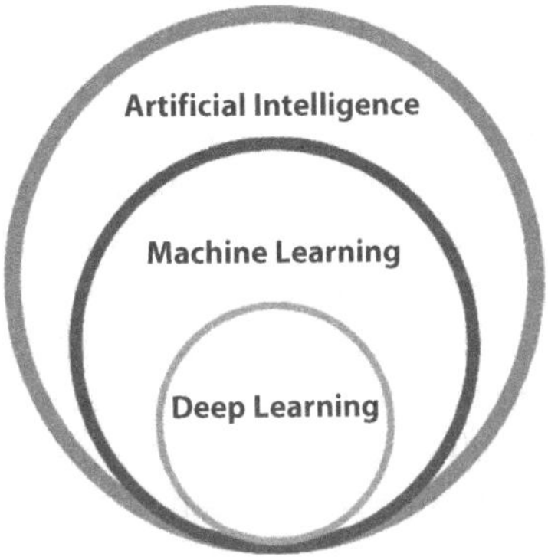

Figure 1.1 Relation between AI, ML, and DL.

is a subset of ML, which itself falls under the broader domain of AI—forms the foundation for many modern advancements in the field (Figure 1.1).

One of the main goals of AI is to develop algorithms and systems that can analyze large amounts of data, identify hidden patterns, and give outcomes based on that information. This has enabled the creation of applications that are transforming industries and everyday life. Examples include virtual assistants like chatbots capable of understanding and responding to human queries and recommendation systems on streaming platforms that suggest content based on user preferences. In essence, AI represents the intersection of computational power, data analysis, and human ingenuity. By simulating aspects of human intelligence, AI systems have the tendency to enhance productivity, drive innovation, and address complex challenges in ways previously unimaginable. From health care to education, transportation to entertainment, AI is poised to shape the future of society profoundly.

1.2.2 Machine Learning

ML [5] is a subset of AI where machines are trained to improve their performance over time using data rather than being explicitly programmed for specific tasks. Instead of telling a computer exactly how to recognize a cat, for example, you provide it with thousands of cat pictures and let it learn patterns that distinguish cats from other objects. ML relies heavily on algorithms like decision trees, support vector machines (SVMs), etc. It finds applications in various real-world scenarios, such as predicting weather, diagnosing diseases, and detecting fraudulent transactions in banking. ML makes AI more adaptable and data-driven. Traditional machine learning techniques have found numerous valuable applications in astronomy, contributing significantly to our understanding of the cosmos. Techniques like support vector machines (SVMs) and random forests are employed for tasks such as classifying stars and galaxies based on their spectral characteristics, identifying exoplanets from subtle dips in stellar brightness (transit detection), and estimating the physical properties of celestial objects like temperature, luminosity, and metallicity. These algorithms excel at handling tabular data derived from astronomical surveys, enabling astronomers to efficiently process vast datasets and discover rare or unusual objects. For instance, SVMs can effectively separate stars from quasars based on their colors and magnitudes, while random forests can be trained to predict the redshift of galaxies, providing crucial information about their distance and the expansion of the universe. Furthermore, traditional ML methods often require less computational resources compared to deep learning approaches, making them accessible for a greater range of astronomical studies and suitable for analyzing data from older surveys or smaller telescopes.

1.2.3 Deep Learning

DL [6] marks a major breakthrough in the realm of ML. Inspired from the tangled morphology of the human brain, DL utilizes artificial neural networks to model complex relationships within data. These networks are composed of numerous interconnected layers, enabling them to automatically extract salient features directly from raw data. This ability removes the necessity for manual feature engineering, which is typically required in traditional machine learning algorithms. As a result, DL excels in tasks involving unstructured and structured data like text, audio, and images, powering applications like voice assistants (e.g., Siri, ChatGPT, etc.) and facial recognition systems.

1.3 Fundamentals of Deep Learning

1.3.1 Building Blocks of a Neural Network

- **Neuron:** It is the core building block of a neural network. Analogous to its biological counterpart, a neuron receives input, processes it through a mathematical operation, and transmits the resulting output to the subsequent layer.

Table 1.1 Activation function, mathematical expression, and their range of output.

Activation function	Mathematical expression	Range of output
ReLU (Rectified Linear Unit)	$f(x) = \max(0, x)$	$[0,\infty)$
Sigmoid	$f(x) = \frac{1}{(1+e^{-x})}$	(0,1)
Tanh (Hyperbolic Tangent)	$f(x) = \frac{(e^{x} - e^{-x})}{(e^{x} + e^{-x})}$	(-1,1)
Leaky ReLU	$f(x) = \max(0.01x, x)$	$(-\infty,\infty)$

- **Activation Function:** It adds nonlinearity to the output of a neuron, enabling the network to capture and learn intricate patterns. Some of the popular activation functions are given in Table 1.1.
- **Layer:** A layer in a neural network comprises multiple neurons. Layers are categorized based on their position within the network:
  1. **Input Layer:** Takes the raw data.
  2. **Hidden Layers:** It does intermediate computations and feature extraction.
  3. **Output Layer:** Provides final prediction.

- **Weights:** Weights are adjustable parameters associated with each connection between neurons. They determine the influence of an input on a neuron's output and are optimized during the training process.
- **Bias:** A bias is a parameter within a neuron that shifts the activation function, enhancing the model's flexibility in fitting the data.

1.3.2 Training and Optimization

The process of training a DL model involves iteratively adjusting its parameters to minimize prediction errors [7]. Key concepts in this process include:

- **Batch:** A batch is a subset of the entire training data used in a single forward and backward pass during training. Utilizing batches reduces computational overhead.
- **Epoch:** It symbolizes one complete iteration through the entire training dataset. Multiple epochs are typically employed to increase the model's performance.
- **Forward Propagation:** It refers to propagating the input through the layers of the neural networks to produce predictions.
- **Loss Function:** It quantifies the discrepancy between the predicted output and the actual target values, serving as a guide for the optimization process.
- **Back Propagation:** Back propagation is the mechanism for calculating gradients for each weight by propagating the loss backward through the network that indicates the direction and magnitude of weight adjustments needed to minimize the loss.
- **Optimization:** It is the process of optimizing/updating model parameters (weights and biases) to achieve the lowest possible loss function value. Popular optimization algorithms include gradient descent and Adam.
  1. **Gradient Descent:** An iterative optimization algorithm that updates weights by moving in the direction of the negative gradient of the loss function. Mathematically, it is expressed as:

$$\theta_{t+1} = \theta_t - \eta \nabla J(\theta_t) \tag{1.1}$$

     where t represents the parameter at iteration t, J(t) is the gradient of the loss function J with respect to and (learning rate) controls the step size of the update.
  2. **Hyperparameters:** Hyperparameters are settings defined before training that influence the model's behavior and performance, for example, learning rate, batch size, and the number of layers.

1.3.3 Addressing Challenges in Deep Learning

Deep learning models are susceptible to certain challenges [8] that can hinder their performance. Understanding and mitigating these issues are crucial for successful model development.

- **Underfitting:** This occurs when a model is too simple, preventing it from capturing underlying data patterns, which results in poor performance on training and testing datasets.
- **Overfitting:** Overfitting happens when a model captures noise and intricate details from the training data, causing it to perform poorly on new unseen data.
- **Gradient:** A gradient represents the rate of change of a function with the change in its inputs. In neural networks, it indicates how much each weight should be adjusted to minimize the loss.
- **Regularization:** It includes techniques, such as like dropout and L2 regularization, which are employed to prevent overfitting.
  1. **Dropout:** A regularization technique where neurons are randomly ignored (dropped out) during training to prevent the network to rely too much on a single neuron and improve generalization.

- **Data Splitting:** To evaluate and optimize model performance effectively, data are typically split into three sets:
  1. **Training Data:** Used to train the model to adjust its parameters.
  2. **Validation Data:** Used during training to tune hyperparameters to avoid overfitting.
  3. **Test Data:** Used for testing the model's performance on unknown data.

1.3.4 Training a Neural Network

Let us start by imagining you are teaching a dog a new trick like fetching a specific ball. You would not just say "fetch" once and expect the dog to get it perfectly, right? You would show the dog many times, correct it when it makes mistakes, and reward it when it gets closer to the right ball. A neural network learns [9] in a similar way, but instead of a dog, it is a computer program, and instead of balls, they are data like pictures, sounds, or text. So, this is how it starts:

- **Input:** Just like you show the dog a ball, you give the neural network some input. This could be a picture of a cat, a recording of someone speaking, a sentence of text, or a picture of a galaxy.
- **The Hidden Layers (Brain Work):** Now, the neural network has a bunch of "thinking layers" called hidden layers. Imagine these as the dog's brain trying to figure out what you want. Each layer is made up of many simple "brain cells" called neurons. Each neuron looks at a small part of the input and tries to make sense of it. For example, one neuron might look for edges in a picture, another for colors, and so on. These neurons are connected to each other like a network of friends whispering clues to each other. They pass information back and forth, each trying to understand a bit more of the input. The connections between these neurons have different "strengths" like some friends are more trusted than others. These strengths are called weights, and they are the key to learning.
- **The Output (The Guess):** After all the brain work in the hidden layers, the network makes a guess or an output. If you showed it a picture of a cat, the output might be "cat." If you asked it a question, the output might be an answer. Initially, this guess is likely to be wrong just like the dog might fetch the wrong ball at first or might just spin around when you tell it to fetch.
- **Learning from Mistakes (Backpropagation):** This is where the magic happens! We tell the network how wrong it was. Did it say "dog" when it was a cat? We measure the error—how far off its guess was. Then, the network uses a clever process called backpropagation to adjust the "strengths" (weights) of the connections between the neurons. It is like saying, "Okay, neurons that thought this was a dog, listen less next time! Neurons that had a hunch it was a cat, listen more!" This happens over and over again, with many examples. The network keeps adjusting its weights, slowly getting better at making the right guesses.

- **Practice Makes Perfect (Training):** We show the network tons of examples, correcting it each time. This is called training. The more examples it sees, the better it gets, just like the dog gets better at fetching with practice. Eventually, the network learns the important patterns present in the data and can make better predictions even on things it has never seen before.

In short, a neural network is like a really complex guessing machine that learns from its mistakes by adjusting the strengths of its internal connections, kind of like how a brain learns. It takes in information, crunches it, makes a guess, and then adjusts its thinking based on how right or wrong it was (Figure 1.2 represents a fully connected neural network). So, you see, an AI model is not very smart! It is hardworking and learning from its mistakes. On a side note, the training process adjusts the parameters (weights and biases) to minimize the loss (mistakes).

The trained model is what can do our task easily. The best sets of weights and biases are thus not created out of nowhere. They always exist! We are just finding them in our space of weights and biases while we are training the model. These sets of weights and biases are not unique, multiple such combinations exist. One can dive deeper into the mathematics of neural networks to learn more about it.

To get the brain some data, we need different sense organs, right? Eyes, nose, tongue, etc. For looking around us, we need eyes; for smell, we need

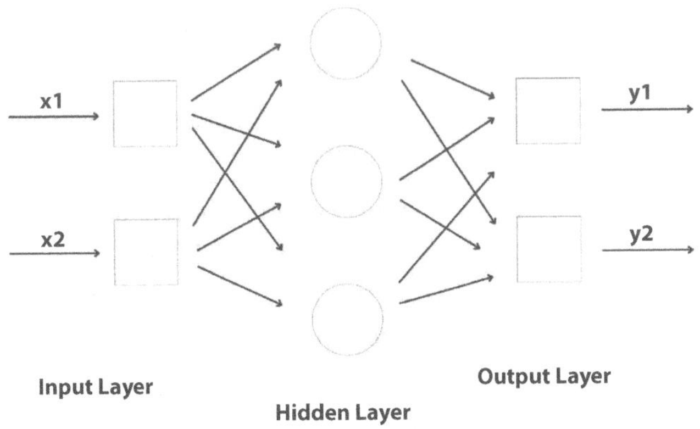

Figure 1.2 Pictorial representation of a fully connected artificial neural network with one hidden layer.

a nose; so, there is a specific organ for a specific data type. Neural networks have something kind of similar for sensing although not really. With advancements in computing capabilities, new methods were devised to feed in various data types of neural networks; these emerged as specific networks designed for specific tasks like CNNs for image data, RNNs for time series, and many others.

1.4 AI Algorithms Shaping Astronomical Research

1.4.1 Convolutional Neural Network

CNNs [10] are a DL architecture primarily made for processing seeable data. They can automatically and adaptively learn features from images, such as edges, textures, and shapes. CNNs are composed of several layers, including convolutional layers, pooling layers, and fully connected layers (Table 1.2). The CNN layers apply filters (kernels) to scan the image and extract patterns, while pooling layers reduce the image's dimensionality to improve computational efficiency. Figure 1.3 illustrates a sample CNN architecture. In astronomy, CNNs are extensively applied to tasks

Table 1.2 CNN layers and their purpose.

Layer type	Purpose
Convolutional Layer	Feature extraction by applying filters/kernels over the input
Pooling Layer	Reduces spatial dimensions while retaining important features
Fully Connected Layer	Connects neurons to all previous activations for final classification
Dropout Layer	Prevents overfitting by randomly dropping neurons during training
Batch Normalization Layer	Normalizes activations to stabilize learning and speed up convergence
Activation Layer	Introduces nonlinearity to help learn complex patterns
Flatten Layer	Converts multidimensional feature maps into a 1D vector

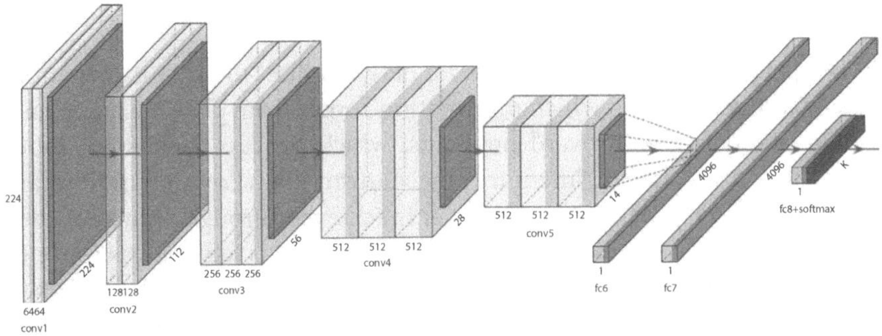

Figure 1.3 Pictorial representation of a CNN (VGG16 architecture).

like image classification and object detection. For instance, CNNs can categorize various types of galaxies from large-scale sky survey images, distinguishing between spiral, elliptical, and irregular galaxies. Additionally, they are used to detect exoplanets (Figure 1.4) by analyzing telescope light curves, automatically identifying slight dips in brightness caused by planets transiting in front of their host stars.

Figure 1.4 Picture of TRAPPIST-1 (An earth sized exoplanet). (Image credit: https://science.nasa.gov/image-detail/amf-pia22098/).

1.4.2 Recurrent Neural Network

RNNs [11] are a unique type of neural network built to process sequential data where sequence in which information comes from plays a crucial role. Unlike standard feedforward networks, which process each input independently, RNNs have a "memory" mechanism that allows them to retain knowledge from previous time steps. It is achieved through recurrent connections, where a neuron's output at one time step is fed back as input to the same neuron in the next step. This looping structure enables RNNs to capture temporal relationships and contextual dependencies in time series data. As a result, RNNs are particularly effective for applications likes natural language processing (e.g., sentiment analysis), speech recognition, and forecasting future values in sequential data. However, standard RNNs can struggle with long sequences due to vanishing gradients; more advanced architectures like LSTMs and GRUs have been developed to address this limitation.

In an RNN, the hidden state h_t depends both on the current input x_t and on the previous hidden state h_{t-1}, allowing the network to maintain a form of memory over time. And the model learns for W_h, weight matrix for the hidden state from the previous time step, and W_x, weight matrix for the input at time step t.

RNNs are particularly useful in a time-series analysis within astronomy. For instance, they are used to analyze light curves from variable stars, supernovae, or transient events. These light curves contain sequences of brightness measurements over time, and RNNs can learn the underlying patterns to predict future behavior or detect unusual events. RNNs are also applied in forecasting solar activity, such as predicting solar flares or coronal mass ejections, based on past data.

Mathematically, RNNs are represented as:

$$h_t = f(W_h h_t - 1 + W_x x_t + b_h) \tag{1.2}$$

1.4.3 Long Short-Term Memory

LSTM networks [12] are a variant of RNNs designed for processing sequential data and effectively capture long-term dependencies. LSTMs overcome this problem through their unique architecture, which contains memory cells and specialized gating mechanisms: the input gate, forget gate, and output gate. These gates work together to regulate how information is stored, updated, or discarded, enabling the network to focus on important patterns while ignoring irrelevant details. This makes LSTMs particularly well-suited for tasks that include sequential data from natural language to genomics, where understanding sequential data is critical. Because of their ability to retain context and adapt to varying sequence lengths, LSTMs excel in applications like machine translation, predictive modeling, and video captioning. They are especially useful in scenarios where identifying relationships across time is essential, such as weather forecasting, financial market prediction, and sequence-to-sequence tasks. By addressing the limitations of traditional RNNs, LSTMs provide a powerful and flexible tool for analyzing and interpreting sequential data, making them an indispensable component in modern deep learning.

1.4.4 Reinforcement Learning

RL is an effective ML approach in which the agent understands decision-making by interacting with its environment. RL agents learn through trial and error, receiving rewards for correct steps taken or penalties for wrong steps taken. The objective is to identify an optimal policy—a strategy that maximizes the total rewards over time. This is achieved by exploring the environment, taking actions, and observing the consequences. The agent uses these experiences to update its understanding of the environment and improve its decision-making process. RL is inspired by behavioral psychology and finds applications in areas such as robotics, game playing (e.g., AlphaGo), etc., where agents must learn to navigate complex situations and optimize their behavior to achieve specific goals. The core challenge in RL lies in balancing exploration and exploitation (using the current knowledge to maximize rewards). To understand it in simpler terms, imagine you are teaching a dog a new trick, but instead of showing it what to do, you let it try things on its own and give it treats when it does something good. That is basically how reinforcement learning (RL) works. The "dog" is called an agent, and it lives in an environment like a room or a game. The agent can take different actions like moving left or right, and the environment gives it a reward or a penalty depending on how good or bad the action was. Think of a treat as a reward and a scolding as a penalty. The primary aim of the agent is to learn a strategy, called a policy, that gets it the most rewards over time [13].

The agent learns by trying different activities or steps and seeing what happens. At first, it makes a lot of random choices like a puppy stumbling around. But every time it gets a reward, it remembers what it did and tries to do similar things in the future. If it gets a penalty, it tries to avoid doing that again. Over time, the agent figures out which actions lead to the most rewards and starts to make better decisions. It is like the dog learning that sitting when you say "sit" gets it a tasty treat! This learning process is all about trial and error, and eventually, the agent becomes really good at achieving its goal whether it is navigating a maze, winning a game, or controlling a robot. While still a nascent field in astronomy, reinforcement learning (RL) is emerging as a promising tool for tackling complex control and optimization problems. One key application lies in the operation of telescopes and observatories where RL agents can learn to autonomously schedule observations, optimizing for factors such as target visibility, weather conditions, and scientific priorities. This can lead to more effective use of valuable telescope time and enable rapid follow-up observations of transient events like supernovae or gamma-ray bursts. RL is also being explored for adaptive optics, training algorithms to dynamically correct for atmospheric distortions and improve image quality in real time. Further potential applications include controlling robotic telescopes, optimizing the pointing of antennas for radio astronomy, and even guiding spacecraft for autonomous exploration of the solar system. As astronomical datasets grow larger and instruments become more complex, RL offers a powerful framework for developing intelligent systems that can enhance our ability to observe and understand the universe.

1.5 Revolutionizing Data Analysis: AI in Astronomical Surveys

The exciting advancements in AI and astronomy [14] are not limited to seasoned researchers in large institutions. Thanks to open-source software, publicly available datasets, and a thriving online community, it is easier than ever for anyone with a passion for the cosmos to begin exploring the universe with AI. A brief overview of some key tools, datasets, and resources is described below to get you started.

1.5.1 Popular Machine Learning Libraries

At the heart of most AI projects are powerful machine learning libraries. These provide pre-built functions and algorithms, making it simpler to develop and deploy AI models without having to code everything from scratch. Here are a few of the most popular choices:

- **Scikit-learn:** This Python library is a great starting point for beginners to machine learning. It provides a wide range of classical ML algorithms and tools for data preprocessing, model selection, and evaluation. Its strength lies in its ease of use and comprehensive documentation, making it ideal for learning fundamental concepts [15].
- **TensorFlow:** It is an open-source library developed by Google LLC for numerical computation and large-scale ML, particularly deep learning. It gives a flexible framework for building and training neural networks, with support for various hardware platforms, including CPUs, GPUs, and TPUs. TensorFlow is used in astronomy for tasks like image classification, time-series analysis, and complex simulations [16].
- **PyTorch:** Another popular deep learning framework, PyTorch is known for its dynamic computation graphs and ease of use, making it particularly favored by researchers. It offers similar capabilities to TensorFlow, with a strong emphasis on flexibility and rapid prototyping. PyTorch is gaining popularity in astronomy for its versatility and growing community support [17].

1.5.2 Astronomical Software and APIs

To work with astronomical data, you will need tools to access, process, and analyze them. A variety of specialized software packages and application programming interfaces (APIs) are available:

- **Astropy:** A core Python package for astronomy, Astropy provides fundamental tools for working with astronomical data formats (like FITS files), celestial coordinates, units, and more. It is an essential package for almost any AI project in astronomy [18].
- **Astroquery:** This Python package simplifies the process of querying astronomical databases and catalogs, allowing you to easily download data from various sources like the Virtual Observatory, the Sloan Digital Sky Survey, and much more [19].

- **SAOImage DS9:** A powerful image visualization and analysis tool, DS9 is widely used by astronomers to inspect astronomical images, create overlays, and perform basic measurements [20].
- **Specific Observatory/Telescope APIs:** Many major observatories and telescopes provide APIs for programmatic access to their data. For example, the Hubble Space Telescope API allows you to query and download data directly from its archive. Consult the documentation for the specific telescope or survey you are interested in.

1.5.3 Public Datasets and Repositories

One of the greatest advantages of working in modern astronomy is the abundance of publicly available data. Numerous surveys and missions have made their data freely available to the scientific community, opening up opportunities for anyone to explore the universe. Here are a few notable resources:

- **Sloan Digital Sky Survey (SDSS):** A massive survey that has mapped a large portion of the sky, providing data on millions of galaxies, stars, and quasars [21].
- **Gaia Mission:** An ESA mission that is creating a precise three-dimensional map of our galaxy, the Milky Way, measuring the positions, distances, and motions of billions of stars [22].
- **Zwicky Transient Facility (ZTF):** A wide-field survey that scans the sky every night, looking for transient events like supernovae, variable stars, and moving objects [23].
- **The NASA/IPAC Extragalactic Database (NED):** A comprehensive database of information on objects outside our galaxy [24].
- **The Mikulski Archive for Space Telescopes (MAST):** A repository of data from numerous space-based telescopes, including the James Webb Space Telescope (JWST), and many others [25].

1.6 Machine Learning Models for Celestial Object Classification

1.6.1 Random Forest Model for Classifying Celestial Objects Into Three Categories: STAR, GALAXY, Or Quasi-Stellar Object

- **Dataset Description:** The SDSS Galaxy Colors dataset contains features derived from the photometric measurements of celestial objects across 5 bands (u, g, r, i, z) (Table 1.3).
- **Redshift:** It is the measure of stretch of light due to the expansion of the universe.

1. **Data Loading and Preprocessing:** The dataset is loaded using the astroML library, and class labels are mapped to numeric values for compatibility with machine learning algorithms.
2. **Model Training:** A random forest classifier is trained on the dataset. This model is robust to overfitting and performs well on small- to medium-sized datasets.
3. **Evaluation and Prediction:** The model after getting trained achieves a high accuracy of approximately 96% on the test set. It can also predict the class of new objects based on their photometric features. (Table 1.4 for model performance metrics other than accuracy).

Table 1.3 Color bands along with their short description.

Color bands	Short description
u-g	Color index between u (ultraviolet) and g (green) bands
g-r	Color index between g and r (red) bands
r-i	Color index between r and i (infrared) bands
i-z	Color index between i and z bands

- **Python code used for classifying celestial objects using a Random Forest model:**

```
import numpy as np
import pandas as pd
from sklearn.model_selection import train_test_split
from sklearn.ensemble import RandomForestClassifier
from sklearn.metrics import classification_report, accuracy_score
from astroML.datasets import fetch_sdss_galaxy_colors  # Fetch astronomical
dataset

# Load the SDSS Galaxy Colors dataset
data = fetch_sdss_galaxy_colors()

# Convert the data into a pandas DataFrame for easier manipulation
df = pd.DataFrame(data, columns=['u-g', 'g-r', 'r-i', 'i-z', 'redshift',
'class'])

# Map class names ('STAR', 'GALAXY', 'QSO') to integer labels (0, 1, 2)
class_mapping = {'STAR': 0, 'GALAXY': 1, 'QSO': 2}
df['class'] = df['class'].map(class_mapping)

# Separate features (input) and target labels (output)
X = df[['u-g', 'g-r', 'r-i', 'i-z', 'redshift']].values  # Input features
y = df['class'].values  # Target labels

# Split the data into training and testing sets
X_train, X_test, y_train, y_test = train_test_split(X, y, test_size=0.3,
random_state=42)

# Initialize the Random Forest Classifier with 100 decision trees
model = RandomForestClassifier(n_estimators=100, random_state=42)

# Train the model
model.fit(X_train, y_train)

# Use the trained model to make predictions
y_pred = model.predict(X_test)

# Evaluate the model's performance using a classification report
print("\nClassification Report:")
print(classification_report(y_test, y_pred))

# Calculate and display the accuracy of the model
accuracy = accuracy_score(y_test, y_pred)
print(f"Model Accuracy: {accuracy:.2f}")

# Example: Predict the class of a new celestial object based on its features
new_object = [[0.95, 0.6, 0.1, 0.02, 0.1]]  # Example feature values
prediction = model.predict(new_object)  # Make prediction
class_name = {0: 'STAR', 1: 'GALAXY', 2: 'QSO'}  # Mapping integers back to
class names
print(f"\nPrediction for the new object: {class_name[prediction[0]]}")
```

Table 1.4 Classification report for each class in order: STAR, GALAXY, and QSO.

Precision	Recall	F1 score
0.98	0.97	0.98
0.96	0.96	0.96
0.92	0.95	0.94

1.6.2 k-Nearest Neighbors for Classification of Celestial Objects Into Distinguishing AGNs from Stars and Galaxies

In the paper [26], the authors investigated the application of the kNN algorithm for the automated classification of celestial objects, leveraging multiwavelength data. The paper emphasizes the growing complexity of astronomical data due to large-scale sky surveys across multiple spectral bands, necessitating efficient classification techniques. The study utilizes data from multiple sources, including ROSAT All-Sky Survey (X-ray), USNO-A2.0 (optical), and 2MASS (infrared), to distinguish active galactic nuclei (AGNs) from stars and normal galaxies. The study concludes that kNN is a practical tool for handling large astronomical datasets and can be integrated into virtual observatories for real-time classification of celestial objects.

1.6.3 Convolutional Neural Network for Classifying Galaxies

This study in [27] presents a DL framework for automated galaxy classification using images from the Sloan Digital Sky Survey and optimized it for FPGAs. Traditional classification methods, including manual efforts by astronomers and large-scale citizen science projects like Galaxy Zoo [28], have significantly contributed to morphological studies but are inherently time-intensive and limited by human subjectivity. Recent advances in ML, particularly CNNs, provide a promising avenue for automating this process with improved efficiency and consistency.

The authors propose a CNN-based model that classifies galaxies into 10 morphological categories (Figure 1.5 shows for example images from

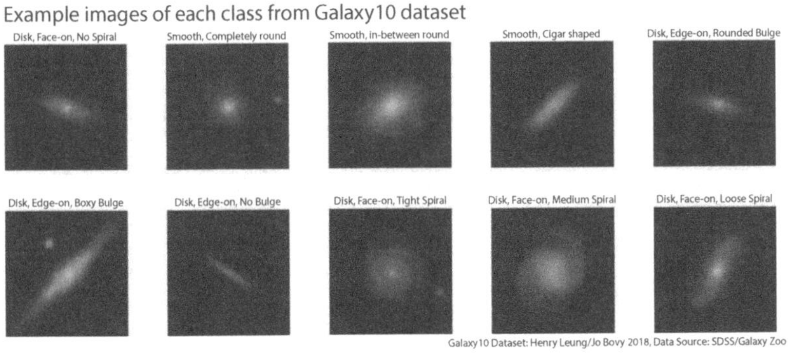

Figure 1.5 Example image from each class of SDSS galaxy10 dataset (Image credit:https://astronn.readthedocs.io/en/latest/galaxy10sdss.html).

each category), an extension of the Hubble Tuning Fork classification [29]. Unlike previous studies that focused on broad classifications such as spiral, elliptical, and irregular galaxies, this model incorporates finer distinctions, including face-on and edge-on disk galaxies with varying bulge prominence, as well as different spiral arm structures. By leveraging convolutional layers for feature extraction and additional fully connected layers for classification, the model effectively captures intricate morphological details from the galaxy images.

This research underscores the growing role of DL in astronomy, demonstrating its potential to handle vast datasets and extract meaningful insights with minimal human intervention. The proposed automated classification system provides a scalable solution for future large-scale sky surveys, aiding in the systematic study of galaxy evolution & structure and dips in brightness caused by planets transiting in front of their host stars.

1.7 AI in Observational Astronomy: Transforming Telescopic Data

With the exponential increase in the quantity and complexity of data generated by modern telescopes (Figure 1.6), traditional methods of analysis are no longer sufficient. AI, especially through deep learning and machine

Figure 1.6 Hubble telescope (Image credit: https://science.nasa.gov/mission/hubble/).

learning, has become a game changer—enabling astronomers to efficiently process vast datasets, uncover hidden patterns, and automate the discovery process. From real-time analysis of cosmic events to the classification of galaxies and exoplanets, AI is revolutionizing how telescopic data are interpreted.

1.7.1 AI-Powered Image Processing in Astronomy

Telescopic images often suffer from noise, atmospheric distortions, and instrumental limitations. AI-based techniques, such as CNNs and generative adversarial networks (GANs), are enhancing image resolution, removing noise, and reconstructing missing data. One of the prominent works in this domain is highlighted in [30]. The paper introduces a novel AI-based method, *Noise2Astro*, designed to improvise the quality of astronomical images. The authors used convolutional neural networks (CNNs) to develop a model that can efficiently "denoise" images. The network learns to distinguish between real astronomical features and noise by processing large datasets of simulated noisy images. The result is an improvement in image quality, where faint celestial objects, such as distant galaxies or stars, become clearer and more identifiable. This AI approach demonstrates a significant advancement in astronomical image processing, offering a powerful tool to enhance the analysis of telescopic data, especially when high-quality images are critical for further research like in the search for exoplanets or deep space exploration.

1.7.2 Real-Time Data Processing and Automated Observations

With the advent of large-scale surveys such as the LSST [31], SKA, and others, real-time data processing has become essential. AI algorithms enable automated telescope scheduling, rapid identification of transient events, and adaptive optics corrections. In the paper [32], the researchers introduce the ***StarWhisper Telescope (SWT)***, an AI-powered system to manage telescope array. The SWT employs large language models (LLMs) to automate the scheduling and execution of astronomical observations. The system generates observation lists tailored to different sites and strategies, ensuring full coverage of the night sky. Observed images are then analyzed promptly, with transient events communicated to the observer for immediate follow-up. Then, an AI agent converts detected transients into real-time follow-up observation proposals, facilitating swift responses to new discoveries. By integrating AI into the observation process, the SWT system streamlines the management of large-scale astronomical surveys.

It reduces the manual workload of astronomers, accelerates the identification of transient events, and enhances the overall efficiency of telescopic observations.

This approach exemplifies how AI can transform real-time data processing and automated observations in astronomy, enabling rapid responses to transient events and improving the effectiveness of large-scale surveys.

1.7.3 AI in Telescope Data Compression and Storage

Astronomical telescopes generate massive volumes of data, particularly with modern surveys and high-resolution imaging systems. The sheer size of these data can overwhelm traditional storage systems, making it difficult to efficiently process and store. AI methodologies, such as DL and unsupervised learning algorithms, are increasingly being used to compress these data without losing essential information. These methods can intelligently identify and retain the most valuable data, removing redundancies and compressing images and time-series data for storage. AI also helps in classifying and tagging data automatically, ensuring that critical observations are flagged for further analysis, reducing the burden on human astronomers and enabling more efficient data management.

The work carried out by the researchers in the paper [33] presents a DL-based approach for compressing astronomical survey data, overcoming the challenges posed by the massive volumes of data generated by modern telescopes. The authors propose using autoencoders, a type of neural network, to reduce the size of large image datasets while preserving key features required for scientific analysis. They demonstrate that deep learning models can efficiently compress images by learning the underlying patterns and structures, achieving high compression rates with minimal loss of information. The model outperforms traditional compression methods, making it highly suitable for large-scale astronomical surveys like the LSST. This approach not only facilitates better storage but also accelerates data transmission and reduces computational overhead for further analysis.

1.8 Harnessing AI for Space Exploration and Planetary Science

AI has become an essential tool in the rapidly advancing field of space exploration and planetary science. From the search for habitable exoplanets to the study of celestial bodies within our own solar system, AI techniques like machine learning, neural networks, and data analytics are revolutionizing how we explore and understand the universe. Some of the research in the given field are discussed below:

1.8.1 AI-Driven Autonomous Navigation for Deep Space Missions

One of the most significant applications of AI in space exploration is the use of autonomous systems for spacecraft navigation. With deep space missions requiring real-time decision-making, AI enables spacecraft to navigate through complex environments without human intervention. AI algorithms are used to process sensor data, detect obstacles, and optimize trajectory planning. This allows spacecraft to adjust their paths or even avoid hazardous situations, such as asteroid fields or debris. One such example is described in [34].

The paper explores the role of artificial intelligence in enhancing spacecraft autonomy for deep space missions. Traditionally, spacecraft rely on ground-based instructions for navigation, but the increasing distance of missions, such as those to Jupiter's moons or interstellar space, makes real-time human control infeasible due to signal delays. The researchers propose an AI-based system that integrates reinforcement learning and deep neural networks to enable spacecraft to make real-time navigation decisions without human intervention.

The paper discusses how AI processes input from onboard sensors, such as star trackers, LiDAR, and optical cameras, to detect celestial landmarks and calculate the best possible trajectory. The authors present case studies where AI algorithms successfully optimize fuel efficiency, minimize collision risks with asteroids, and even adjust flight paths in response to unexpected space weather conditions. The research highlights a successful simulation of an AI-controlled spacecraft avoiding space debris using predictive modeling.

Moreover, the study demonstrates the potential of AI-powered navigation for future missions to Saturn's moon Titan and the asteroid belt, where rapid decision-making is crucial. The authors emphasize that AI-based navigation systems can significantly reduce mission costs and improve overall mission success rates by increasing spacecraft autonomy.

1.8.2 AI for Exoplanet Discovery and Characterization

The discovery of exoplanets has increased severely in recent years, thanks to the power of AI. ML techniques are applied to analyze enormous amounts of data from telescopes and space observatories. AI aids in identifying potential exoplanets by analyzing light curves, detecting subtle variations in star brightness caused by transits. Furthermore, AI helps in characterizing these planets, such as estimating their size, temperature, and atmospheric composition, by interpreting data from spectrographs and imaging systems.

The researchers in the paper [35] highlight how AI has become an indispensable tool in exoplanet detection and characterization. Traditional methods, such as the transit method and radial velocity measurements, generate heaps of data that require extensive human analysis. The authors introduce machine learning models trained on datasets from NASA's Kepler and TESS missions, enabling faster and more accurate identification of exoplanet candidates. The work highlights the use of CNNs to analyze light curves and detect periodic dips in a star's brightness, which indicate the presence of orbiting planets. The AI model significantly reduces false positives caused by stellar variability and instrumental noise. Furthermore, the paper discusses how deep learning algorithms improve the classification of exoplanets based on size, orbital distance, and atmospheric properties. Beyond detection, the study explores AI applications in exoplanet characterization, particularly in atmospheric analysis. The authors describe a neural network trained to interpret spectral data, identifying the presence of gases such as oxygen, methane, and carbon dioxide. This capability is crucial for assessing whether a planet's atmosphere might support life. The AI models tested in this research demonstrate a 30% improvement in detection accuracy compared to traditional statistical methods.

The findings suggest that AI can revolutionize the search for habitable exoplanets by automating and refining data analysis, allowing astronomers to focus on the most promising candidates for further study.

1.8.3 AI in Planetary Surface Exploration and Robotic Operations

Planetary rovers play a crucial role in exploring the surfaces of Mars, the Moon, and other celestial bodies, but they face significant challenges such as rough terrains, unpredictable environments, and communication delays with Earth. The paper [36] investigates how AI-driven autonomy enhances robotic exploration, enabling rovers to operate with minimal human intervention. The researchers detail the integration of AI techniques, including deep RL and computer vision, into robotic systems like NASA's Perseverance rover. The AI framework allows rovers to analyze terrain features, select optimal paths, and avoid hazards such as large boulders, cliffs, or unstable ground. The study presents an AI model trained on simulated Martian landscapes, demonstrating how the rover successfully navigated previously uncharted terrains.

In addition to mobility, the paper explores AI's role in scientific exploration. The authors describe how AI-powered image recognition enables rovers to classify rocks, identify potential signs of past water activity, and prioritize interesting geological formations for sampling. The research also highlights the use of AI in autonomous drilling operations, where the system determines the best locations to extract soil samples without waiting for instructions from mission control.

One of the key findings of this study is that AI-driven autonomy increases mission efficiency by 40%, as rovers can complete more scientific tasks in a given time frame. The authors conclude that AI-powered robotic exploration will be essential for future missions, such as those to Europa and Enceladus, where real-time human intervention is impossible. They emphasize that AI-driven systems will play a vital role in the search for extraterrestrial life by enabling rovers to explore remote and extreme environments with unprecedented precision.

1.9 AI-Driven Discoveries: Case Studies in Astronomy

Artificial intelligence (AI) has significantly advanced astronomical research, enabling discoveries that were previously challenging or impossible. In this section, we will see three such examples.

1.9.1 AI-Enhanced Gravitational Wave Detection

In the groundbreaking study [37], researchers introduced a new AI-driven approach to optimize the design of gravitational wave detectors (Figure 1.7), which are sensitive instruments that capture minute ripples in space-time caused by catastrophic cosmic events such as black hole mergers. Traditional detector designs are limited by hardware and noise constraints. By applying machine learning algorithms, researchers were able to simulate various configurations for detector designs, including innovative setups that were previously unexplored. AI helped to fine-tune detector sensitivity, leading to new recommendations for the future design of interferometers, the primary instruments used to detect gravitational waves. This research holds the potential to improve the accuracy and sensitivity of future gravitational wave observations, enabling the detection of weaker

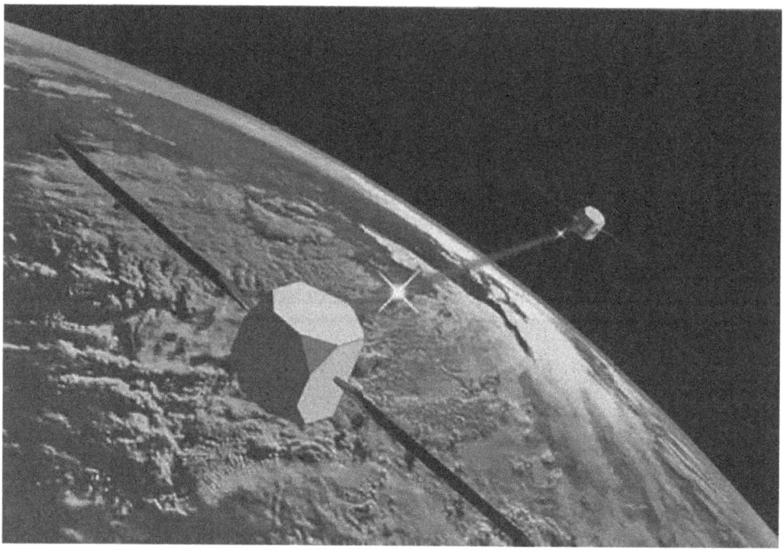

Figure 1.7 Space-based gravity wave detector (Image credit: https://www.nasa.gov/general/a-gravitational-wave-detector-based-on-an-atom-interferometer/).

signals from more distant or faint cosmic events. The application of AI in this context underscores its ability to push the boundaries of technological advancement in observational astronomy.

1.9.2 ML for Anomaly Detection in Astronomical Data

The paper [38] presents an innovative application of unsupervised ML algorithms to detect abnormalities in large astronomical datasets. Given the sheer volume of data generated by telescopes and space observatories, detecting rare or unusual astronomical events can be a daunting task for traditional methods. The researchers employed a clustering-based approach using density-based spatial clustering, which automatically identifies anomalies by grouping similar data points together and flagging those that do not fit the typical patterns. This approach led to the discovery of unusual phenomena such as the unpredictable variations in the brightness of Boyajian's Star [39], an enigmatic star system whose behavior has puzzled astronomers for years. The ability to apply AI-driven anomaly detection to massive datasets has the potential to revolutionize the search for new and rare cosmic events like fast radio bursts or transient supernovae. This method improves both the speed and efficiency of astronomical discovery, enabling researchers to focus on the most promising leads in a sea of data.

1.9.3 AI-Driven Meteor Shower Mapping

In the paper [40], the authors describe the development of an AI-enhanced pipeline designed to process and map meteor data, significantly improving how meteor showers (Figure 1.8) are observed and cataloged. Meteor showers, while relatively common events, are often difficult to monitor due to their sporadic nature and the vast areas of sky they cover. This research introduced an automated system that uses machine learning techniques to classify and track meteors captured by cameras and sensors around the globe. The AI system not only was able to validate existing meteor showers but also uncovered more than 200 new meteor showers, expanding the meteor catalog and offering new insights into the dynamics of meteoroid streams. A key feature of the project was its use of citizen science, encouraging the public to contribute to the monitoring process, which helped amass a vast amount of data quickly. This approach showcases the power of AI to facilitate large-scale collaborative efforts in astronomy, combining the strengths of automated data processing with the enthusiasm of public involvement. It demonstrates how AI can assist in the discovery of

Figure 1.8 Perseid meteor shower - 2015 (Image credit: https://science.nasa.gov/resource/perseid-meteor-shower-2015/).

previously unknown celestial events while improving the accuracy of existing observations.

These AI-driven discoveries are reshaping the way astronomers observe and analyze the cosmos, providing new tools to explore the universe at an unprecedented scale.

1.10 Challenges and Limitations of AI in Astronomy

The integration of AI in astronomy holds great promise for advancing our understanding of the cosmos, from uncovering hidden patterns in vast datasets to enabling real-time analysis of dynamic astronomical events. However, this technological transformation is not without its challenges [41]. It is crucial to address these challenges thoughtfully, responsibly, and ethically, ensuring that AI is deployed in ways that enhance, rather than compromise, scientific integrity.

One of the most significant challenges arises from the inherent biases in astronomical datasets. These biases stem from multiple sources, including the design of instruments, the strategies employed in conducting surveys, and even the subjective decisions made during data processing and labeling. Left unaddressed, these biases can distort AI-driven insights, potentially leading to inaccurate or misleading interpretations of celestial phenomena. Recognizing and mitigating these biases are fundamental to ensuring the reliability and robustness of AI applications in astronomy.

1.10.1 Sources of Bias in Astronomical Data

Astronomical data, like any dataset, can reflect the limitations and assumptions embedded within the processes used to collect and analyze them. Common sources of bias include:

- **Selection Bias:** Datasets often prioritize specific types of objects or regions of the sky, potentially excluding underrepresented populations. For example, a dataset focused on nearby galaxies may not adequately capture the diversity of distant or faint galaxies, leading to models that fail to generalize across the broader cosmic population.
- **Instrumental Bias:** Characteristics inherent to telescopes or detectors, such as sensitivity variations or calibration errors, can introduce systematic inaccuracies. These biases may

disproportionately affect certain types of data, skewing the results of AI analyses.

- **Human Bias:** During data labeling or annotation, subjective decisions made by researchers can introduce unintended biases. For instance, the categorization of galaxy morphologies might vary between individuals, influencing the training of supervised learning models in subtle but impactful ways [42].

1.10.2 Mitigating Bias in AI Applications

To address these biases, a multipronged approach is required:

- **Comprehensive Data Validation:** Rigorous validation techniques should be employed to identify and quantify potential biases in datasets. This includes cross-checking results against independent datasets and conducting sensitivity analyses.
- **Bias Mitigation Techniques:** Advanced statistical methods and algorithmic approaches, such as re-weighting or re-sampling, can help reduce the impact of biases. Additionally, incorporating domain knowledge into AI models can improve their skill to account for known limitations in the data.
- **Transparency and Interpretability:** Transparent AI systems that allow researchers to understand how models make predictions are essential. This requires focusing on interpretability and explainability to build trust and ensure scientific accuracy.

1.10.3 The Importance of Interpretability: Understanding How AI Models Make Decisions

DL models, those using neural networks, are commonly described as "black boxes" because of their intricate and unclear decision-making processes. While these models can achieve remarkable accuracy, understanding the "*why*" behind their predictions is just as important as the predictions themselves, particularly in scientific domains like astronomy. Interpretable AI ensures that the insights derived from models are not only accurate but also scientifically meaningful.

1.10.4 Key Approaches to Interpretability

- **Feature Importance:** Identifying which features in the data most significantly influence a model's predictions. For instance, in a galaxy classification task, determining whether the model prioritizes morphological features, spectral characteristics, or other parameters can reveal critical insights.
- **Explainable AI (XAI):** Leveraging techniques that provide explanations for individual predictions or the overall behavior of the model. For example, SHapley Additive exPlanations or Local Interpretable Model-agnostic Explanations can elucidate the relationships between input data and outputs [43].
- **Visualization Techniques:** Tools like saliency maps, heatmaps, and dimensionality reduction visualizations (e.g., t-SNE, UMAP) can help researchers explore how models process data, identify patterns, and connect features with outcomes.
- **Model Simplification:** Simplifying complex models through distillation or creating interpretable surrogate models can provide approximate but understandable representations of decision-making processes.

1.11 The Future of AI in Astronomy: Opportunities and Horizons

Based on the current scenario and challenges faced by astronomers and scientists in applying AI in the field of astronomy, we can point a few directions where extensive development will be necessary and is already ongoing [44]:

- **Explainable Models:** As already discussed, explainable AI is not just a desirable feature but a crucial requirement for the future of AI in astronomy. If we cannot understand how our models are arriving at their conclusions, we cannot fully trust them, nor can we use them to advance our scientific understanding. Building truly explainable models is an active area of research, with several promising avenues being explored:

1. **Physics-Informed Models:** One approach is to integrate known physical laws and principles directly into the AI models. For example, when classifying galaxies, a physics-informed model could incorporate constraints based on our understanding of galactic dynamics and stellar populations. This not only helps to improve the accuracy of the model but also makes its predictions more physically meaningful and interpretable. In practice, this could involve adding terms to the loss function of a neural network that penalize deviations from known physical laws or designing network architectures that explicitly encode physical symmetries and conservation principles. For instance, in exoplanet atmospheric studies, radiative transfer equations and chemical kinetic models can be incorporated within the machine learning framework to constrain atmospheric parameters and ensure physically plausible interpretations.
2. **Symbolic Model Extractors:** Another strategy involves extracting symbolic representations or rules from complex AI models. This is often referred to as "rule extraction" or "model distillation." The goal is to approximate the behavior of a black-box model with a simpler, more transparent model, such as a decision tree or a set of logical rules. For example, a symbolic model extractor might reveal that a galaxy classification model is primarily relying on features related to the galaxy's shape, color, and surface brightness profile. In a time-series analysis for variable stars, symbolic extractors can distill patterns that correlate with physical processes within the star such as pulsations or binary interactions. This allows for the generation of human-readable rules describing the light curve behaviors that may not be immediately obvious through direct observation [45].
3. **Human Feedback-Based Learning:** Incorporating human expertise into the AI training process can also enhance explainability. In this approach, astronomers or other domain experts provide feedback on the model's predictions, highlighting areas where the model is performing well or poorly and providing explanations for their judgments. This feedback can be used to refine the model and to make it more aligned with human understanding. For

instance, a model might initially misclassify a rare type of astronomical object. Through interactive feedback, an astronomer could correct the misclassification and provide additional information about the distinguishing features of that object. This creates an iterative refinement cycle, leading not only to more accurate but, more importantly, more interpretable models. This is particularly valuable in cases where the training data are limited or noisy and where expert knowledge is essential for interpreting subtle patterns [46].

- **Machine-based Literature Review:** In the rapidly evolving landscape of astronomical research, the volume of published literature grows exponentially each year. This deluge of information presents a significant challenge for astronomers trying to stay abreast of the latest discoveries, identify relevant prior work, and synthesize knowledge across different subfields. Traditional literature review methods, while valuable, are becoming increasingly difficult to scale with this ever-expanding body of knowledge. This is where the power of AI, particularly LLMs like ChatGPT and other similar models, comes into play. These tools offer the potential to revolutionize how astronomers conduct literature reviews, enabling more comprehensive, efficient, and insightful exploration of the vast astronomical literature. LLMs offer a powerful set of capabilities that can address these challenges and transform how astronomers conduct literature reviews. Here are some key ways LLMs are being used:
    1. **Semantic Search:** Unlike traditional keyword searches, LLMs can understand the meaning and context of words and phrases, leading to more relevant search results. They can identify papers that are conceptually related to a research question even if they do not use the exact same keywords.
    2. **Topic Summarization:** LLMs can summarize the key findings and contributions of a large collection of papers, providing researchers with a high-level overview of a specific research area. This allows them to quickly identify the most important papers and understand the current state of knowledge without having to read every paper in detail.

3. **Identifying Trends and Connections:** LLMs can analyze the relationships between different papers, identifying emerging research trends, areas of consensus and disagreement, and potential connections between seemingly disparate subfields.
4. **Automated Reference Extraction and Organization:** LLMs can automate the tedious process of extracting and organizing references, saving researchers significant time and effort.
5. **Personalized Recommendations:** LLMs can learn a researcher's individual interests and recommend relevant papers that they might otherwise have missed.
6. **Generating Research Questions:** By analyzing existing literature, LLMs can even help generate new research questions and identify promising areas for future investigation.
7. **Multilingual Literature Access:** LLMs are increasingly capable of processing and understanding text in multiple languages, breaking down language barriers and making it easier for researchers to access global astronomical literature.

While LLMs can make the future a lot easier, the current scene is not so clean and challenges such as model hallucination and latency prohibit their use in actual research, techniques such as retrieval-augmented generation (RAG) [47], model fine-tuning, model pruning, and model quantization are opening the doors for mess-free and trustworthy LLM uses in astronomy research.

- **Multimodal Large Astronomical Models (MLAMs):** The future of astronomical research is poised to be revolutionized by multimodal large astronomical models (MLAMs) [48], a new generation of AI systems capable of processing and integrating diverse data types from across the electromagnetic spectrum and beyond. Unlike single-modality models, MLAMs can simultaneously analyze images, spectra, and time-series data and even catalog information, along with gravitational wave and neutrino data, creating a holistic understanding of celestial objects and phenomena. By fusing these disparate data streams, MLAMs can overcome the limitations of individual instruments and observation techniques, identifying subtle correlations and patterns that would otherwise remain hidden. For instance, an MLAM could combine optical, infrared, and radio data to characterize a galaxy's star formation history while simultaneously analyzing its morphology and incorporating its redshift from spectroscopic observations. Moreover, these models can leverage the principles of contrastive learning to identify subtle differences and similarities across modalities, enhancing their ability to classify, interpret, and make predictions about various astronomical events. As the volume and complexity of astronomical data continue to grow with next-generation telescopes and multi-messenger facilities, MLAMs, trained on massive multimodal datasets and equipped with advanced architectures like transformers, will play an increasingly crucial role in unlocking the universe's deepest secrets, driving discoveries in areas ranging from exoplanet characterization to the understanding of dark matter and dark energy, and fundamentally transforming our approach to cosmic exploration. The development and deployment of MLAMs represent a paradigm shift in astronomical data analysis, promising a new era of data-driven discovery and a more comprehensive understanding of the universe in all its multifaceted splendor.

- **Generative Models for Astrophysical Simulations:** Generative AI models are emerging as powerful tools for advancing astrophysical simulations, offering the ability to create remarkably realistic synthetic universes that mirror the complexity of observed cosmic phenomena. Techniques like GANs and variational autoencoders (VAEs) [49] have demonstrated significant potential in this domain. GANs, with their generator-discriminator interplay, excel at producing highly detailed and realistic samples, while VAEs offer a probabilistic framework for modeling the underlying data distribution, allowing for controlled generation and interpolation within the latent space of astrophysical parameters. The models can be trained on vast datasets of observational data or outputs from existing simulations to learn the intricate relationships governing the formation and evolution of cosmic structures. Once trained, they can generate novel simulations of astrophysical processes, such as the intricate dance of galaxy formation, the dynamic life cycle of stars from birth to death, or even the extreme physics surrounding black holes and their accretion disks. This capability allows astronomers to explore a wider range of scenarios than traditional computational methods, probe parameter spaces that are difficult to access observationally, and test hypotheses about the underlying physics driving these events. Moreover, these models have shown remarkable ability to create synthetic datasets that can supplement real astronomical data, proving invaluable in scenarios where observational data are sparse, incomplete, or prohibitively expensive to acquire. For example, generating synthetic images of distant galaxies or rare transient events can augment training sets for other AI models focused on classification, object detection, or anomaly detection, ultimately boosting their performance and robustness. This ability to generate synthetic data "on demand" can be particularly helpful for training models that require substantial amounts of labeled data, effectively democratizing access to high-quality training resources for a wider range of astronomical research. In essence, generative models are not just replicating existing data, they are becoming virtual laboratories for exploring the universe, giving a new vision through which to study the cosmos and potentially uncovering new physical insights that might otherwise remain hidden.

- **AI-Powered Education and Outreach:** AI is poised to transform astronomy education and public outreach, fostering a more engaging and personalized learning experience for individuals of all ages and backgrounds. By leveraging the capabilities of advanced AI, particularly in the realm of natural language processing and generative models, educational platforms can move beyond static textbooks and presentations toward dynamic, interactive environments. Imagine AI-powered simulations that allow students to not just passively observe but actively manipulate celestial objects, witnessing the effects of gravity, altering the parameters of stellar evolution, or even designing their own galaxies, all in real time. Such interactive experiences can demystify complex astrophysical concepts, making them far more intuitive and memorable.

Furthermore, AI can act as a personalized tutor, generating customized learning materials tailored to an individual's pace and learning style. It can create practice problems, provide hints and explanations based on specific questions, and adapt the difficulty level as the learner progresses. For example, a student struggling with the concept of redshift could receive targeted explanations using analogies, visualizations, and interactive exercises designed to address his or her specific misunderstandings. AI can even curate personalized content feeds, recommending articles, videos, and activities aligned with a learner's particular interests, whether exoplanets, black holes, or the history of space exploration. In public outreach, AI-powered chatbots can engage in dialogues with museum visitors or online learners, answering questions about astronomical phenomena, providing up-to-date information on space missions, and even offering guidance on stargazing. By making astronomy more accessible, interactive, and tailored to individual needs, AI has the potential to ignite a passion for science in a broader audience, fostering scientific literacy and inspiring the next generation of astronomers, astrophysicists, and space explorers, ultimately cultivating a more scientifically informed and engaged public. The ability of AI to personalize and adapt the educational experience will democratize learning and break down barriers that have traditionally limited participation in STEM fields, ultimately guiding in a new era of astronomical engagement.

- **Quantum Machine Learning for Astronomy:** Being still in its nascent stages, quantum ML (QML) holds the tantalizing prospect of revolutionizing astronomical data analysis by harnessing the unique capabilities of quantum computers [50]. These machines, which leverage the principles of quantum mechanics, have the potential to outperform classical computers on specific computational tasks. In the context of astronomy, where datasets are rapidly growing in size and complexity, QML algorithms could unlock new avenues for discovery by dramatically accelerating ML tasks or enabling entirely new approaches that are intractable for classical systems. One area of promise lies in quantum-enhanced optimization algorithms. Many ML problems, including those common in astronomy like parameter estimation and model fitting, boil down to finding the optimal solution within a vast search space. Quantum algorithms like the quantum approximate optimization algorithm (QAOA) or quantum versions of gradient descent could potentially navigate these complex landscapes much faster than their classical counterparts, leading to more efficient model training. Another exciting avenue is the application of quantum ML models themselves. For instance, quantum neural networks (QNNs), which utilize quantum circuits as their building blocks, might be able to identify complicated patterns and correlations in astronomical data that elude classical neural networks due to their inherent ability to explore higher-dimensional feature spaces. Moreover, quantum feature mapping techniques could transform classical astronomical data into quantum states, potentially revealing hidden structures and relationships when processed using quantum algorithms. Imagine applying these techniques to analyze the cosmic microwave background for subtle signatures of early universe physics or using quantum clustering algorithms to identify new faint structures in galaxy surveys. While the realization of large-scale fault-tolerant quantum computers is still on the horizon, research into QML for astronomy is already underway. Scientists are exploring the application of existing quantum hardware and simulators to tackle specific astronomical problems, developing new quantum algorithms tailored to astrophysical datasets, and investigating the theoretical foundations of QML. This includes studying the potential speedups offered by quantum algorithms for tasks like spectral analysis, image processing, and gravitational wave data analysis. Although significant challenges remain in building and programming quantum computers, as well as in developing quantum-resistant ML models, the potential benefits of QML for astronomy are too significant to ignore. This burgeoning field promises to guide in a new era of cosmic discovery, where the marriage of quantum mechanics and artificial intelligence could unlock some of the universe's most profound secrets.

1.12 Conclusion

The fusion of artificial intelligence and astronomy is not just an incremental advancement—it is a paradigm shift in our quest to understand the universe. As AI-driven methodologies continue to evolve, they are redefining the very fabric of astronomical exploration, unlocking mysteries that were once beyond the reach of human capability. With ML, DL, and RL at the helm, researchers are now able to sift through vast cosmic datasets with unmatched precision, uncovering celestial phenomena that would otherwise remain hidden in the depths of space.

The impact of AI in astronomy is nothing short of transformative. From classifying galaxies and detecting exoplanets to analyzing gravitational waves and simulating cosmic evolution, AI has become an indispensable tool in modern astrophysics. AI-powered systems are enhancing the efficiency of astronomical surveys, automating the discovery of transient events, and even optimizing space mission strategies. The recent advancements in DL models, particularly CNNs and RNNs, have enabled astronomers to decipher complex spatial and temporal patterns within astronomical observations, pushing the boundaries of discovery. AI's ability to detect anomalies in massive datasets has also led to unexpected discoveries, such as identifying new types of celestial bodies and phenomena that challenge existing astrophysical theories.

However, as with any technological revolution, challenges persist. The accuracy and fairness of AI models hinge upon the quality and diversity of training datasets. Biases in data can lead to skewed results, making it imperative to develop robust methodologies that mitigate such risks. Interpretability remains a crucial hurdle—scientific progress demands not only predictive accuracy but also a clear understanding of why an AI system makes specific decisions. Computational constraints also pose a challenge, especially given the exponentially growing volume of astronomical data. Addressing these limitations will require not only improved algorithms but also more powerful computational infrastructures and interdisciplinary collaborations between astrophysicists and AI researchers. Ethical considerations also come into play, as reliance on AI models necessitates responsible development to avoid unintended consequences in research and space exploration.

Looking ahead, the future of AI in astronomy brims with boundless potential. As next-generation observatories like the Vera C. Rubin Observatory and the Square Kilometer Array come online, they will generate unprecedented amounts of data, necessitating real-time AI processing to extract meaningful insights. The integration of AI with quantum computing and neuromorphic hardware could further propel astronomical research into uncharted territories, enabling simulations and analyses at scales previously unimaginable. AI-powered robotic telescopes and automated spacecraft navigation systems are set to revolutionize space exploration, ensuring efficient utilization of resources and enhanced scientific return from interplanetary missions. Moreover, AI-driven citizen science platforms could democratize astronomy, allowing enthusiasts and researchers alike to contribute to groundbreaking discoveries, fostering a more inclusive and collaborative scientific environment.

In many ways, the rise of AI in astronomy mirrors the technological leaps that defined previous eras of scientific progress—akin to the invention of the telescope or the development of space probes. It represents a new frontier where computational intelligence meets cosmic curiosity, bridging the gap between theoretical physics and empirical observations. As humanity continues to explore the vastness of space with AI as its guide, one thing is certain: the mysteries of the universe are far from fully unraveled, and the best is yet to come. AI's potential to revolutionize our understanding of dark matter, dark energy, and the origins of the universe itself makes it one of the most promising tools in the history of astrophysical research. By embracing innovation, fostering interdisciplinary collaborations, and addressing challenges with ethical and scientific rigor, AI will remain at the forefront of astronomical exploration, inspiring generations to delve deeper into the cosmos and uncover its secrets.

References

1. Barbieri, C. and Bertini, I., *Fundamentals of Astronomy (2nd ed.)*, CRC Press, Boca Raton, 2020.
2. Sheikh, H., Prins, C., Schrijvers, E., Artificial Intelligence: Definition and Background, in: *Mission AI*. Research for Policy, Springer, Cham, 2023.
3. Petroff, E., Hessels, J.W.T., Lorimer, D.R., Fast radio bursts. *Astron. Astrophys. Rev.*, 27, 4, 2019.
4. Grossi, E. and Buscema, M., Introduction to Artificial Neural Networks. *Eur. J. Gastroenterol. Hepatol.*, 19, 1046–1054, 2008.
5. Sarker, I.H., Machine Learning: Algorithms, Real-World Applications, and Research Directions. *SN Comput. Sci.*, 2, 3, 160, 2021.
6. LeCun, Y., Bengio, Y., Hinton, G., Deep Learning. *Nature*, 521, 436–444, 2015.
7. El Naqa, I. and Murphy, M.J., What Is Machine Learning?, in: *Machine Learning in Radiation Oncology*, Springer, Cham, 2015.
8. Najafabadi, M.M., Villanustre, F., Khoshgoftaar, T.M., Seliya, N., Wald, R., Muharemagic, E., Deep learning applications and challenges in big data analytics. *J. Big Data*, 2, 1, 2015.
9. da Silva, I.N., Hernane Spatti, D., Andrade Flauzino, R., Liboni, L.H.B., dos Reis, A., Artificial Neural Network Architectures and Training Processes, Springer, Cham, 2017.

10. O'Shea, K. and Nash, R., An Introduction to Convolutional Neural Networks, *arXiv:1511.08458* [cs.NE], 2015.
11. Sherstinsky, A., Fundamentals of Recurrent Neural Network (RNN) and Long Short-Term Memory (LSTM) Network. *Physica D*, 404, 132306, 2020.
12. Staudemeyer, R.C. and Morris, E.R., Understanding LSTM – A Tutorial into Long Short-Term Memory Recurrent Neural Networks, *arXiv:1909.09586* [cs.NE], 2019.
13. Buffet, O., Pietquin, O., Weng, P., Reinforcement Learning. *arXiv:2005.14419* [cs.LG], 2020.
14. Fluke, C.J. and Colin, J., Surveying the reach and maturity of machine learning and artificial intelligence in astronomy. *Wiley Interdiscip. Rev.: Data Min. Knowl. Discov.*, 10, 2, e1349, 2020.
15. Pedregosa, F., Varoquaux, G., Gramfort, A., Michel, V., Thirion, B., Grisel, O., Blondel, M., Prettenhofer, P., Weiss, R., Dubourg, V., Vanderplas, J., Passos, A., Cournapeau, D., Brucher, M., Perrot, M., Duchesnay, E., Scikit-Learn: Machine Learning in Python. *J. Mach. Learn. Res.*, 12, 2825–2830, 2011.
16. Abadi, M., Barham, P., Chen, J., Chen, Z., Davis, A., Dean, J., Devin, M., Ghemawat, S., Irving, G., Isard, M., Kudlur, M., Levenberg, J., Monga, R., Moore, S., Murray, D.G., Steiner, B., Tucker, P., Vasudevan, V., Warden, P., Wicke, M., Yu, Y., Zheng, X., TensorFlow: A System for Large-Scale Machine Learning. *Proc. 12th USENIX Symp. Oper. Syst. Des. Implement (OSDI)*, pp. 265–283, 2016.
17. Paszke, A., Gross, S., Massa, F., Lerer, A., Bradbury, J., Chanan, G., Killeen, T., Lin, Z., Gimelshein, N., Antiga, L., Desmaison, A., Kopf, A., Yang, E., DeVito, Z., Raison, M., Tejani, A., Chilamkurthy, S., Steiner, B., Fang, L., Bai, J., Chintala, S., PyTorch: An Imperative Style, High-Performance Deep Learning Library. *Adv. Neural Inf. Process. Syst.*, 32, 8024–8035, 2019.
18. Astropy Collaboration, Robitaille, T.P., Tollerud, E.J., Greenfield, P., Droettboom, M., Bray, E., Aldcroft, T., Davis, M., Ginsburg, A., Price-Whelan, A.M., Kerzendorf, W.E., Conley, A., Crighton, N., Barbary, K., Muna, D., Ferguson, H., Grollier, F., Parikh, M.M., Nair, P.H., Günther, H.M., Streicher, O., The Astropy Project: Building an Open-science Project and Status of the v2.0 Core Package. *Astron. J.*, 156, 3, 123, 2018.
19. Ginsburg, A., Sipőcz, B.M., Brasseur, C.E., Cowperthwaite, P.S., Craig, M.W., Deil, C., Guillochon, J., Guzman, G., Liedtke, S., Lim, P.L., Lockhart, K.E., Mommert, M., Morris, B.M., Norman, H., Parikh, M., Persson, M.V., Robitaille, T.P., Segovia, J.-C., Singer, L.P., Tollerud, E.J., de Val-Borro, M., Valtchanov, I., Woillez, J., The Astroquery Collaboration, and a subset of the Astropy Collaboration, Astroquery: An Astronomical Web-Querying Package in Python. *Astron. J.*, 157, 98, 2019.
20. Joye, W.A. and Mandel, E., SAOImage DS9: A Reference Guide. Astron. Data Anal. Softw. Syst. (ADASS). *ASP Conf. Ser.*, vol. 238, 2001.
21. York, D.G., Adelman, J., Anderson, J.E., Anderson, S.F., Annis, J., Bahcall, N.A., Bakken, J.A., Barkhouser, R., Bastian, S., Berman, E., Boroski, W.N.,

Bracker, S., Branly, R., Carey, L., Carr, M.A., *et al.*, The Sloan Digital Sky Survey: Technical Summary. *Astron. J.*, 120, 1579–1587, 2000.

22. Gaia Collaboration, Prusti, T., de Bruijne, J.H.J., Brown, A.G.A., Vallenari, A., Babusiaux, C., Bailer-Jones, C.A.L., *et al.*, The Gaia Mission. *Astron. Astrophys.*, 595, A1, 2016.
23. Bellm, E.C., Kulkarni, S.R., Graham, M.J., Dekany, R., Smith, R.M., Riddle, R., Masci, F.J., *et al.*, The Zwicky Transient Facility: System Overview, Performance, and First Results. *Publ. Astron. Soc. Pac.*, 131, 995, 018002, 2019.
24. Helou, G., Madore, B.F., Schmitz, M., Bicay, M.D., Wu, X., Bennett, J., Gelles, J., The NASA/IPAC Extragalactic Database. *Publ. Astron. Soc. Pac.*, 103, 668, 1033–1068, 1991.
25. Space Telescope Science Institute (STScI), Mikulski Archive for Space Telescopes (MAST) Portal, https://mast.stsci.edu/portal/Mashup/Clients/Mast/Portal.html.
26. Li, L., Zhang, Y., Zhao, Y., k-Nearest Neighbors for Automated Classification of Celestial Objects. *Sci. China Ser. G-Phys. Mech. Astron.*, 51, 916–922, 2008.
27. Barnwal, R. and S., K., Morphological Galaxy Classification Using Convolutional Neural Networks on FPGA. *2024 IEEE Space, Aerospace and Defence Conference (SPACE)*, Bangalore, India, 2024.
28. Walmsley, M., Lintott, C., Géron, T., Kruk, S., Krawczyk, C., Willett, K.W., Bamford, S., Kelvin, L.S., Fortson, L., Gal, Y., Keel, W., Masters, K.L., Mehta, V., Simmons, B.D., Smethurst, R., Smith, L., Baeten, E.M., Macmillan, C., Galaxy Zoo DECaLS: Detailed Visual Morphology Measurements from Volunteers and Deep Learning for 314,000 Galaxies. *Mon. Not. R. Astron. Soc.*, 509, 3, 3966–3988, 2022.
29. ESA/Hubble, NGC 4414 – A Typical Spiral Galaxy, Eur. Space Agency, 1999, URL: https://esahubble.org/images/heic9902o/.
30. Zhang, Y., Nord, B., Pagul, A., Lepori, M., Noise2Astro: Astronomical Image Denoising with Self-supervised Neural Networks. *Res. Notes AAS*, 6, 9, 187, 2022.
31. Ivezić, Ž., Kahn, S.M., Tyson, J.A., *et al.*, LSST: From Science Drivers to Reference Design and Anticipated Data Products. *Astrophys. J.*, 873, 2, 111, 2019.
32. Wang, C., Hu, X., Zhang, Y., *et al.*, StarWhisper Telescope: Agent-Based Observation Assistant System to Approach AI Astrophysicist, *arXiv preprint, arXiv:2412.06412*, 2024.
33. Liu, X., Liu, Y., Yang, L., Wu, S., Jiang, R., Xiang, Y., AstroDLLC: Efficiently Reducing Storage and Transmission Costs for Massive Solar Observation Data *via* Deep Learning-based Lossless Compression. *Publ. Astron. Soc. Pac.*, 136, 075001, 2024.
34. Habib, T.M.A., Artificial intelligence for spacecraft guidance, navigation, and control: A state-of-the-art. *AS*, **5**, 503–521, 2022.

35. Jara-Maldonado, M., Alarcon-Aquino, V., Rosas-Romero, R., *et al.*, Transiting exoplanet discovery using machine learning techniques: A survey. *Earth Sci. Inform.*, 13, 573–600, 2020.
36. Russo, A. and Lax, G., Using artificial intelligence for space challenges: A survey. *Appl. Sci.*, 12, 5106, 2022.
37. Krenn, M., Drori, Y., Adhikari, R.X., Digital Discovery of Interferometric Gravitational Wave Detectors, *arXiv preprint, arXiv:2312.04258*, 2023.
38. Giles, D. and Walkowicz, L., Systematic serendipity: a test of unsupervised machine learning as a method for anomaly detection. *Mon. Not. R. Astron. Soc.*, 484, 834–849, 2018.
39. Woods, P., Melting moons make Tabby's Star murky. *Nat. Astron*, 3, 11, 961, 2019.
40. Ganju, S., Hatua, A., Jenniskens, P., *et al.*, AI-Enhanced Data Processing and Discovery Crowd Sourcing for Meteor Shower Mapping, *arXiv preprint, arXiv:2308.02664*, 2023.
41. M., J.M., Challenges and Approaches for Mining Astronomical Data and Complex Models. *2019 IEEE International Conference on Big Data, Cloud Computing, Data Science & Engineering (BCD)*, 2019.
42. F., X., v. N., T., Such, J.M., Coté, M., C., N., Bias and Discrimination in AI: A Cross-Disciplinary Perspective. *IEEE Technol. Soc. Mag.*, 40, 2, 72–80, 2021.
43. Xu, F., Uszkoreit, H., Du, Y., Fan, W., Zhao, D., Zhu, J., Explainable AI: A Brief Survey on History, Research Areas, Approaches and Challenges. *Natural Language Processing and Chinese Computing. NLPCC 2019.* Lecture Notes in Computer Science, vol. 11839, Springer, Cham, 2019.
44. Huang, K., Hu, T., Cai, J., Pan, X., Hou, Y., Xu, L., Wang, H., Zhang, Y., Cui, X., Artificial Intelligence in Astronomical Optical Telescopes: Present Status and Future Perspectives. *Universe*, 10, 5, 210, 2024.
45. Weng, J., Symbolic Models and Emergent Models: A Review. *IEEE Trans. Auton. Ment. Dev.*, 4, 1, 29–53, 2012.
46. MacGlashan, J., Ho, M.K., Loftin, R., Peng, B., Wang, G., Roberts, D.L., Taylor, M.E., Littman, M.L., Proceedings of the 34th International Conference on Machine Learning. *PMLR*, vol. 70, pp. 2285–2294, 2017.
47. Fan, W., Ding, Y., Ning, L., Wang, S., Li, H., Yin, D., Chua, T.-S., Li, Q., A Survey on RAG Meeting LLMs: Towards Retrieval-Augmented Large Language Models. *Proc. 30th ACM SIGKDD Conf. on Knowledge Discovery and Data Mining (KDD '24)*, Association for Computing Machinery, New York, NY, USA, pp. 6491–6501, 2024.
48. Rizhko, M. and Bloom, J.S., Self-supervised Multimodal Model for Astronomy. *NeurIPS 2024 Workshop: Foundation Models for Science – Progress, Opportunities, and Challenges*, 2024.
49. Ravichandran, P., Pushing Boundaries with Deep Generative Models: Innovations and Applications of VAEs and GANs. *Adv. Deep Learn. Tech.*, 2, 1, 37–48, 2022.
50. Kordzanganeh, M., Utting, A., Scaife, A., Quantum Machine Learning for Radio Astronomy, *arXiv, 2112.02655*, 2021.

2

Data Mining and Machine Learning in Astrophysics

Gissmol Saji[1]* and Sanjay Singh Bisht[2]

[1]*Department of Electrical Engineering and Information Technology, Ruhr University Bochum, Bochum, Germany*
[2]*Department of Physics, Radhe Hari Govt. PG College, Kashipur (U S Nagar), Uttarakhand, India*

Abstract

In recent years, data mining and machine learning (ML) have revolutionized numerous fields, including astrophysics, by efficiently analyzing the vast and complex datasets that are collected from telescopes, satellites, other observational devices, and simulations. These technologies have proven beneficial in automating and accelerating tasks that are traditionally labor-intensive and time-consuming for humans. Applications in astrophysics span a wide range of areas, including exoplanet detection, gravitational wave analysis, galaxy classification, and transient event identification. By employing advanced techniques such as deep learning, supervised learning, clustering, and anomaly detection, researchers can extract meaningful patterns from large and often noisy datasets, gaining critical insights into the dynamic and diverse nature of celestial phenomena. Machine learning techniques play a pivotal role in addressing challenges associated with the analysis of big data in astronomy, for instance, supervised learning aids in classifying stellar types while clustering algorithms group similar galaxies based on their properties. Additionally, anomaly detection algorithms help identify rare or unusual phenomena, such as certain types of supernovae or fast radio bursts. Major advancements have been made in image processing and pattern recognition by the use of deep neural networks (DNNs) in astronomical datasets. These models excel at identifying faint exoplanetary signals in light curves, classifying galaxies based on their morphology, and isolating gravitational wave signals from detector noise. Despite these advancements, challenges remain in the field. The sheer volume of data generated by modern observatories, often in the order of terabytes per night,

**Corresponding author*: Gissmol.Saji@edu.ruhr-uni-bochum.de

Yogesh Chandra, Manjuleshwar Panda and Mahesh Chandra Mathpal (eds.) The Intelligent Universe: AI's Role in Astronomy, (47–86) © 2025 Scrivener Publishing LLC

requires scalable computational frameworks and high-performance computing resources. Observational noise, imbalanced datasets, and the interpretability of complex ML models further complicate the analytical process. A solution to these obstacles demands powerful algorithms plus new feature methods with better pre-processing methods to achieve dependable and precise outcomes. Looking ahead, the integration of data mining and ML with theoretical astrophysical model might bridge the gap between observations and simulations. The development of autonomous observatories powered by ML algorithms could further transform the field, enabling telescopes to make real-time decisions about what to observe and analyze. In general, improved simulations of cosmic phenomena, such as galaxy formation and dark matter distribution, combined with predictive models hold the potential to uncover fundamental truths about the universe. This chapter highlights the transformative impact of data mining and machine learning in astrophysics and showcases their role in better understanding the universe. By leveraging these cutting-edge technologies, researchers are not only enhancing our understanding of the universe but also paving the way for groundbreaking discoveries in cosmology, astroinformatics, and astrobiology.

Keywords: Data mining, machine learning, deep learning, exoplanet detection, gravitational waves, galaxy classification, transient event identification, autonomous observatories

2.1 Introduction

The concept of intelligent machines has its roots in ancient myths and early philosophical ideas. In the 20th century, Alan Turing established the foundational principles with his Turing Test (1950), where he famously posed the question, "Can machines think?" Artificial intelligence (AI) is modeled after the human brain, which includes various methods, from traditional rule-based systems to advanced algorithms. John McCarthy, a computer and cognitive scientist, originally used the term "artificial intelligence" in 1956 as part of the Dartmouth Summer Research Project on Artificial Intelligence [1]. The idea of machine learning was first introduced by Arthur Samuel in 1959 when he developed a computer program that gradually became more proficient in checkers. Samuel proved that a computer could be taught to "learn" from experience and adhere to preset rules, eventually surpassing the programmer. His work represents a significant milestone in teaching machines to improve through experience, and he coined the term "machine learning." In the 1990s and 2000s, the rise of machine learning and big data helped AI make major strides in areas like speech recognition, image processing, and pattern recognition [1].

The development in deep learning and neural networks has led to significant advancements in the fields of image recognition, face detection, natural language processing, and automation. Figure 2.1 illustrates the evolution of AI/ML, balancing progress with concerns about ethics, including issues like bias, privacy, and safety, making them truly understand, reason, and adapt, bringing it closer to human-like intelligence. Implementing AI/ML in astrophysics has made a rapid expansion in revolutionizing the field in astronomical data interpretation, driven by advances in observational technology, and has necessitated new approaches to data analysis. The vast and abstract data generated by telescopes, satellites, and simulations are growing at an unprecedented rate, presenting challenges in storage, processing, and interpretation. Traditional manual methods of analysis are no longer sufficient for handling such large datasets efficiently. Before AI/ML, astrophysicists manually cleaned and processed data from telescopes, adjusting for errors and imperfections. They used statistical methods to interpret the data, fitting models to observations and analyzing periodic patterns. Simulations helped predict phenomena that could not be directly observed. Pattern recognition largely relied on human expertise, with scientists visually inspecting data for anomalies. Data visualization tools helped make complex results easier to interpret and share. As datasets grew larger, AI began to assist by automating tasks and revealing patterns faster, marking a new era in astrophysics. Moreover, data mining and ML have

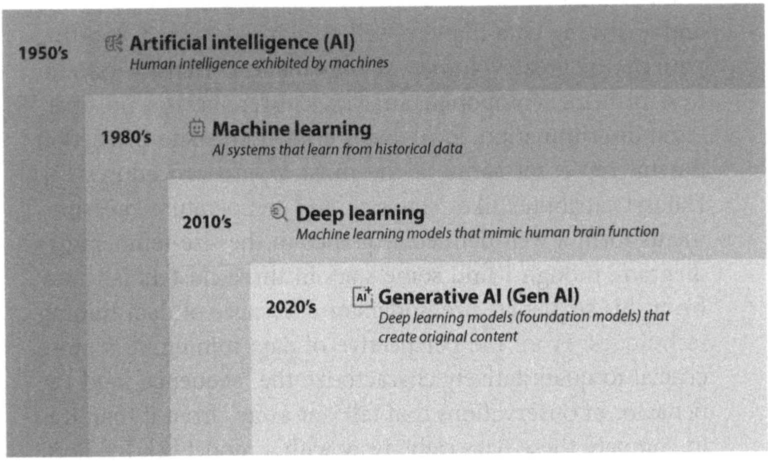

Figure 2.1 Evolution of artificial intelligence (Image credit: International Business Machines (IBM) Corporation).

emerged as powerful solutions, enabling astronomers to uncover hidden patterns, classify celestial objects, and make precise predictions about cosmic phenomena. ML algorithms, particularly deep learning models, can analyze complex datasets with remarkable accuracy, automating tasks that were once labor-intensive and time-consuming.

This chapter explores the role of data mining (DM) and machine learning (ML) in astrophysics, covering fundamental techniques, applications, and future directions. It aims to deliver a thorough overview of how these advanced computational tools are transforming our understanding of the universe.

2.2 Foundations of Data Mining and Machine Learning

2.2.1 Data Science and Its Components

Data science is an interdisciplinary method of deriving meaning from structured and unstructured data through statistical methods, computational methods, and knowledge from the discipline. Data science in astrophysics allows the systematic investigation of large datasets in an attempt to find meaningful patterns and trends. The key components of data science include:

- **Data mining** represents a collection of methods for modeling and analyzing data that are well-structured, such as locating patterns in great volumes of information. Density estimation, principal component analysis, cluster analysis, unsupervised discrimination, local linear embedding, and projection pursuit represent some of the most typical procedures [2]. Dataset attributes like "star size and temperature measurements form a well-defined sequence in the size-temperature diagram, though I find some stars in three clusters far away from this sequence" are understood because of data mining techniques. From the perspective of data mining, it is more crucial to quantitatively characterize the "sequence" and the behavior of observations that fall "far away" from it than it is to compare these data right away with a model [2]. In short, data mining is about what data are telling us themselves. Figure 2.2 demonstrates an outline of a data mining process involved in any possible scenario.

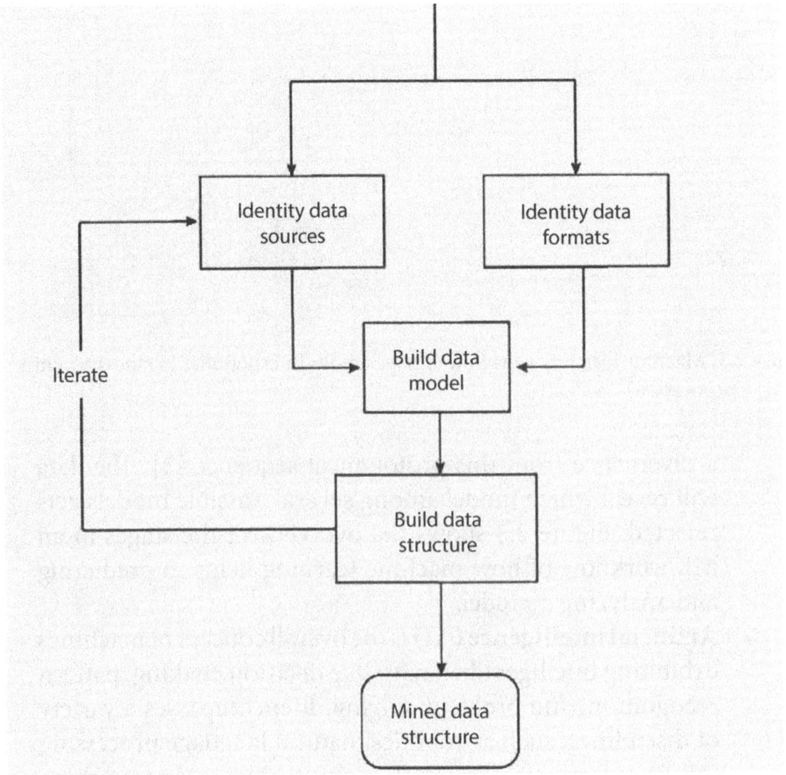

Figure 2.2 Outline of a data mining process (Image credit: International Business Machines (IBM) Corporation).

- **Machine learning (ML)** is a subset of AI that enables computers to learn from data and make predictions. By comparing machine learning models to models for data behavior, such as different regression techniques, supervised classification techniques, maximum likelihood estimators, and the Bayesian approach, ML models gradually increase their accuracy as they are exposed to more data. They are frequently referred to as data-based statistical inferences, inference procedures, or simply "fitting." A physical stellar structure model enables scientists to predict both the position and shape of the main sequence stars in size-temperature diagrams, as demonstrated in the data mining example above. By combining with galaxy formation models, the model predicts the distribution pattern of stars that shows

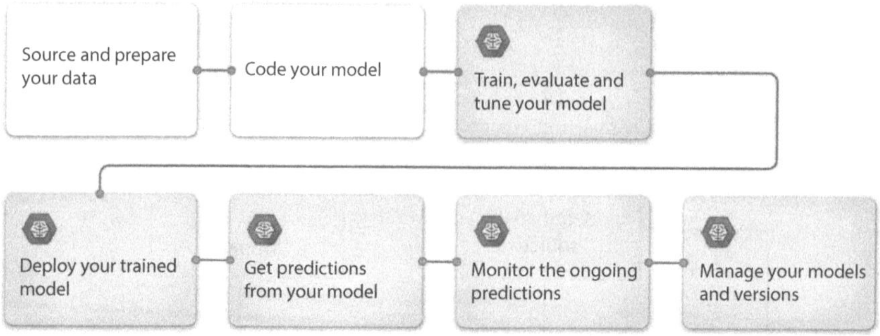

Figure 2.3 Machine learning workflow (Image credit: International Business Machines (IBM) Corporation).

a divergence from this prototypical sequence [2]. The data will reveal which model among several possible models gets rejected. Figure 2.3 shows the overview of the stages in an ML workflow of how machine learning helps in predicting and analyzing a model.

- **Artificial intelligence (AI)** is the overall concept of machines exhibiting intelligent behavior like decision-making, pattern recognition, and problem-solving. It encompasses a variety of disciplines, such as robotics, natural language processing (NLP), information retrieval, computer vision, and machine learning [2].
- **Big data analytics** is the management and processing of massive astronomical datasets, often requiring high-performance computing and cloud-based infrastructures. With the vast volume and complexity of data from different wavelengths, traditional methods are no longer sufficient. Machine learning is often used to identify patterns, classify objects, and detect anomalies, such as discovering new exoplanets or analyzing galaxy structures. Big data also enables real-time analysis, allowing scientists to respond quickly to events like supernovae or gamma-ray bursts. Overall, big data helps astrophysicists make faster discoveries and gain deeper insights into the universe.

2.2.2 Classical Machine Learning Versus Deep Learning

Both classical machine learning (ML) and deep learning (DL) are powerful approaches in data-driven analysis. Classical ML is often preferred

Table 2.1 Key features of classical machine learning versus deep learning.

Features	**Classical machine learning**	**Deep learning**
Feature Extraction	Requires manual feature engineering	Learns features automatically
Data Requirements	Works well with small- to medium-sized datasets	Requires large labeled datasets
Computation Power	Can run on standard CPUs	Needs GPUs/TPUs for efficiency
Interpretability	Easier to interpret and explain	Functions as a "black box"
Example Algorithms	Decision trees, SVM, random forest	CNNs, RNNs, GANs

for structured data and small datasets, while deep learning shines in tasks involving large-scale unstructured data, such as image processing and speech recognition. The following Table 2.1 summarizes the key differences between classical machine learning and deep learning:

Although DL, ANN, ML, and artificial intelligence (AI) are related technologies, the phrases are commonly used interchangeably, which frequently confuses their distinctions. The most comprehensive system is AI. A subset of AI is machine learning (ML). Deep learning (DL) is a branch of machine learning (ML), and the foundation of deep learning techniques is neural networks. A deep learning algorithm requires more than three node layers, or depth, which is what separates a single neural network from a deep learning algorithm [3].

Classical machine learning (ML) and deep learning (DL) differ in data interpretation.

- **Classical ML** requires manually extracted features and includes algorithms like decision trees, support vector machines (SVMs), and random forests. These techniques rely on domain expertise to define features for classification and prediction tasks [4].
- **Deep learning:** Deep learning uses multilayered neural networks to learn automatic representations from data and therefore is most suited for image classification, time-series

analysis, and astrophysical spectral classification [5]. Deep learning is capable of directly handling raw and high-dimensional data with minimal preprocessing. Deep learning is very efficient at finding complex patterns, e.g., detection of weak exoplanet signals or morphological galaxy classification [6]. Additionally, it automates feature extraction, reducing human effort and minimizing potential biases. However, deep learning also comes with challenges. It requires large labeled datasets for effective training, and its high computational demands necessitate specialized hardware like GPUs and TPUs. Another issue is interpretability, as deep learning models function as a "black box," making it hard to know how they arrive at conclusions.

2.2.3 Data Mining Software and Tools

Several software tools (Table 2.2) and libraries have been developed to facilitate data mining and ML in astrophysics. These tools help researchers analyze large datasets efficiently, uncovering new insights into celestial phenomena. The table below provides a comparison of commonly used data mining tools in astrophysics [7]:

Table 2.2 Data mining tools used in astrophysics.

Tool	Description	Key features
AstroML	Python library for ML in astronomy	Includes clustering, classification, and regression algorithms
Weka	Open-source ML software	Offers data visualization and various ML algorithms
StatCodes	Statistical tools for astrophysical applications	Helps in spectral classification and redshift estimation
VOStat	Statistical analysis tool for the virtual observatory	Supports advanced statistical modeling
DAME	Cloud-based data mining framework	Designed for handling large astronomical datasets

2.3 Machine Learning Applications in Astrophysics

2.3.1 Supervised Learning Techniques

Supervised learning algorithms work on labeled datasets to train models for classification and regression tasks [8]. In astrophysics, supervised learning plays a crucial role in:

- **Classification:** Assigning labels to celestial objects (e.g., stars, galaxies, quasars) based on their spectral and photometric properties. Techniques like support vector machines, random forests, and artificial neural networks (ANNs) are commonly used.
- **Regression:** Predicting continuous values such as stellar mass, temperature, or redshift using algorithms like K-nearest neighbors, decision trees, and Gaussian process regression. These models help astronomers estimate key parameters of celestial objects with high precision.

Table 2.3 below highlights common supervised learning techniques used in astrophysics:

2.3.2 Unsupervised Learning Techniques

Unsupervised learning techniques help identify patterns in unlabeled datasets, making them useful for exploratory astrophysical studies [8]. Table 2.4 below presents common unsupervised learning techniques and their astrophysical applications:

Table 2.3 Supervised learning techniques and their applications.

Technique	**Application**
Support Vector Machines (SVMs)	Classifying stars, galaxies, and quasars
Random Forest	Predicting stellar properties
Decision Trees	Estimating redshifts
K-Nearest Neighbors	Identifying galaxy clusters
Neural Networks	Detecting exoplanet signals

Table 2.4 Unsupervised learning techniques and their applications.

Technique	Application
K-Means Clustering	Grouping galaxies based on spectral properties
DBSCAN	Identifying structures in large-scale cosmic data
Principal Component Analysis (PCA)	Reducing dimensionality in spectral classification
Isolation Forest	Detecting anomalies such as supernovae
Gaussian Mixture Models	Understanding the distribution of galaxy populations

- **Clustering:** Algorithms like K-Means and DBSCAN group similar celestial objects based on properties such as brightness and morphology, aiding in the study of the evolution of galaxy and structure formation at a large scale [9].
- **Anomaly Detection:** Techniques like One-Class SVM and isolation forest identify rare astronomical events, such as supernovae, fast radio bursts (FRBs), and gravitational wave signals, enabling their timely detection and classification [10].

2.3.3 Structure of Training Data: Supervised and Unsupervised

The fundamental difference between supervised and unsupervised learning is that there is a presence of labeled data in supervised learning and their absence in unsupervised learning. Figure 2.4 demonstrates the difference between supervised versus unsupervised learning by looking at the structure of the training data. In supervised learning, the dataset contains predefined output labels. This model's goal is to accurately map inputs to outputs. In contrast, unsupervised learning deals with datasets that lack labeled outputs. This model's objective is to discover possible patterns within the data [11].

The process of supervised learning involves several key steps. First, during the training phase, the model is fed labeled data. That implies that each input variable is associated with a known output variable. The model

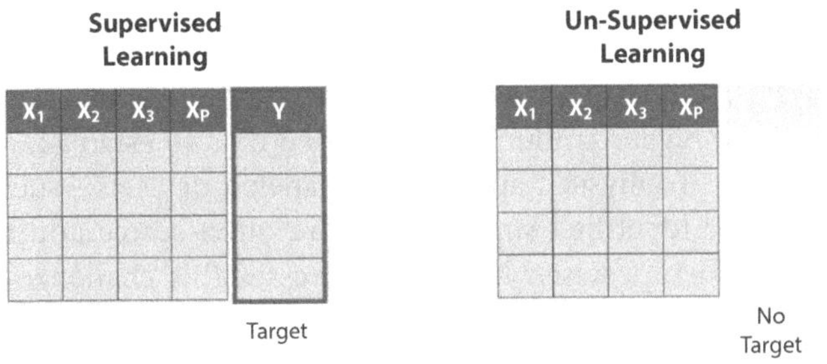

Figure 2.4 Difference between supervised and unsupervised learning. (Image credit: Sharp Sight Labs).

attempts to learn the relationship between the inputs and the outputs. Once training is complete, the model moves to the prediction phase, where it applies the learned function to new unseen inputs to generate predictions. The accuracy of these predictions is then evaluated by comparing them to the actual target values. The term "supervised learning" originates from the fact that the presence of output labels essentially acts as a supervisor during the training process. The model continuously compares its predictions to the correct outputs and refines itself to minimize errors.

Whereas in an unsupervised learning process, the model is given raw input data and must analyze them to uncover hidden patterns or groupings. Unlike supervised learning, there is no external guidance or feedback from predefined labels. The model explores the data independently and extracts meaningful insights. Unsupervised learning is primarily used for clustering and dimensionality reduction [9]. Clustering seeks to gather data points that share common intrinsic features through their inherent characteristics. The objective of dimensionality reduction consists of diminishing input features without losing important data points. Such a method proves beneficial for working with extensive datasets containing multiple variables during image processing operations.

The term "unsupervised learning" reflects the fact that there are no output labels (no target) to guide the learning process. The model operates independently, searching for patterns within the input data [9]. Although supervised learning can be very difficult when it comes to classifying large data, the outcomes are very reliable and accurate. Unsupervised learning, on the other hand, is capable of processing massive amounts of data in real time. However, there is a greater chance of errors in outcomes and a lack of transparency regarding the clustering of data. Herein lies the role of semi-supervised learning.

2.3.4 Semi-Supervised Learning Techniques

In astrophysics, semi-supervised learning utilizes both smaller amounts of labeled data and bigger amounts of unlabeled data to establish predictive models [8]. In astrophysical applications, labeled datasets—such as classified galaxies or identified supernovae—are often scarce and expensive to produce. Semi-supervised learning addresses this challenge by leveraging the abundance of unlabeled data, such as images or spectra from telescopes, while utilizing a smaller set of labeled examples to guide the learning process. This approach is particularly valuable in tasks like object classification, anomaly detection, and feature extraction, where it can reduce the reliance on manual annotation. For instance, it can enhance the classification of galaxy morphologies or aid in the identification of rare events, such as gravitational wave signals, even when labeled examples are limited. By utilizing both labeled and unlabeled data, semi-supervised learning enables astrophysicists to optimize the use of vast datasets from modern observatories, facilitating more accurate models and accelerating the pace of discovery.

2.4 Role of Machine Learning in Key Astrophysical Research Areas

2.4.1 Exoplanet Detection

Since the discovery of the first exoplanets, their detection has become a key focus of astrophysical research, with over 4,000 exoplanets identified using methods such as radial velocity, astrometry, direct imaging, and gravitational microlensing. Modern scientists confirm most exoplanets through the transit method because this method reveals periodic brightness reductions when a planet crosses in front of its star [12]. Despite its effectiveness, this method has limitations, including the need for nearly perfect orbital alignment and a high rate of false positives. Exoplanet identification involves detecting threshold crossing events (TCEs), which are manually and automatically vetted to eliminate noise and false positives. Traditional approaches, such as the box-fitting least squares (BLS) algorithm, have inherent limitations, prompting an increasing shift toward machine learning for automated vetting [13].

Early models relied on tree-based methods like Robovetter and Autovetter, but deep learning approaches, such as Astronet, have since gained prominence due to their superior accuracy [14]. Figure 2.5 illustrates

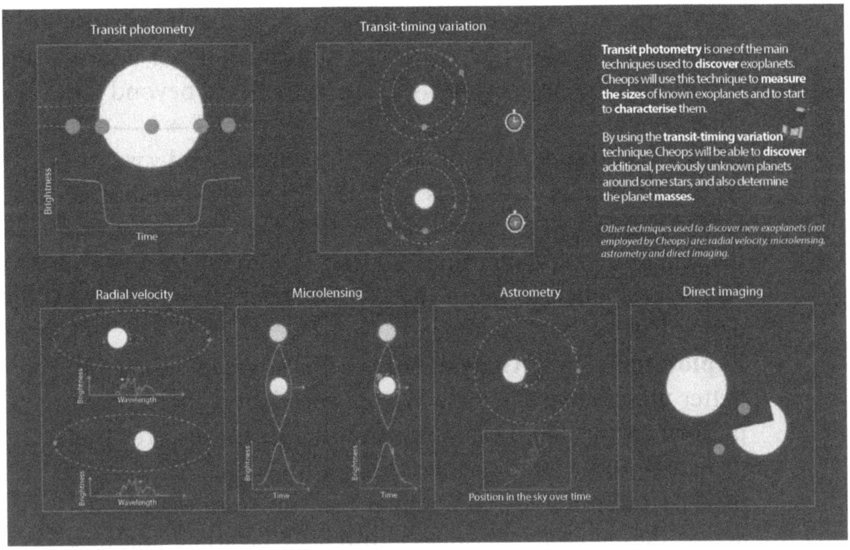

Figure 2.5 Exoplanet detection methods. (Image credit: European Space Agency (ESA)).

various conventional methods used for exoplanet detection. Using AI/ML for exoplanet detection includes simple steps for data processing such as dataset description, data cleaning, exoplanetary data analysis, and feature selection [15].

- **Dataset Description:** The Kepler Dataset was collected by NASA's Kepler Space Telescope, which was launched in 2009 for the search of exoplanets in the nearby galaxy and stars. Over its 9-year mission, Kepler identified 9,564 potential exoplanets. However, not all of these are real planets, so scientists classify them into three groups. The first group consists of confirmed exoplanets, which includes 2,358 planets that have been verified as real. The second group, called candidates, contains 2,366 planets that might be real but require further verification. The last group, labeled false positives, includes 4,840 objects that were initially mistaken for planets but later determined not to be. Scientists use this dataset to train machine learning models to distinguish between real and false exoplanets.

  Another important dataset is the Confirmed Exoplanets Dataset, which provides detailed information on 4,375 exoplanets confirmed by Kepler and other telescopes. This dataset includes measurements such as planetary size, mass, and temperature. Researchers analyze these data to classify exoplanets into different categories and gain a better understanding of the diverse types of planets that exist beyond our solar system.

- **Data Cleaning:** Before analysis, the dataset underwent a cleaning process. Columns with a significant number of missing values were removed to ensure data reliability. For the remaining missing values, appropriate imputation methods were applied to fill in the gaps without compromising data integrity.
- **Exploratory Data Analysis (EDA):** EDA was conducted to better understand the data. First, a correlation analysis was performed to identify and remove redundant features that provided similar information. Next, univariate analysis was used to examine how individual features were distributed across the dataset. This was followed by bivariate analysis, which explored relationships between key variables, including a visual representation of planetary positions on a star map and an investigation into how many planets exist per system [16].
- **Feature Selection:** To improve model performance and avoid collinearity, features with a high correlation were removed. The most essential features, such as planetary mass, radius, orbital period, and stellar distance, were retained to provide meaningful insights into exoplanet classification and characterization. Table 2.5 below summarizes machine learning applications in exoplanet detection.

Table 2.5 Machine learning methods for exoplanet detection.

Method	Purpose	Example use case
Transit Method	Identifies periodic dips in a star's brightness	Kepler and TESS missions
Convolutional Neural Networks (CNNs)	Detects planetary signals in light curves	Differentiating exoplanets from false positives
Random Forest	Classifies potential exoplanet candidates	Improving detection accuracy
Decision Trees	Helps refine exoplanet selection	Reducing false alarm rates

- The **Transit Method** identifies periodic dips in a star's brightness caused by orbiting planets, a technique enhanced by deep learning models that automate transit signal detection. The transit method shown in Eqn. 2.1 was used for the measurement of an exoplanet's radius [17], assuming the stellar disc is of uniform brightness and neglecting any flux from the planet. The ratio of the observed change in flux, ΔF.

$$\Delta F = \pi \frac{R_p^2}{R_*^2} \tag{2.1}$$

where ΔF is the fractional decrease in flux, Rp is the radius of the planet, R_* is the radius of the star, and $\frac{R_p^2}{R_*^2}$ is the true transit depth [18].

- **Kepler and TESS Datasets** provide extensive light curve data analyzed using ML techniques, improving the discovery of exoplanets with higher accuracy. NASA's Kepler mission and its follow-up K2 survey contributed significantly to exoplanet cataloging, shifting focus from individual system analysis to statistical population studies. Machine learning and exoplanet detection have been combined in certain research based on the Kepler dataset or other comparable datasets [19]. The astrophysical signals were processed by researchers using a convolutional neural network, and new exoplanets were confirmed to be in line with CNN predictions. To determine the likelihood of an exoplanet candidate's existence, they also applied various techniques such as decision trees, logistic regression, and neural networks to the Kepler's dataset. Comparable exoplanet research was summed up as transit finding using machine learning methods. These connected studies give us valuable information about the processes of data preprocessing, model selection, training, and inference. The Transiting Exoplanet Survey Satellite (TESS), which was launched in 2018, expanded the search area and improved the detection of exoplanets around brighter stars, making follow-up observations easier.
- **CNNs for Exoplanet Identification** use pattern recognition to differentiate between real planetary signals and

false positives, significantly improving detection rates. Convolutional neural networks (CNNs) are vital in exoplanet detection due to their ability to analyze large and abstract datasets, such as light curves and transit signals. They excel in pattern recognition, identifying the periodic dips in brightness caused by exoplanet transits with high accuracy. CNNs have proven to outperform traditional methods in classifying exoplanets, offering better precision and recall. Exoplanet detection becomes automated through the use of this method, which reduces the requirement for human inspection, thus accelerating discovery procedures. The ability of CNNs to generate hierarchical data representations automatically distinguishes them from conventional signal recognition techniques since they learn from raw data patterns to distinguish true exoplanets from false detections. The generalization ability of CNNs allows them to operate with new datasets, which enables consistent exoplanet detection across Kepler and TESS space missions, thus improving their reliability. The application of CNNs provides astronomers with an efficient tool that helps identify new exoplanets with enhanced accuracy levels [20].

2.4.2 Gravitational Wave Analysis

Gravitational waves (GWs) are ripples in space-time that are caused by black hole mergers as we detected in LIGO, neutron star collisions (heavy objects can curve the space-time), and events that involve massive interactions [20]. Detecting these waves requires highly sensitive interferometric detectors like LIGO, which measure tiny fluctuations in the change of space between masses using lasers. However, the sensitivity of these detectors is limited by other sources, which introduces errors in the system by producing noise, which must be identified and removed to improve signal clarity. Traditional methods rely on manual selection and linear filtering, but machine learning offers a more automated and efficient approach. However, the sensitivity of these gravitational wave detectors can be enhanced by systematically identifying and removing noise from the main gravitational wave signal; for this purpose, data mining and machine learning can help a lot. Virgo and LIGO detectors' sensitivity can be heavily impacted by the location of the source in the sky. When a linearly polarized gravitational wave (GW) propagates orthogonally to the plane of the interferometer arms, the interferometer's response is maximized. This is because the

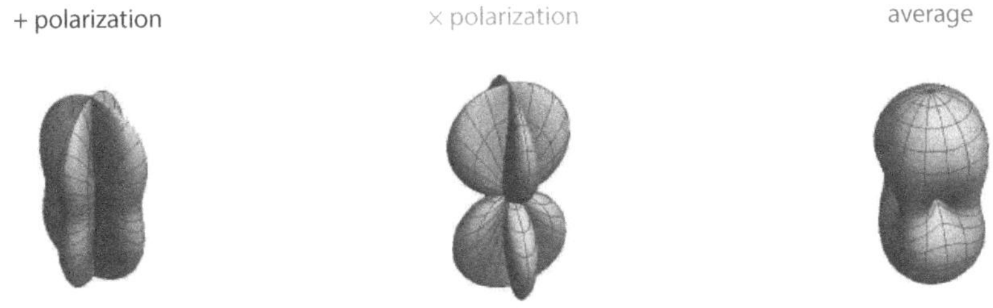

Figure 2.6 Antenna response pattern for a LIGO GW detector. (Image credit: WOLFRAM).

GW's oscillating strain pattern is best matched by the interferometer's two orthogonal arms in this configuration [20], whereas the response is diminished for any other angle of incidence or polarization. The sensitivity of the detector can be calculated in relation to the GW's direction. This sensitivity is displayed in Figure 2.6 and is referred to as the detector's antenna pattern. We can clearly see in this figure that the antenna response pattern for gravitational wave detectors has a very peculiar peanut shape, indicating that LIGO gravitational wave detectors are more sensitive to waves coming directly from above or below them [21].

Gravitational wave detectors operate as systems with numerous auxiliary monitoring channels yet generate only one output signal that represents the GW strain. The auxiliary monitoring channels assess both environmental and instrumental noise that could affect the main signal. A gravitational wave travels through the detection area of an interferometric GW detector [22]. After the calibration, the signal output is a strain s(t) and is given by Eqn. 2.2:

$$s(t) = n(t) + h(t) \tag{2.2}$$

where *n(t)* is the noise and *h(t)* is the strain of the gravitational wave. In case of an elliptically polarized wave, h(t) is the sum of both polarizations h_+ and $h_\times$ [22]. However, the interferometer response is different for these two polarizations, depending on the orientation of the arms of the interferometer. The strain response of the detector is expressed in Eqn. 2.3:

$$h(t) = F_+ h_+(t) + F_\times h_\times(t) \tag{2.3}$$

where F_+ and $F_\times$ are the antenna patterns shown in Figure 2.6, left and middle panels, respectively. The detector noise shows no correlation with the signal according to our assumptions. The mathematical requirement demands setting the correlation between n(t) and h(t) to zero during multiple noise realization averages [22]. The correlation between the two signals is shown in Eqn. 2.4:

$$R_{xy} = R\big(x(t), y(t)\big) = x(t) \bigotimes y(t) = \int_{-\infty}^{+\infty} x(\tau)y(t+\tau)d\tau \quad (2.4)$$

where R_{xy} is the correlation and τ is called displacement or lag. The condition for which n(t) and h(t) are not correlated is therefore shown in Eqn. 2.5:

$$\langle n(t) \bigotimes h(t) \rangle = 0 \quad (2.5)$$

where $\langle . \rangle$ denotes the average over multiple noise realizations.

Noise sources in GW detectors fall into two categories: fundamental noise that includes quantum noise (uncertainties in laser photons) and thermal noise (random molecular vibrations) and are inherent to the physics of detectors. Second is technical noise that includes seismic activity, scattered light, and electromagnetic interference, which can be mitigated with proper noise subtraction techniques. Given that the output signal is a mixture of GW signals and noise, improving detection requires identifying which auxiliary channels contribute noise and how they interact with the main output channel [23].

- **Data Mining for Noise Identification** enhances signal clarity by filtering out background noise in LIGO/Virgo data. This is a brute-force data mining technique that systematically analyzes the thousands of auxiliary channels recorded by LIGO. The goal is to find channels that show strong correlations with the main GW output, indicating that they contribute noise. To do this, the frequency-domain coherence between each auxiliary channel is computed with the GW output. The coherence function measures how strongly two signals are related at different frequencies. If a channel exhibits high coherence with the GW signal at a specific frequency, it is a major noise contributor. This method is highly efficient, as it automatically scans thousands of channels, reducing reliance on human intuition [24]. By quantifying noise contributions, it enables the identification of optimal candidates for noise subtraction. Furthermore, it prioritizes the most problematic frequency bands, ensuring that the most significant noise sources are effectively addressed.

- **Real-Time Signal Detection** using ML enables quick identification of gravitational wave events, critical for astrophysical follow-up observations. Unlike traditional methods that rely on manually selected features, machine learning can automatically detect patterns in large datasets. Deep learning and parametric modeling are used to improve noise subtraction. Neural networks receive historical LIGO data to learn noise source characteristics that help them predict their effects on GW signals. Hence, machine learning offers several key advantages in data analysis and signal processing, particularly real-time gravitational wave (GW) detection. One of its most significant benefits is automated feature selection, which allows the model to identify important noise sources without requiring manual intervention. This reduces human bias and improves efficiency in isolating relevant data [24]. Another crucial advantage is its ability to perform nonlinear modeling, enabling the capture of complex relationships between auxiliary channels and the GW output that traditional linear methods might miss. Additionally, machine learning enhances real-time adaptability, allowing systems to adjust dynamically to changing noise conditions during observations, ensuring more reliable and accurate detections. These capabilities make machine learning an invaluable tool in improving the precision and efficiency of astrophysical data analysis [25]. Figure 2.7 shows the flowchart depicting the outline used for identifying the glitches that are found in the data collection.

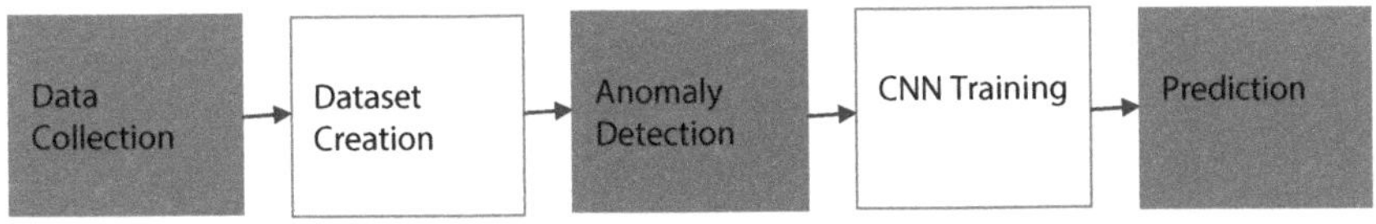

Figure 2.7 Outline for identification of glitches in the data.

2.4.3 Galaxy Classification

- **Traditional Methods versus ML-Based Classification:** Researchers use galaxy classification as a basic process to study how galaxies form along with their structural development and pathway of evolution. The traditional approach for classifying galaxies uses their morphological characteristics to identify spiral, elliptical, and irregular types as shown in Figure 2.8. The Sloan Digital Sky Survey (SDSS) along with its vast data collection has made manual classification of galaxies impossible. The automated pattern learning technology in machine learning allows the analysis of labeled datasets to perform galaxy classification efficiently. CNN-based models have replaced manual classification because these models provide efficient galaxy categorization [26].

Data processing and feature engineering involve three fundamental stages such as feature extraction, data preprocessing, and model training optimization [27].

- **Feature Extraction**
  At the heart of astrophysical machine learning lies the challenge of identifying meaningful patterns within galactic images and spectroscopic data. The success of such models hinges on the careful selection of features—measurable properties that encapsulate the defining characteristics of astronomical objects. The morphology of a galaxy, whether

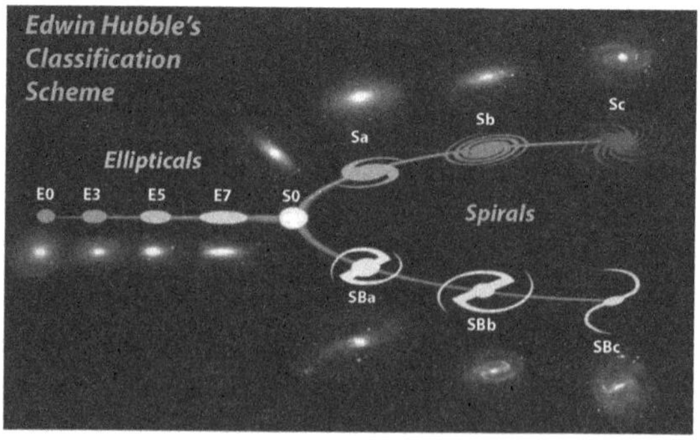

Figure 2.8 Hubble tuning fork scheme for galaxy classification. (Image credit: NASA & ESA).

it be spiral, elliptical, or irregular, serves as a fundamental classification parameter. Traditional methods of morphological classification, once the domain of human astronomers, are now supplanted by machine learning algorithms trained on vast image datasets. The intensity of a galaxy's emitted light, often quantified in absolute magnitudes, provides insight into its stellar population and evolutionary state. This property is crucial for distinguishing between dwarf galaxies and massive clusters. The color of a galaxy, determined by its spectral energy distribution, reveals information about its age and composition. A blue-shifted spectrum is indicative of young hot stars, whereas a redshifted spectrum suggests an older stellar population dominated by cooler evolved stars. Metrics such as the Sérsic index, concentration, and asymmetry coefficients allow for a deeper understanding of a galaxy's evolutionary path [28]. These parameters encapsulate dynamical processes, including mergers, tidal interactions, and star formation activity.

- **Data Preprocessing**
  Raw astronomical data, despite their richness, often contain inconsistencies, missing values, or noise introduced by instrumental limitations and observational constraints. The preprocessing phase is crucial for ensuring that only high-quality standardized data are used in machine learning models, thereby enhancing predictive accuracy and generalizability. Observational limitations, including instrumental sensitivity and atmospheric distortions, frequently lead to missing or unreliable measurements. Several techniques are employed to address this issue:
  - **Interpolation:** Missing values are estimated using adjacent data points within a well-defined parameter space.
  - **Statistical Imputation:** Absent values are replaced with the mean, median, or mode of the feature distribution.
  - **Predictive Estimation:** Regression models are utilized to infer missing attributes based on correlated parameters.

- **Model Training Optimization**
  Standardization and normalization prove vital for managing various galactic property scales because stellar magnitudes operate through logarithmic scales and distances span across millions of light-years. The normalization technique called

min-max scaling transforms features by rescaling them to a specified range from 0 to 1 while maintaining proportional data relationships. The transformation of data through Z-score standardization produces Gaussian distributions with zero mean and unit variance that improves numerical stability especially when using gradient-based optimization algorithms. High-dimensional datasets naturally face the dimensional curse. Machine learning models trained on excessive features may suffer from overfitting, wherein they memorize noise rather than learning fundamental relationships. Principal component analysis (PCA) and autoencoders serve as powerful tools for distilling high-dimensional data into their most informative components. Once the data have been meticulously prepared, they are introduced into the machine learning model. The training process involves iterative refinement, where the model learns underlying astrophysical correlations through exposure to labeled datasets. However, the effectiveness of this learning process is contingent upon algorithm selection and hyperparameter optimization [29, 30].

- **Algorithmic Selection: Matching the Problem to the Solution:** Different astrophysical inquiries require distinct machine learning paradigms. Supervised learning is employed for classification tasks, such as predicting galaxy morphology, where labeled training data guide the model. Unsupervised learning is applied to clustering problems, enabling the identification of novel galaxy populations or the discovery of hidden structures in large-scale surveys. The processing power of CNNs in deep learning enables its ability to extract complex hierarchical features from astronomical pictures like those trained astrophysicists do [26].
- **Hyperparameter Optimization: Fine-tuning the Machinery:** Machine learning models rely on adjustable hyperparameters that govern their learning dynamics, significantly impacting overall performance. The learning rate determines the speed of convergence in gradient-based optimizers; an excessively high rate can cause instability, while an overly low rate results in slow learning. The regularization techniques L1/L2 regularization

and dropout help prevent overfitting through their mechanism of penalizing complex models. The deep learning network architecture controls feature abstraction capabilities through adjustments to its network depth and width, which represent the number of layers and neurons [31].

To ensure robustness, models undergo cross-validation, where data are partitioned into training and validation subsets. Machine learning and data mining significantly enhance the automated processing and classification of millions of galaxy images. This approach improves efficiency by reducing the need for manual classification, thereby saving astronomers valuable time. It also ensures scalability, enabling the handling of increasingly large datasets generated by ongoing astronomical surveys. Moreover, it enhances classification accuracy by minimizing human error.

- **PCA and Clustering Techniques:** The recursive procedure of principal component analysis (PCA) fits a plane (principal component, PC) to previous iteration residuals by choosing a PC that minimizes point-to-PC average squared distance. The procedure continues until all remaining residuals disappear [32]. The contribution measurement of each principal component becomes more efficient when analyzing the projected data covariance matrix in the new coordinate system. Orthogonality requirements for each PC make the covariance matrix diagonal. The natural ordering emerges because smaller residual differences in successive iterations make subsequent PCs explain less and less variance. The PCs along with their corresponding variances become available through singular value decomposition (SVD) applied to the data point covariance matrix C. SVD allows us to factorize it in the form given below in Eqn. 2.6:

$$C = V \pounds V^{T} \tag{2.6}$$

where **Σ** is a diagonal matrix with the eigenvalues (variances) and **V** contains the eigenvectors (principal components) in the respective columns [33].

PCA enables researchers to decrease data dimensions, and clustering techniques establish classifications for galaxies based upon their morphological and spectral properties. The PCA scores enable us to conduct clustering analysis using Gaussian mixture models or k-means to define galaxy types automatically or detect potential unusual morphologies of galaxies. The clustering algorithm functions to create groups of similar events. By analyzing multidimensional features of transient events, DM algorithms can separate them into distinct classes based on shared attributes. A common approach for this assignment relies on DBSCAN (Density-Based Spatial Clustering of Applications with Noise) to locate data clusters representing important events while distinguishing noise from actual astrophysical manifestations [34].

2.4.4 Transient Event Identification

Transient astronomical sources show a brief appearance and disappearance behavior that happens within short time ranges. The phenomena occur together with relativistic compact objects and relate to highly energetic astrophysical occurrences that include supernovae (SNe), fast radio bursts (FRBs), and gamma-ray bursts (GRBs) [35]. Extremely rapid and intense energy discharge during these events produces severe physical environments that are vital to understanding high-energy astrophysical phenomena. Research into astronomy today centers mainly on transient sources. Research on transient events received significant benefit from the progression of multiwavelength and multi-messenger astronomy. Extensive research into extreme astrophysical phenomena depends on multiple space signals including neutrinos together with cosmic particles and gravitational waves to enrich traditional electromagnetic detection methods. The fast nature of these sources requires sophisticated equipment for their detection, and their full analysis demands efficient real-time alert systems. Observational systems should possess quick abilities to find astrophysical events within high-energy events before sharing findings worldwide to the astronomical research community. The system allows researchers to execute synchronized follow-up observations at different wavelengths and messengers through multiple observational facilities. Over time, multiwavelength along with multi-messenger observations have become fundamental to modeling astrophysical research using next-generation astronomical observatories. The identification of short-lived celestial events with fast temporal changes becomes possible through the use of ML. Consider an example of gamma-ray bursts (GRBs) [36]. ML plays a significant role in analyzing and classifying gamma-ray bursts (GRBs), providing tools to automate data analysis, recognize patterns, and predict future events. Unlike traditional data mining, which extracts useful patterns, ML enables adaptive learning and predictive modeling, making it a valuable asset in astrophysical research. Some of the applications of ML in GRB analysis include pattern recognition, predictive modeling, query optimization and automated exploration, and adaptive learning.

- **Pattern Recognition:** ML algorithms, particularly supervised learning models, can recognize patterns in GRB event data based on labeled datasets. These models classify new unseen data into relevant categories by analyzing spectral characteristics such as energy distribution across different wavelengths and temporal features like rise/fall time and pulse width. By training on diverse datasets, ML models can identify novel GRB types or detect unusual events [36]. Fuzzy logic, decision trees, and support vector machines (SVMs) help classify transient events and recognize subtle distinctions between different GRB types and other astrophysical phenomena.
- **Predictive Modeling:** The ability of ML to forecast transient astronomical events is very advantageous. By analyzing past GRB light curves, ML models can predict when and where a new GRB might occur and whether an observed event belongs to a specific GRB classification.

In photometric observations [37], the flux of an astronomical object is measured over a specified time interval, known as the exposure time. The true flux *F(t)* observed at time *t* in each filter is the integral of the spectral energy distribution $S(t,\nu)$ and the filter transmission curve *T(v)*, as shown in Eqn. 2.7:

$$F(t) = \int_{-\infty}^{+\infty} S(t,\nu).T(\nu)d\nu \tag{2.7}$$

The magnitude *m(t)* at time *t*, which is defined using a reference source, such as Vega for Vega magnitudes or a hypothetical constant source for AB magnitudes, is related to the flux, as shown in Eqn. 2.8:

$$m(t) = -2.5 \log_{10}\left(\frac{F(t)}{F_0}\right) \tag{2.8}$$

where F_0 is the flux zero-point. However, in practice, we cannot measure the true flux directly, as any observation requires a finite exposure time. Therefore, the time-averaged flux $\langle F \rangle_{[t_1,t_2]}$ over a time interval $[t_1,t_2]$ with exposure time $\Delta t = t_2 - t_1$ shown in Eqn. 2.9 is used:

$$\langle F \rangle_{[t_1,t_2]} = \frac{1}{\Delta t}\int_{t_1}^{t_2} F(t)dt \tag{2.9}$$

This time-averaged flux is what we observe, and it can differ from the true flux due to the finite exposure time. The difference between the time-averaged flux and the true flux is referred to as the exposure time bias, ΔF, and is shown in Eqn. 2.10:

$$\Delta F = \langle F \rangle_{[t_1,t_2]} - F_{true} \tag{2.10}$$

When the exposure time Δt approaches zero, the time-averaged flux becomes nearly equal to the true flux, provided that the timescale of the flux variations is much larger than the exposure time and is shown in Eqn. 2.11:

$$\lim_{\Delta t \to 0} \langle F \rangle_{[t_1,t_2]} \approx F_{true} \tag{2.11}$$

The approximation applies when the timescale of the light curve extends far beyond the duration of the exposure. The exposure time bias can be measured precisely by understanding the shape of the light curve. A convex light curve shape leads to time-averaged flux values that exceed the true flux values by the Hermite–Hadamard inequality. A light curve that is concave shape produces an averaged flux that falls below the actual flux value. A linear light curve indicates that the time-averaged flux matches exactly with the true flux.

Light curves in real observations tend to exhibit nonlinearity, so time-averaged flux measurements become systematically incorrect regarding the true flux. The peak position of the light curve becomes affected by this bias. A light curve with fast rising and slow decaying phases will cause the peak position to move rightward in the time-averaged light curve. The peak moves toward the left when a light curve rises slowly while it decays rapidly. The peak position of a symmetric light curve shows no change. The difference in peak positions, $|t_{avg.peak} - t_{true\ peak}|$, is always non-zero [37]. This exposure time bias is particularly important for identifying and characterizing transient events such as supernovae (SN), gamma-ray bursts (GRBs), and flaring stars, which are often characterized by rapid flux variations. In these cases, the timescales of variability can be comparable to or even shorter than the exposure time, leading to significant biases in the observed light curve. If the exposure time is not short enough compared to the duration of the transient, the time-averaged flux $\langle F \rangle_{[t_1,t_2]}$ may not accurately reflect the true flux during the event, especially for light curves that are convex or concave. For convex, it shows rapid rise and slow decay, and slow rise and rapid decay in case of concave. For example, in the case of a convex light curve, the time-averaged flux will overestimate the true flux, leading to a brighter observed magnitude than the actual peak brightness. Similarly, for a concave light curve, the time-averaged flux will underestimate the true flux, which could cause the event to appear dimmer than it is. These errors can significantly affect the detection and characterization of the transient event, including its apparent timing, peak brightness, and duration. Moreover, exposure time bias can lead to shifts in the peak position of the light curve. If the transient has a fast rise followed by a slow decay, the peak of the time-averaged light curve will appear later than the true peak. Conversely, if the event has a slow rise and fast decay, the observed peak will occur earlier. This shift in the peak timing is important for accurately determining the time of peak emission, which is often critical for identifying the nature of the transient, such as distinguishing between different types of astrophysical events. Thus, exposure time bias must be carefully considered when identifying and studying transient events [37]. To mitigate these effects, it is crucial to use shorter exposure times relative to the timescales of the transient events being observed or to apply correction methods that account for the bias when constructing light curves. This ensures that the true properties of the transient event are accurately captured, allowing for reliable identification and detailed analysis.

- **Query Optimization and Automated Exploration:** Astronomical datasets are vast and multidimensional, making manual querying cumbersome. ML automates data exploration, optimizes queries within large astrophysical archives, and utilizes online analytical processing (OLAP)

techniques to create multidimensional data representations for efficient querying and analysis.
- **Adaptive Learning:** With continuous data input from space missions, ML models can be retrained periodically, ensuring they remain effective in classifying new GRB events and adapting to changes in GRB characteristics or newly discovered transient phenomena.

Benefits of ML in GRB is that automation reduces manual classification efforts, allowing researchers to focus on interpretation. ML scales to process vast GRB archives, handling large datasets efficiently. By identifying subtle patterns, ML improves classification accuracy and aids in the discovery of new GRB types [36]. Predictive modeling helps forecast future transient events, assisting in observational planning. Clustering and anomaly detection techniques reveal hidden patterns and correlations, leading to new discoveries. An emerging field known as astroinformatics merges astronomy, statistics, and computer science to manage and analyze massive datasets from modern sky surveys [38]. As astronomical data enters the petascale regime, traditional methods become inadequate, necessitating ML-based automation. Real-time transient detection minimizes false alarms by rapidly classifying events like supernovae and GRBs. Time-domain astronomy generates massive heterogeneous data streams requiring real-time classification. As data complexity increases, scalable ML models and standardized procedures for reproducibility become crucial. Key features for GRB classification include color indices (differences in magnitudes across various bands), light curve behavior following a power-law decay, and spectral behavior exhibiting a power-law spectrum [36]. SVMs execute classification through hyperplane optimization by implementing RBF kernel functions. The random forest model consists of multiple decision trees that improves performance by countering overfitting with Gini impurity as its splitting measure for tree branches. Neural networks (NNs) with multilayer perceptrons (MLPs) require gradient descent to train them while using loss functions such as cross-entropy. The development of real-time ML pipelines seeks to achieve immediate classification of transient events. Anomaly detection algorithms enhanced by new improvements will reveal the occurrence of rare astronomical events [39]. Scalable ML architectures will be essential for handling Exascale datasets. Scalability remains a challenge as ML models must efficiently process high-dimensional data. Standardized ML procedures are needed to ensure reproducibility across studies. Collaboration between astronomers, data scientists, and statisticians will drive the field forward.

2.5 Challenges in the Era of Big Data

Machine learning has revolutionized astrophysics by enabling automated data analysis, uncovering hidden patterns, and improving predictive modeling. However, several challenges still limit its full potential. These challenges include managing vast data volumes, dealing with observational noise, handling imbalanced datasets, and improving model interpretability [40]. Addressing these issues requires scalable solutions, innovative preprocessing techniques, and hybrid learning approaches.

2.5.1 Managing Vast Data Volumes

Modern astronomical observatories produce data at such overwhelming volumes that this serves as the primary challenge in modern astrophysics. The James Webb Space Telescope (JWST) and Vera C. Rubin Observatory and Square Kilometers Array (SKA) generate petabyte-sized data annually at an accelerating speed [41]. Each night of observations results in terabytes of information, making data storage, processing, and real-time analysis increasingly complex. To handle this enormous data influx, scalable machine learning models must leverage high-performance computing (HPC), cloud-based solutions, and distributed computing architectures. Efficient data compression, indexing, and parallel-processing algorithms play a crucial role in managing large datasets while maintaining accuracy and efficiency. Researchers are also developing advanced streaming methods to process and analyze data in real time, ensuring timely insights and addressal into astrophysical phenomena.

2.5.2 Addressing Observational Noise and Data Imbalance

Another major challenge in astrophysical data analysis is the presence of significant observational noise. Noise arises from various factors, including instrumental artifacts, calibration errors, and detector sensitivity variations. Ground-based telescopes face additional issues such as atmospheric distortions, which blur images and introduce seeing variations, while cosmic radiation and celestial interference further contaminate observations. All of these factors make it difficult for data analysis and hamper the knowledge drawn from machine learning models. To cope with these problems, researchers are devising sophisticated preprocessing methodologies, such as adaptive filtering, wavelet signal creation and systems, domain adaptation type, etc., to improve signal-to-noise ratio. Moreover, self-supervised learning and generative models are key in the pasting and denoising of sparse datasets that lead to a better overall model sample in performance. In addition to noise, imbalanced datasets pose a significant challenge, particularly when studying rare astrophysical events such as supernovae, fast radio bursts (FRBs), and gravitational wave detections [42]. These rare phenomena are vastly outnumbered by common objects like stars and galaxies, leading to models that struggle to detect them accurately. Techniques such as synthetic data generation using generative adversarial networks (GANs) and variational autoencoders (VAEs), anomaly detection algorithms, and cost-sensitive learning are being employed to address this issue. Additionally, resampling methods like Synthetic Minority Oversampling Technique (SMOTE) help balance datasets and improve model generalization.

2.5.3 Enhancing Model Interpretability

One major challenge in astrophysical machine learning is interpretability. Deep learning methods, especially neural networks, have had tremendous success in object classification in the sky, forecasting cosmic events, and simulation of astrophysical processes. Nevertheless, the models tend to be "black boxes," meaning they do not give much insight into their decisions. This aspect of lack of transparency complicates the ability of scientists to cross-check machine learning outputs against prevailing physical theories. For enhanced interpretability, scientists are combining explainable AI (XAI) methods like SHAP (Shapley Additive Explanations), LIME (Local Interpretable Model-agnostic Explanations), and attention-based approaches that draw emphasis on crucial features that shape the predictions. Additionally, physics-informed machine learning methods, which incorporate physical constraints and domain knowledge, ensure that model outputs align with astrophysical principles. Hybrid approaches combining machine learning with traditional techniques like Bayesian inference and Markov Chain Monte Carlo (MCMC) further enhance both predictive power and interpretability. To fully harness the potential of machine learning in astrophysics, future research is focusing on developing more efficient model architectures tailored for sparse high-dimensional data. Efforts are also being directed toward automating feature extraction to reduce reliance on manual intervention and improving cross-survey data compatibility through robust pipelines. Cloud-based platforms and federated learning frameworks are being explored to enable distributed data analysis across multiple institutions. By overcoming these challenges, machine learning will continue to advance astrophysical research, providing more accurate, efficient, and interpretable insights that drive new discoveries about the universe.

2.6 Bridging Observations and Theory

The integration of data-driven techniques, particularly machine learning, with theoretical astrophysical models marks one of the most transformative advancements in modern astrophysics. Traditionally, astrophysics has relied on theoretical frameworks to describe and predict cosmic phenomena, employing complex mathematical equations and simulations to model galaxy formation, dark matter behavior, and the universe's expansion. However, these models have inherent limitations due to simplifying assumptions, observational constraints, and computational challenges. Machine learning is now bridging the gap between theory and observation in unprecedented ways, unlocking new pathways for understanding the cosmos.

2.6.1 Enhancing Theoretical Simulations with Observational Data

Machine learning is being deployed more and more to improve theory-driven simulations using massive amounts of observational data. Researchers can tighten up hypotheses on the universe's nature by training forecasting models from large datasets such as those available from the European Space Agency's Gaia mission or the Sloan Digital Sky Survey. These models contrast observational data with predictions of cosmic evolution from theory, enabling scientists to detect inconsistencies, refine assumptions, and reveal new avenues for investigation. One particularly influential use of machine learning is in the testing of cosmological theories. By analyzing galaxy clusters, the cosmic microwave background (CMB), and gravitational wave signals, machine learning models help validate theories on the structure and expansion of the universe. These techniques improve our understanding of large-scale cosmic structures, dark energy dynamics, and galaxy evolution. Additionally, machine learning excels at detecting subtle patterns in extensive datasets—patterns that traditional methods might overlook—providing fresh insights into the fundamental physics governing cosmic phenomena [43].

2.6.2 Revolutionizing Astrophysical Simulations

Beyond improving observational analysis, machine learning is revolutionizing astrophysical simulations. Simulating complex events, such as star formation, black hole mergers, and supernova explosions, is computationally intensive due to the high-dimensional parameter spaces involved. Machine learning algorithms, particularly reinforcement learning and surrogate modeling, significantly enhance simulation efficiency by approximating complex calculations and optimizing parameter exploration. Neural networks and Gaussian processes, for example, can emulate expensive simulations, enabling researchers to explore a wider range of physical conditions without requiring massive computational resources. Machine learning has also reshaped our understanding of stellar and galactic evolution. Star formation simulations, which involve intricate fluid dynamics, magnetic fields, and thermodynamics, benefit from the machine learning's ability to uncover hidden relationships within the data. This allows researchers to identify key factors influencing star formation rates, turbulence in the interstellar medium, and stellar feedback mechanisms. Similarly, black hole merger simulations—requiring relativistic physics and high-energy modeling—are increasingly enhanced with machine learning, leading to more precise gravitational wave predictions and deeper insights into black hole dynamics.

2.6.3 Processing and Analyzing Vast Astrophysical Datasets

A further important field in which machine learning is crucial is the processing and analysis of huge astrophysical datasets. New-generation observatories, for example, the Large Hadron Collider (LHC), the Event Horizon Telescope (EHT), and the Laser Interferometer Gravitational-Wave Observatory (LIGO), produce gigantic amounts of data. Machine learning performs automated event detection, anomaly detection, and data categorization, optimizing the search for transient phenomena such as supernovae, gravitational wave signals, and unusual cosmic events. Deep learning models, for instance, can efficiently classify galaxies, detect faint signals buried in noise, and prioritize the most scientifically valuable data. Furthermore, machine learning enhances the simulation and interpretation of complex astrophysical processes that are computationally challenging to model accurately. A prime example is cosmic reionization—the epoch when the first stars and galaxies ionized intergalactic hydrogen. Simulating this event requires modeling radiation transport, gas dynamics, and large-scale structure formation, all of which demand significant computational power. Machine learning models trained on high-redshift observations refine these simulations, leading to more accurate predictions of early galaxy formation and their impact on the intergalactic medium [44].

2.6.4 Discovering New Astrophysical Phenomena

Machine learning (ML) plays an important role in discovering new astrophysical phenomena. Algorithms that analyze extensive datasets can uncover previously unknown variable stars, gravitational wave sources, and other celestial objects, leading to groundbreaking discoveries. These findings can then be incorporated into theoretical models to refine our understanding of cosmic evolution. Additionally, machine learning facilitates the integration of multiwavelength observations from different telescopes—combining optical, infrared, radio, and X-ray data—to provide a more comprehensive view of astrophysical systems, shedding light on phenomena such as active galactic nuclei and dark matter behavior. By merging machine learning with theoretical modeling and observational data, astrophysicists are addressing some of the most profound questions in the field, from the nature of dark matter and black hole physics to the origins of the universe. This interdisciplinary approach is driving a new era of discovery, where data-driven insights refine theoretical models, which leads to more accurate predictions and a deeper understanding of the cosmos. The synergy between machine learning, observational astronomy, and theoretical physics holds immense promise for unraveling the universe's greatest mysteries and advancing the frontiers of astrophysical research.

2.7 The Future: Autonomous Observatories and Predictive Models

The future of astrophysics is poised for a revolutionary transformation with the integration of autonomous observatories powered by real-time machine learning. These intelligent systems will not only streamline data collection but also enhance our ability to dynamically respond to the ever-changing universe. Equipped with advanced machine learning algorithms, autonomous observatories will process vast streams of observational data in real time, enabling them to swiftly identify and classify celestial events such as supernovae (SN), gamma-ray bursts (GRBs), and fast radio bursts (FRB).

2.7.1 Real-Time Event Detection and Response

Traditional observatories often rely on human intervention for data analysis and prioritization, leading to delays in detecting transient events that evolve rapidly [45]. This can result in missed opportunities for discovery. By leveraging real-time machine learning, observatories can continuously analyze incoming data, instantly flag significant events, and adjust observational priorities accordingly. This real-time responsiveness ensures that astronomers can capture crucial cosmic phenomena as they unfold, allowing for timely follow-up observations that are essential for understanding transient events in greater depth. One of the most promising applications of machine learning in autonomous observatories is the detection and study of transient astronomical events. Supernovae, for instance, are explosive star deaths that emit tremendous energy but only for a short time. GRBs and FRBs are also rare and short-lived and need to be detected quickly for any meaningful investigation. Machine learning allows observatories to detect such events in real time, to prioritize them for investigation, and to dynamically reallocate observation tactics according to their properties.

2.7.2 Predictive Modeling in Astrophysics

Beyond event detection, machine learning-driven simulations and predictive models are set to revolutionize various subfields of astrophysics. In cosmology, for instance, deep learning models are enhancing our understanding of dark matter distribution—one of the universe's most enigmatic components. Traditional simulations of cosmic structure formation are constrained by computational limitations and the lack of direct dark matter observations. Machine learning can optimize these simulations, providing more accurate predictions of dark matter's role in galaxy formation and evolution. In astrobiology, machine learning is becoming a crucial tool in the search for extraterrestrial life. The discovery of exoplanets—planets beyond our solar system—has been a landmark achievement, but identifying potentially habitable Earth-like planets remains challenging. Given the vast number of exoplanets and the subtlety of their signals, machine-learning algorithms are being deployed to analyze extensive datasets from missions like Kepler and the James Webb Space Telescope (JWST) [42]. These models can automatically identify exoplanets with conditions conducive to life, such as those within the habitable zone of their stars.

2.7.3 A New Era of Astrophysical Discovery

As these advances develop, machine learning will not only make data analysis more powerful but also spur new theoretical advances. The combination of real-time event detection, autonomous data processing, and predictive modeling will enable astrophysicists to probe the universe with greater efficiency and depth than ever before. The collaboration between autonomous observatories and advanced machine learning algorithms marks a new era of astrophysical discovery, leading us closer to solving some of the greatest scientific mysteries, such as the nature of dark matter, the origins of galaxies, and the hunt for extraterrestrial life.

2.8 Conclusion

Machine learning (ML) integrated with data mining techniques into astrophysics has fundamentally transformed the field by providing powerful tools for handling the vast and complex datasets generated by modern telescopes, satellites, and simulations. As astronomical data grow exponentially, driven by advances in observational technology and computational resources, traditional methods of data analysis are no longer sufficient. This is where ML and data mining come into play, offering efficient and automated solutions that significantly improve both the speed and accuracy of astrophysical research. Machine learning methods, and specifically deep learning, have found their way into many aspects of astrophysics, from exoplanet discovery to gravitational wave analysis, galaxy classification, and transient event detection. Through automated processing of enormous datasets, ML models are capable of rapidly discovering patterns, categorizing celestial objects, and predicting cosmic events that would otherwise prove too complex or time-consuming to analyze with classical approaches. For example, deep neural networks excel at processing light curves to detect faint exoplanets, classifying galaxies based on their morphology, and filtering out noise from gravitational wave signals. These applications highlight how ML techniques enable astronomers to make groundbreaking discoveries more efficiently, uncovering hidden structures in the universe and identifying rare phenomena such as fast radio bursts or supernovae. One of the most significant benefits of applying ML and data mining to astrophysical research is the ability to handle "big data" in real time. The large volumes of data generated by modern observatories require scalable computational frameworks, high-performance computing, and advanced data storage solutions. Machine learning, particularly in the context of big data analytics, allows astronomers to process these datasets efficiently and extract meaningful insights, enabling faster responses to events such as supernovae, gamma-ray bursts, and gravitational waves. Real-time data analysis is also crucial for applications like autonomous observatories, which rely on ML algorithms to make decisions about what to observe and analyze, optimizing the use of available resources. Although great strides have been made in the use of machine learning in astrophysics, there are still challenges. The "black box" approach of deep learning models, for instance, makes results difficult to interpret, with researchers unsure of how the model comes to its conclusions. Moreover, the need for large labeled datasets on which to train deep learning models can be restrictive, especially when labeled data are in short supply. These issues are pushing the evolution of new methods, including explainable AI (XAI), which is trying to enhance machine learning model transparency, as well as semi-supervised learning, which utilizes labeled and unlabeled data to bypass data limitation. In the future, the contribution of machine learning to astrophysics will keep increasing. The combination of ML with theoretical models will enable more precise simulations of cosmological phenomena and enhance the analysis of observational data. Furthermore, the continued development of autonomous observatories, powered by AI and ML, has the potential to revolutionize the way we collect and analyze data, enabling real-time decision-making and paving the way for new discoveries. As these technologies evolve, they will not only enhance our understanding of the universe but also open new frontiers in cosmology, astroinformatics, and astrobiology. In conclusion, machine learning and data mining are reshaping the landscape of astrophysics, offering new ways to analyze complex data, uncover hidden patterns, and accelerate scientific discovery. By leveraging these cutting-edge technologies, researchers are making significant strides in our understanding of the cosmos and laying the groundwork for groundbreaking discoveries in the future.

Data Availability

The images are taken from various open resources of different sites like European Space Agency, International Business Machines (IBM) Corporation, etc.

Acknowledgement

The authors sincerely thank book's editors and reviewers for their valuable suggestions and insightful reviews.

References

1. Mucci, T., The History of Artificial Intelligence, IBM, 21 oct. 2024, https://www.ibm.com/think/topics.
2. Ivezić, V., and Z. Ivezić. Statistics, Data Mining, and Machine Learning in Astronomy: A Practical Python Guide for the Analysis of Survey Data. *Vol. 1, 08540, Princeton University Press, Princeton, New Jersey, 2014.*
3. IBM Data and AI Team, AI vs. machine learning vs. deep learning vs. neural networks: What's the difference?, IBM, 6 July 2023, https://www.ibm.com/think/topics/ai-vs-machine-learning-vs-deep-learning-vs-neural-networks.
4. Han, T., Jiang, D., Zhao, Q., Wang, L., Yin, K., Comparison of random forest, artificial neural networks and support vector machine for intelligent diagnosis of rotating machinery. *Trans. Inst. Meas. Control*, 40, 8, 2681–2693, 2018.
5. Meher, S.K. and Panda, G., Deep learning in astronomy: a tutorial perspective. *Eur. Phys. J. Spec. Top.*, 230, 10, 2285–2317, 2021.
6. Cao, J., Xu, T., Deng, Y., Deng, L., Yang, M., Liu, Z., Zhou, W., Galaxy morphology classification based on Convolutional vision Transformer (CvT). *Astron. Astrophys.*, 683, 3, 2024.
7. Faaique, M., Overview of big data analytics in modern astronomy. *Int. J. Math. Stat. Comput. Sci.*, 2, 96–113, 2024.
8. Delua, J., Supervised versus unsupervised learning: What's the difference?, IBM, 12 March 2021, https://www.ibm.com/think/topics/supervised-vs-unsupervised-learning.
9. Ebner, J., Supervised vs Unsupervised Learning, Explained, Sharp Sight, 12 April 2021, https://www.sharpsightlabs.com/blog/supervised-vs-unsupervised-learning/.
10. Fotopoulou, S., A review of unsupervised learning in astronomy. *Astron. Comput.*, 48, 9–11, 2024.
11. Malanchev, K.L., Volnova, A.A., Kornilov, M.V., Pruzhinskaya, M.V., Ishida, E.E., Mondon, F., Korolev, V.S., Use of Machine Learning for Anomaly Detection Problem in Large Astronomical Databases. *DAMDID/RCDL*, 2019.
12. Malik, A., Moster, B.P., Obermeier, C., Exoplanet detection using machine learning. *MNRAS*, 513, 5505–5516, 2022.

13. Khan, S. and Khan, M.A.R., A Review of Exoplanet Detection Processes using Artificial Intelligence and Neural Networks. *Int. J. Comput. Appl.*, 186, 975–8887, 2024.
14. Greim, H., Investigations of Improvements to Deep Learning Models for the Detection of Exoplanets in Kepler Data, College of Charleston, 2021.
15. Koumbis, G.A., Examining Deep Learning Artificial Neural Networks with Hyperparameter Optimization and Ensemble Learning in the Search for Habitable Exoplanets: An Experimental Study, Diss., National University, 2024.
16. Camizuli, E. and Carranza, E.J., Exploratory data analysis (EDA), in: *The encyclopedia of archaeological sciences*, pp. 1–7, 2018.
17. Seager, S. and Deming, D., Exoplanet atmospheres. *Annu. Rev. Astron. Astrophys.*, 48, 631–672, 2018.
18. Chakraborty, H., Lendl, M., Akinsanmi, B., Petit dit de la Roche, D.J.M., Deline, A., SAGE: A tool for constraining the impacts of stellar activity on transmission spectroscopy. *Astron. Astrophys.*, 685, 173, 2024.
19. Priyadarshini, I. and Puri, V., A convolutional neural network (CNN) based ensemble model for exoplanet detection. *Earth Sci. Inf.*, 14, 735–747, 2021.
20. What are gravitational Waves?, LIGO Caltech, Dec. 2015, https://www.ligo.caltech.edu/page/what-are-gw.
21. Abbott, B.P., *et al.*, LIGO: the laser interferometer gravitational-wave observatory. *Rep. Prog. Phys.*, 72, 10–11, 2009.
22. Boudart, V., Detection of minute-long Gravitational Wave transients using Deep Learning methods, 2023.
23. Yan, J., Interpretable Machine Learning Architectures for Efficient Signal Detection with Applications to Gravitational Wave Astronomy, Columbia University, 2024.
24. Flanagan, E.E. and Hughes, S.A., Measuring gravitational waves from binary black hole coalescences. II. The waves' information and its extraction, with and without templates. *Phys. Rev. D*, 57, 4566–4587, April 1998.
25. Vajente, G., Data mining and machine learning improve gravitational-wave detector sensitivity. *Phys. Rev. D, 105*, 10, 102005, 2022.
26. Webb, S., *et al.*, Unsupervised machine learning for transient discovery in deeper, wider, faster light curves. *MNRAS*, 498, 3, 3077–3094, 2020.
27. Sen, S., *et al.*, Astronomical big data processing using machine learning: A comprehensive review. *Exp. Astron.*, 53, 1, 1–43, 2022.
28. Csizi, B., *et al.*, Euclid preparation. Deep learning true galaxy morphologies for weak lensing shear bias calibration, *arXiv preprint arXiv:2409.07528*, 2024.
29. Guruprasad, A., Galaxy Classification: A machine learning approach for classifying shapes using numerical data, *arXiv preprint arXiv:2312.00184*, 2023.
30. Reza, M., Galaxy morphology classification using automated machine learning. *Astron. Comput.*, *37*, 100492, 2021.

31. Pravin, P.S., *et al.*, Hyperparameter optimization strategies for machine learning-based stochastic energy efficient scheduling in cyber-physical production systems. *Digit. Chem. Eng.*, 4, 5–6, 2022.
32. Jaadi, Z., Principal Component Analysis (PCA): A Step-by-Step Explanation, Built In, 23 Feb. 2024, https://builtin.com/data-science/step-step-explanation-principal-component-analysis.
33. Çakir, U. and Buck, T., MEGS: Morphological Evaluation of Galactic Structure-Principal component analysis as a galaxy morphology model. *Astron. Astrophys.*, 691, 3–4, 2024.
34. Pérez-Díaz, V.S., *et al.*, Unsupervised machine learning for the classification of astrophysical X-ray sources. *MNRAS*, 528, 3, 4852–4871, 2024.
35. Fang, J.-H., *et al.*, TransientVerse: A Comprehensive Real-Time Alert and Multi-Wavelength Analysis System for Transient Astronomical Events, *arXiv preprint arXiv:2501.04247*, 2025.
36. Garcia-Cifuentes, K., Becerra, R.L., De Colle, F., ClassiPyGRB: Machine Learning-Based Classification and Visualization of Gamma Ray Bursts using t-SNE, *arXiv preprint arXiv:2404.06439*, 2024.
37. Liu, D. and Fan, Z., Photometric light curve studies: potential bias induced by exposure time. *MNRAS*, 537, 1334–1342, 2025.
38. Biehl, M., Bunte, K., Longo, G., Tino, P., Machine learning and data analysis in astroinformatics. *Special Session at ESANN 2018: Machine Learning and Data Analysis in Astroinformatics*, pp. 307–314, 2018.
39. Goode, S., Artificial Intelligence for Fast Data Analysis and Fast Transient Detection Applications. Diss. Swinburne, 2024.
40. Rudin, C., *et al.*, Interpretable machine learning: Fundamental principles and 10 grand challenges. *Stat. Surv.*, 16, 1–85, 2022.
41. An Overview About JWST, NASA, 22 Aug. 2024, https://science.nasa.gov/mission/webb/spacecraftoverview/.
42. Hocking, A., Geach, J.E., Sun, Y., Jarvis, M.J., Bonfield, D.G., Crawford, S.M., Deep learning for multi-object detection in wide-field images of the radio sky. *MNRAS*, 474, 2, 2848–2856, 2018.
43. Youvan, D.C., Simulating AI-Extraterrestrial Communication: Exploring Theoretical Models and Experimental Approaches, 2024.
44. Kremer, J., Stensbo-Smidt, K., Gieseke, F., Pedersen, K.S., Igel, C., Big universe, big data: Machine learning and image analysis for astronomy. *IEEE Intell. Syst.*, 32, 16–22, 2017.
45. Liu, J.J., Boongoen, T., Iam-On, N., Improved detection of transient events in wide area sky survey using convolutional neural networks. *Data Inf. Manage.*, 8, 3, 100035, 2024.

3

The Role of Artificial Intelligence in the Discovery and Characterization of Exoplanets

Shraddha. Biswas[1]*, D. Bisht[1] and Ing-Guey Jiang[2]

[1]*Indian Centre For Space Physics, Barakhola, Netai Nagar, Kolkata, West Bengal, India*

[2]*Department of Physics and Institute of Astronomy, National Tsing-Hua University, Hsinchu, Taiwan*

Abstract

Exoplanets, which are planets orbiting stars outside our solar system, have intrigued both the scientific community and the public due to the possibility of uncovering diverse and distant worlds far beyond our own. As of 17 October 2024, 5,780 exoplanets have been confirmed across 4,314 planetary systems, with 969 of these systems containing multiple planets. In the pursuit of detecting and characterizing these distant objects, astronomers employ a range of advanced techniques and technologies, with the transit method playing a central role. This technique, which has been instrumental in identifying over 90% of known exoplanets, involves detecting periodic reductions in a star's brightness, as a planet transits across its face. In this chapter, titled The Role of Artificial Intelligence in the Discovery and Characterization of Exoplanets, we first provide a comprehensive overview of the different types of exoplanets. We then explore the primary methods used in exoplanet detection, including transit photometry, radial velocity, direct imaging, gravitational microlensing, pulsar timing, and astrometry. In recent years, the integration of artificial intelligence (AI) has significantly enhanced our ability to identify, analyze, and interpret the data associated with exoplanet discovery and characterization. AI, particularly through machine learning (ML) algorithms, has refined the analysis of complex datasets, enabling more efficient identification of exoplanet candidates and uncovering previously elusive planets. Machine learning models, trained on existing data, can recognize the subtle signatures of exoplanets,

**Corresponding author*: hiyabiswas12@gmail.com

Yogesh Chandra, Manjuleshwar Panda and Mahesh Chandra Mathpal (eds.) The Intelligent Universe: AI's Role in Astronomy, (87–128) © 2025 Scrivener Publishing LLC

allowing researchers to rapidly identify potential candidates for further study. Moreover, AI can process vast datasets from space missions such as Kepler and TESS more effectively than traditional methods, accelerating the discovery of new exoplanets. In addition to improving detection capabilities, AI is also pivotal in the characterization of exoplanets. Once an exoplanet is identified, understanding its physical properties, such as its composition, size, atmosphere, and potential for habitability, requires the analysis of complex spectral data and other observational measurements. The AI-driven approaches are particularly valuable in the search for Earth-like exoplanets, where precise predictions of habitability are crucial for directing future missions aimed at detecting signs of life. The application of AI and ML has significantly enhanced detection techniques, enabling more efficient analysis of complex data and revealing previously overlooked planets. Looking to the future, AI is expected to remain a driving force in exoplanet research, especially with the emergence of next-generation space telescopes, such as ESA's PLAnetary Transits and Oscillations of stars (PLATO) space mission and NASA's Nancy Grace Roman Space Telescope (RST). These missions will generate unprecedented amounts of data, further necessitating the use of AI for data processing, analysis, and interpretation. The continued integration of AI into exoplanet science will accelerate the pace of discoveries and open new avenues for exploring the broader cosmic context of exoplanetary systems. This chapter is intended to be accessible to both experts and general readers, illustrating how AI is transforming the field of exoplanet research. As AI technology evolves, it promises to further deepen our understanding of the universe and may ultimately help answer the profound question: Are we alone in the cosmos?

***Keywords*:** Exoplanets, artificial intelligence, machine learning algorithms, transit photometry, radial velocity, direct imaging, space telescopes

3.1 Introduction

It is essential to understand exoplanets, starting with their definition and categories, before analyzing the functions of artificial intelligence in their detection procedures and characterization, as well as examining the different missions launched to discover exoplanets and follow up on previous efforts.

> *"There are an infinite number of worlds, some like this world, others unlike it."*
>
> *Epicurus — in his letter to Herodotus (~300 BC)*

Since ancient times, humans have been intrigued by the question of whether we are the only living beings in the universe or if other star systems, akin to our own, exist. The search for exoplanets—planets orbiting distant stars—has long been driven by the curiosity of whether any of these planets share characteristics with Earth, particularly if they lie within a "habitable zone" where conditions could potentially support life. This enduring quest for extraterrestrial life and habitable planets reflects a fundamental human interest in understanding our place in the cosmos. In his 1686 book *Entretiens sur la pluralite des monde* (Conversations on the Multiplicity of Worlds), Bernard le Bovier de Fontenelle wrote, "It is not probable we are the only fools in the universe" [1], highlighting the historical continuity of speculation regarding other worlds and life beyond our own planet.

Today, this age-old question has been answered with a resounding "yes!" The centuries-old quest to discover Earth-like planets has been revitalized by the discovery of hundreds of exoplanets—planets orbiting stars beyond the Sun. The term "exoplanet" comes from the ancient Greek language "**ἔξω**" (**exo**), meaning "**outside**," and "**πλανήτης**" (**planētēs**), meaning "**wanderer**," referring to a planet that orbits a star outside our solar system. These exoplanets vary widely in terms of size, composition, and distance from their parent stars. The first potential evidence of an exoplanet was observed in 1917, although it was not initially identified as such. This early observation was based on periodic variations in the light coming from a star, which were later interpreted as being caused by the gravitational influence of an orbiting planet. However, at the time, astronomers lacked a comprehensive understanding of the significance of these light fluctuations, and the concept of planets orbiting stars beyond our solar system remained a speculative hypothesis.

3.2 Exoplanet Discovery

In 1992, astrophysicists Aleksander Wolszczan and Dale Frail made the first ever confirmed discovery of exoplanets, marking a significant breakthrough in exoplanetary science [2]. They identified three planets orbiting the pulsar PSR1257+12, a rapidly rotating neutron star located about 2,300 light-years from Earth. This discovery was both extraordinary and unexpected, as pulsars—known for their intense radiation and extreme conditions—were initially believed to be inhospitable to planet formation.

Figure 3.1 Didier Queloz and Michel Mayor are pictured in front of the 3.6 meter telescope installed at the La Silla Observatory in Chile while being operated by the European Southern Observatory (ESO). (Image credit: L. Weinstein/Ciel et espace photos).

The planets they detected were situated in an environment vastly different from the solar system-like conditions that had previously been considered ideal for planet formation. The discovery of exoplanets around a pulsar was groundbreaking for several reasons. First, it marked the first direct confirmation that planets exist beyond our solar system. Second, it broadened the understanding of where planets could form, suggesting that planetary formation might occur even in the harsh conditions surrounding pulsars. This finding not only confirmed the existence of planets orbiting stars beyond our solar system but also marked the beginning of a new era in exoplanet research, which has since led to the identification of thousands of exoplanets. The first detection of an exoplanet orbiting a "normal" star came in 1995, when Swiss astronomers Michel Mayor and Didier Queloz (Figure 3.1) from the University of Geneva discovered 51 Pegasi b [3]. This discovery was made using the 3.6-meter telescope at the European Southern Observatory (ESO), which was equipped with the HARPS instrument, a leading tool for exoplanet detection. Their pioneering work earned them the 2011 BBVA Foundation Frontiers of Knowledge Award in recognition of their contributions to the field of exoplanet science. 51 Pegasi b orbits the 8 billion-year-old main-sequence star 51 Pegasi, located about 42 light-years away in the constellation Pegasus. The mass of 51 Pegasi b amounts to half of Jupiter's, while its orbital cycle takes just over four Earth days due to its proximity of eight times closer to its parent star than Mercury is to the Sun. As a result, its surface temperature is approximately 1,000°C. This groundbreaking discovery marked the beginning of rapid advancements in exoplanetary science, propelling it into one of the most dynamic fields of modern astronomy. Just a year later, the discovery of seven additional exoplanets expanded the scope of research. Notably, one of these exoplanets was found orbiting the star 47 Ursa Major, with a surface temperature ranging from –90°C to –20°C. Another significant discovery involved a planet orbiting 70 Virginis, with an estimated surface temperature between 70°C and 160°C. This exoplanet was the first known to exhibit conditions that could potentially support the presence of liquid water. These findings broadened the potential for habitable worlds beyond our solar system, stimulating further interest and research in the search for exoplanets. Additionally, Ling Xin has discussed the concept of life on other worlds in a recent interview [4]. Based on data from 29 January 2025, the NASA Exoplanet Archive [5] has identified 5,832 confirmed exoplanets present in 4,354 planetary systems, with 976 of these systems containing multiple planets. Many of these exoplanets lie within their stars' habitable zones (the "Goldilocks zone"), where conditions may allow liquid water to exist. These discoveries are of profound scientific interest, as they offer valuable insights into planet formation and evolution, as well as reveal a variety of exotic planetary types not found in our solar system. Furthermore, some of these exoplanets could potentially support life. In recent years, research on Venus and its evolutionary history has led scientists to consider Venus in the context of exoplanet studies [6].

These discoveries have opened up exciting possibilities for the study of potential extraterrestrial life and the nature of other planetary systems. In 2005, YJ Liu demonstrated the progress of exoplanets [7]. We can access information about the newly discovered exoplanets, including their characteristics and their stellar host information, from the following website: The Extra-solar Planets Encyclopedia at http://exoplanets.eu. The International Astronomical Union defines astronomical names and definitions through their status as an international governing organization. While some exoplanets are given informal nicknames (e.g., "Osiris," the first exoplanet known to transit its host star; [8]), most are referred to by their scientific designations. The formal names exist as catalog identifiers that match the database entry of their star host such as the Henry Draper Star Catalogue (HD209458). The official scientific designation of exoplanets starts with their stellar catalog designation then adds a letter next to it (where HD209458 b serves as an example). The naming convention for exoplanets continues historical naming systems of their stellar host constellations where we discover exoplanet 51 Pegasi b as the first exoplanet orbiting a star like our Sun.

3.3 Naming Rules/Nomenclature

The International Astronomical Union (IAU) set some guidelines, and according to this, any proposed names for exoplanets must adhere to the following criteria:

- The name should be 16 characters or fewer in length.
- It should ideally consist of a single word.
- The name must be pronounceable in at least one language.
- It must be non-offensive.

It should not closely resemble the name of any existing astronomical object. To verify whether a name is already in use, one can consult resources such as the Galactic and Extragalactic Names, the General Catalogue, or the IAU-recognized Celestial Names.

Additionally, proposed names should not:

- Be the names of pets.
- Have a commercial nature.
- Be names of individuals, locations, and related events from military or religious spheres.
- Be the name of a living individual.
- Match the name of the host star.

Some examples of exoplanets are HAT-P-14 b, HD 131496 b, WASP 72 b, etc.

3.4 Types of Exoplanets

Just like the planets in our solar system, exoplanets—planets that orbit stars beyond our solar system—display a variety of types based on their size, composition, and orbital characteristics. Exoplanets are mainly divided into 4 categories: Gas Giants, Terrestrial Planets, Super-Earths, and Neptunian Planets (Figure 3.2). Here is an overview of the main categories of exoplanets.

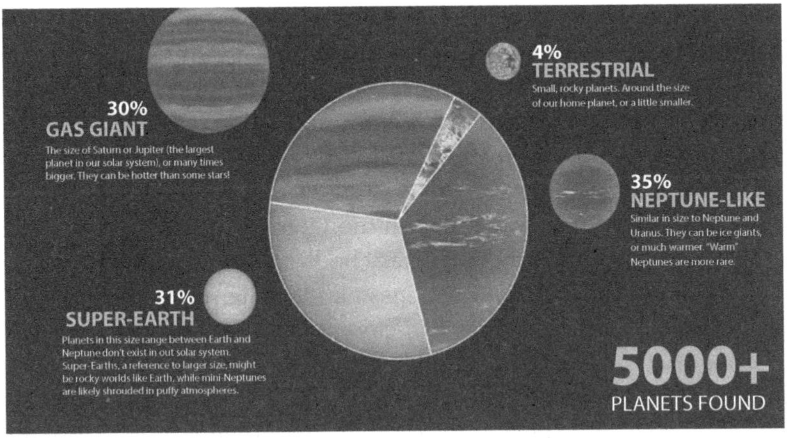

Figure 3.2 A diagram illustrating the classification of exoplanets. (Image credit: NASA/JPL-Caltech).

3.4.1 Gas Giants

Gas giants in exoplanetary systems are planets located outside our solar system that share similarities with Jupiter and Saturn. These planets are predominantly composed of hydrogen and helium, with a presumed small rocky core. They typically have masses greater than 10 Earth masses and possess thick atmospheres that can extend thousands of kilometers, often lacking solid surfaces, which differentiates them from terrestrial planets. Furthermore, some gas giants orbit much closer to their stars than Jupiter and Saturn do to the Sun, leading to extreme temperature and environmental conditions. Due to their large size and mass, gas giants are generally easier to detect than smaller exoplanets, as they induce more significant gravitational effects on their host stars. Approximately 25% of all discovered exoplanets are classified as gas giants.

3.4.2 Hot Jupiters

Hot Jupiters form as a specific type of exoplanets with gas giant characteristics that either formed close to their host stars or started farther out in the protoplanetary disk and later moved inward [9]. The term "hot" refers to the planet's proximity to its star, resulting in extremely high surface temperatures. These planets typically orbit at very close distances, ranging from 0.015 to 0.5 AU, and complete an orbit in just a few Earth days. This leads to surface temperatures often exceeding 1,000°C (1,832°F). The term "Jupiter" in their name indicates that these exoplanets have sizes and masses comparable to Jupiter, although they may be either more massive or smaller than Jupiter in both mass and size. The migration of hot Jupiters is believed to occur due to gravitational interactions with other planets in the system, which exchange energy and momentum, causing the planets' orbits to either shrink or expand. This inward migration likely occurs after the planet has formed beyond the frost line (or snow line), where hydrogen compounds such as water and ammonia exist in solid form, facilitating their accretion into larger planetary bodies. So, hot Jupiters must have originally formed beyond the frost line and then later moved closer to their stars. Because of their large size and close orbits, hot Jupiters are easier to spot using methods like the transit and radial velocity techniques. This is why they are the most commonly discovered type of exoplanet, accounting for approximately 50% of all known exoplanets. Their size and close proximity to their stars make them particularly detectable, contributing to their prevalence in exoplanet surveys.

3.4.3 Terrestrial Exoplanets

Terrestrial exoplanets are Earth-sized worlds located outside our solar system that have solid planetary surfaces primarily composed of heavier materials such as rock, silicates, water, oxygen, and carbon. Unlike gas giants and ice giants, which are primarily composed of gases or ices, the bulk composition of terrestrial exoplanets is largely rock or iron. According to NASA, similar to Earth, as well as Mercury, Venus, and Mars in our solar system, rocky planets can have either solid or liquid surfaces. As Earth is the sole planet where life has been detected, these worlds are prime candidates in the search for extraterrestrial life. Studying them further could provide insights into whether they have atmospheres or oceans that could potentially support life [10]. Typically, smaller and denser than gas giants, terrestrial exoplanets can be found both within and outside the habitable zone of their host stars based on factors like their size, makeup, and how far they are from their star.

3.4.4 Super-Earths

Super-Earths, also known as mini-Neptunes [11], are exoplanets that are more massive than Earth but significantly smaller than ice giants or gas giants. Their masses typically range from about 1.5 to 10 times that of Earth, and they can be composed of a variety of materials, including rock, ice, and gas. Super-Earths with high hydrogen and helium content tend to have lower densities, while those composed primarily of water and silicate materials typically exhibit higher densities. These planets may or may not possess atmospheres, and their characteristics can vary significantly based on their composition.

Super-Earths with radii up to 1.5 times that of Earth are often classified as ocean planets or rocky planets with thin atmospheres. Their larger size and diverse compositions make them some of the most fascinating exoplanets discovered to date. These features make them prime candidates for studying planetary atmospheres, surface conditions, and the potential for habitability. Many super-Earths orbit within the habitable zones of stars while permitting water to stay in its liquid state. However, the actual habitability of these planets depends on a combination of factors, including their atmosphere, composition, and distance from their stars. Consequently, the study of super-Earths remains a key focus in exoplanet research, with the ultimate goal of identifying planets that may be capable of supporting life beyond our solar system.

3.4.5 Neptunian Exoplanets

Neptune-like exoplanets, also known as Neptunian exoplanets [12], are a class of exoplanets that share similarities in size, mass, and composition with Neptune and Uranus, the ice giants in our solar system. These planets typically feature atmospheres rich in hydrogen and helium. Exoplanets of Neptunian type have larger dimensions than Earth including a radius that equals four times Earth's and a mass equivalent to 17 times Earth's, yet they remain smaller than gas giants Jupiter and Saturn. Depending on their specific characteristics, these planets are sometimes categorized as mini-Neptunes or sub-Neptunes.

Neptunian exoplanets are believed to have mixed interiors, with heavy metals likely concentrated in their cores, surrounded by layers of ice, like water, ammonia, and methane, with an outer gas layer made mostly of hydrogen and helium. These planets often have substantial atmospheres and exhibit a range of atmospheric conditions. Studying Neptunian exoplanets remains crucial for understanding planetary development processes along with atmospheric behavior and planetary suitability across the universe. Their diverse compositions and the variety of surface and atmospheric conditions they may possess provide valuable insights into the wide range of planetary types found throughout the universe. Neptunian exoplanets also serve as a critical link between smaller rocky planets and larger gas giants, offering important clues about the transitional stages of planetary evolution. In addition, there are several other subclasses of exoplanets, each defined by specific characteristics such as size, composition, and orbital properties.

3.4.6 Exo-Earths

Exo-Earths are among the most significant discoveries in exoplanet research. They are a subclass of terrestrial exoplanets that share key characteristics with Earth, including similar mass, radius, and temperature. The majority of exo-Earths have solid compositions featuring rocky structures, yet they commonly exist inside the habitable zone surrounding their star system, where suitable water conditions exist. Due to their Earth-like properties and the possibility of supporting life, exo-Earths are primary targets for astrobiological studies. However, despite their significance, only a limited number of exo-Earth candidates have been discovered to date, as they are among the most challenging types of planets to detect.

3.4.7 Water World (Ocean Planets)

Water worlds, also known as ocean planets, are exoplanets that contain sufficient amounts of water to cover their entire surface with oceans. These planets likely formed from icy materials in the outer regions of their protoplanetary disks and migrated inward, where heat from their host stars caused the water to melt, resulting in global oceans. This process, known as planetary migration, is similar to the formation of ice giants like Neptune. Water worlds differ from Earth in that they lack significant landmasses, with their surfaces dominated primarily by liquid water, often under extreme pressures and temperatures. Typically located within their star's habitable zone, these planets may possess subsurface oceans or surface water, making them key targets in the search for extraterrestrial life. Studying water worlds offers valuable insights into planetary formation, climate dynamics, and the conditions necessary for life beyond Earth.

3.4.8 Chthonian Planets

Chthonian planets are rare exoplanets that are thought to be the remnants of gas giants that lost most of their atmospheres due to their close proximity to their host stars. These planets are left with a dense rocky core and possibly a thin atmosphere, having shed much of their volatile material. Some super-Earths may actually be chthonian planets. They likely formed as gas giants, similar to Jupiter or Saturn, but lost their gaseous envelopes through stellar radiation or tidal interactions with their stars. Although chthonian planets are unlikely to be habitable, they offer valuable insights into planetary evolution and the interactions between exoplanets and their stars.

3.4.9 Rogue Exoplanets

Rogue planets [13], also known as *nomad planets* or *orphan planets*, are celestial bodies that exist outside of any stellar system and travel through interstellar space without being gravitationally bound to a star (Figure 3.3). They are also referred to as isolated planetary-mass objects (iPMOs) or free-floating planets (FFPs), as they wander the Milky Way without a parent star. Since they lack a star, rogue planets are dark and challenging to detect, characterized by extreme surface conditions likely resulting from intense geological or atmospheric processes. These planets are thought to have rugged rocky surfaces, with significant features such as large mountain ranges, deep valleys, and possibly volcanic activity. Estimates suggest that between 200 and 400 billion rogue planets may be drifting through the Milky Way. They can vary in size, from small rocky planets to massive gas giants, although the difficulty in detecting them means that all discovered rogue planets to date (such as CFBDSIR 2149-0403) have been large gas giants. Rogue planets can be directly imaged, especially in infrared, or detected through gravitational microlensing. While their harsh environments make them unlikely to support life, they offer valuable insights into

Figure 3.3 An artist's depiction of a bizarre rogue "planet" SIMP j01365663+0933473 with incredible auroras puzzles. (Image credit: Chuck Carter; NRAO/AUI/NSF/Caltech).

the wide variety of planetary environments across the universe and the dynamic processes that shape planetary surfaces. One possible explanation for the existence of rogue planets is that they form within planetary systems but are later ejected into interstellar space typically due to gravitational interactions with other planets or encounters with nearby star systems. A less likely formation mechanism suggests that rogue planets may form independently, outside of established planetary systems.

According to NASA, a study suggests that when the Nancy Grace Roman Space Telescope becomes operational, with its scheduled launch in May 2027, it could detect up to 400 rogue exoplanets with masses similar to Earth using the gravitational microlensing technique.

3.5 Detection Methods

The search for exoplanets is often compared to finding a needle in a haystack, as these planets do not emit detectable light of their own—at least not at levels sufficient for direct observation. Stars are much brighter and more massive than planets, with the light from exoplanets primarily coming

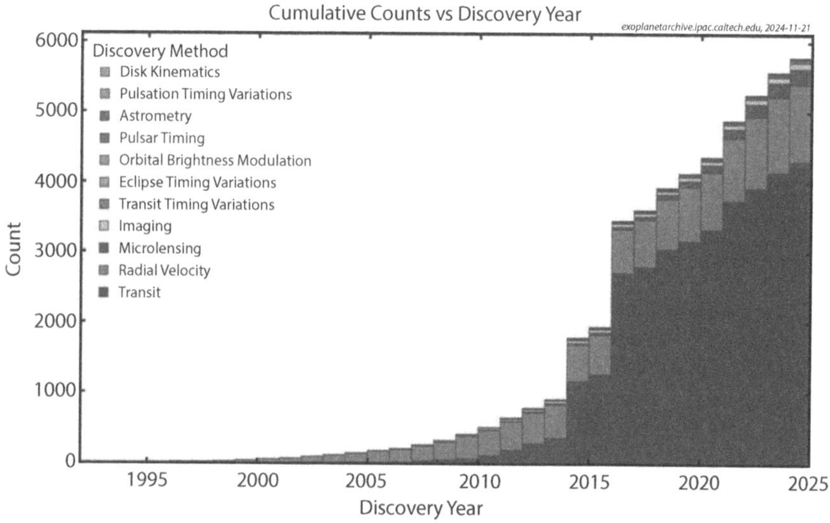

Figure 3.4 Cumulative number of exoplanets discovered using various methods from 1995 to 2025. (Image credit: NASA exoplanet archive).

from the reflection of their host star's light. As a result, the light emitted by a planet is incredibly faint compared to that of its parent star, making direct detection of exoplanets a significant challenge. People attempting to observe light from distant planets face the same challenge of discerning a minimal candle from overwhelming flames of a blazing inferno. Due to this stark difference in luminosity, astronomers have developed indirect methods to detect exoplanets, primarily by studying their effects on their parent stars. The cumulative number of exoplanets discovered each year is shown in Figure 3.4.

Two of the most successful and widely employed methods for exoplanet detection are the transit and radial velocity techniques. In the transit method, astronomers observe the periodic dimming of a star's light as a planet passes in front of it, blocking a portion of the star's radiation. Approximately 90% of known exoplanets have been discovered using this method. Modern technology allows astronomers to measure changes in stellar brightness as small as 1%, and since the amount of light blocked is proportional to the planet's size, this method can provide estimates of the planet's radius. The first exoplanet discovered, **51 Pegasi b**, was identified using the radial velocity method. This technique involves measuring the star's motion caused by the gravitational influence of an orbiting planet. The gravitational pull from the planet causes the star to move during which time its light spectrum changes through a Doppler effect that produces redshift during movement away and blueshift during movement toward Earth. Tangible evidence of this shift depends directly on both the planet's mass and its orbital distance from the star. Scientists calculate planetary masses by measuring how large the shift in a star becomes depending on the size of the planet and its orbital distance.

Other methods for detecting exoplanets include direct imaging, which captures the light that the planet itself emits or reflects, and the gravitational microlensing method, which spots the bending of light caused by a planet's gravitational field. Each of these techniques provides valuable data, contributing to our understanding of the vast diversity of exoplanets. Specifically, combining the transit and radial velocity methods enables astronomers to determine both the density and composition of exoplanets, offering deeper insights into their physical properties and potential for habitability. The number of exoplanets detected using various methods is summarized in Table 3.1.

Below, we provide a detailed explanation of each of these detection methods:

Table 3.1 Data for confirmed exoplanet statistics using various detection methods taken from the NASA exoplanet archive as of 29 January 2025.

Discovery method	Number of planets
Imaging	82
Radial velocity	1,100
Astrometry	3
Pulsar timing	8
Transit	4,342
Gravitational microlensing	235

3.5.1 Direct Detection

Exoplanets are generally difficult to detect due to the minimal light they emit, making them nearly invisible even to advanced telescopes. However, recent technological advancements have enabled astronomers to capture direct images of exoplanets using the direct detection method. This approach involves studying exoplanets by capturing the light they emit or scatter, providing valuable insights into their characteristics.

3.5.1.1 Imaging

Direct imaging [14] of an exoplanet is one of the most challenging detection methods because of the huge difference in brightness between the host star and the planet. Stars are much brighter than their companion planets, which makes it hard to see the planet directly. A successful observation of exoplanets relies on reducing starlight intensity because this enables viewers to see the dimmer signals from planets. The primary challenge in directly imaging exoplanets arises from this contrast, as planets—particularly gas giants like Jupiter—emit much less light than stars. For example, a Jupiter-like planet emits only about one billionth of the light of its host star in the visible spectrum, creating an extreme contrast that makes it nearly impossible to distinguish the planet from the star's intense brightness using conventional visible light imaging techniques.

To address this issue, astronomers turn to infrared radiation where the contrast between the planet and star is reduced to just a few thousandths of the star's brightness. This method is particularly effective for young planets, which are still contracting and emitting excess heat. Another technique involves using a coronagraph, and the usage of coronagraphs as optical devices helps block stellar light to unveil previously impossible-to-see star coronal rings. This enables the detection of faint light emitted by any nearby planets. The first direct detection of an exoplanet was achieved by capturing a single pixel in an image where a coronagraph blocked starlight to enable researchers to observe the planet through extended exposure. This groundbreaking detection involved Fomalhaut b, which orbits the bright star Fomalhaut. Direct imaging is the only method that enables the direct assessment of specific physical parameters of exoplanets, such as surface water content and potential biospheric properties. In 2004, Chauvin *et al.* [15] achieved the first detection of the exoplanet 2M1207b using direct imaging with the NACO adaptive optics system on the Very Large Telescope (VLT) at the Paranal Observatory in Chile, with the CCD frame of the observation shown in Figure 3.5. The E-ELT, European Extremely Large Telescope, is scheduled to begin operations as its first light appears in 2028, and it will significantly enhance exoplanet detection through direct imaging, utilizing its superior optical resolution to obtain detailed observations.

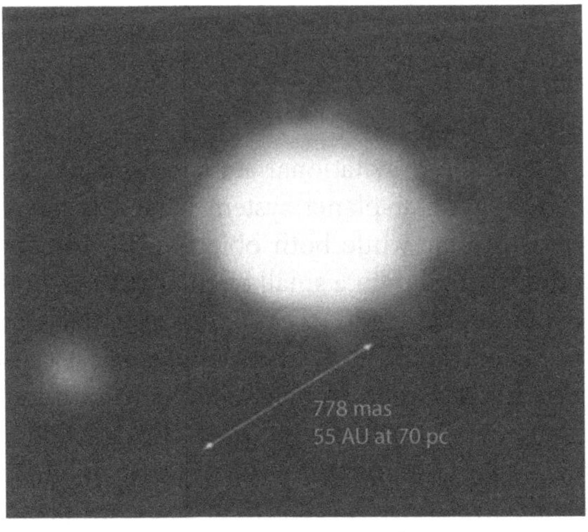

Figure 3.5 The first direct image of 2M1207 b (marked by the red spot) using NACO at the VLT through the direct method. Planetary orbit occurs around the brown dwarf object called 2M1207a (blue spot). (Image credit: European Southern Observatory, ESO).

Direct detection involves observing the planet itself, bypassing the need to account for stellar influences, which is required by methods such as transits or radial velocity. This approach simplifies data interpretation and reduces uncertainties caused by stellar noise.

By tracking a planet's motion over its orbital period, accurate measurements of its mass, orbital radius, and orbital shape can be obtained. Direct imaging is also capable of detecting rogue planets and those in wide orbits, which enhances our understanding of exoplanetary diversity. However, this technique has certain limitations. The large difference in brightness between an orbiting planet and its parent star hinders attempts to detect the planet's light, necessitating advanced methods such as coronagraphs or starshades. Direct imaging generally requires high-resolution telescopes with narrow fields of view, which limits the number of planets that can be observed simultaneously. Additionally, long observation periods are often needed to collect sufficient data, and the resources required for such missions can be substantial.

3.5.2 Indirect Detection

The *indirect detection* method depends on methods that help identify exoplanets by establishing their presence without directly observing the planet. Since exoplanets are typically too distant and faint to be seen directly, these methods rely on detecting an exoplanet's effects on its star or surroundings.

3.5.2.1 Radial Velocity Tracking

The radial velocity method served as an original exoplanet detection technique that measures the gravitational starspot wobble caused by orbiting planetary bodies. A star-planet system places its barycenter slightly off-center toward the star while both objects orbit the common center. Consequently, the star moves in a small circular path. Spectral lines of the star exhibit periodic shifts due to its orbital motion that can be measured by the Doppler effect (Figure 3.6). The Doppler effect causes light to be compressed (blueshifted) when an object moves toward the observer and stretched (redshifted) when it moves away. This method was explained by Dai Ziqi in 2021 [16]. As the star "wobbles," astronomers can measure these Doppler shifts in the star's spectrum to detect the presence of an exoplanet. The periodicity and amplitude of the shifts offer insights into the mass of the planet and its orbit, as well as the orientation of its orbital plane. This technique is particularly effective for detecting large planets in

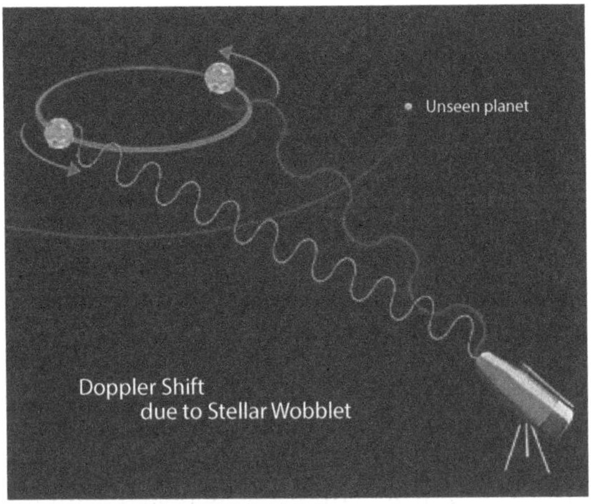

Figure 3.6 Doppler shift method. (Image credit: NASA/JPL-Caltech).

close orbits, as their stronger gravitational influence results in more pronounced stellar wobbles.

The HARPS spectrograph operates as one of the best tools to find low-mass exoplanets through radial velocity measurements at the ESO's 3.6-meter telescope located in Chile. The high precision radial velocity measurements enable astronomers to determine the minimum mass of the planet. This technique gives only minimum mass estimates, since scientists cannot determine the orbit inclination angle of the planet. The radial velocity method enables scientists to calculate eccentricity and semimajor axis measurements in binary star systems along with determining orbital components (Figure 3.7). The measurement technique produces ambiguous results because a large distant planet could create identical radial velocity changes as a smaller planet orienting in the same location.

Scientists utilize stellar wobble observation as their leading tool to search for exoplanets by measuring planetary orbits' impact on stars. Research teams rely on the radial velocity method as a core technique for discovering exoplanets in current scientific studies. This method achieves excellent results for big exoplanet detection, since it allows scientists to detect the wobbles of stellar bodies that originate from hot Jupiter and other short-orbit exoplanets. Researchers use changes in measured stellar speed during orbits to determine planetary properties together with orbital characteristics as well as planet radius values. Scientific exploration of new exoplanets happens by using current technology and equipment to detect multiple exoplanets. The technique reveals no details regarding

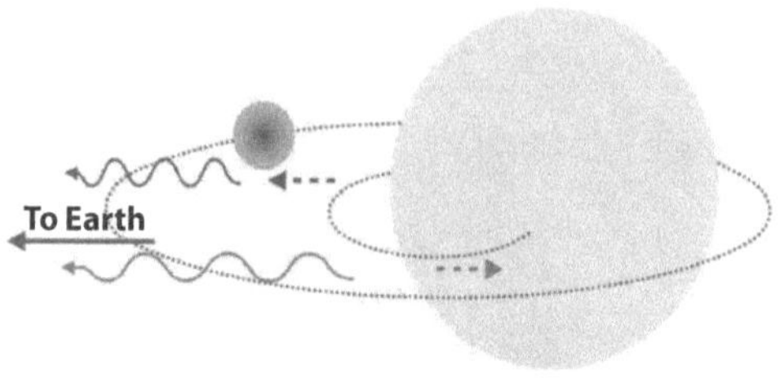

Figure 3.7 Schematic view of the radial velocity method. (Image credit: Yamila Miguel, Leiden Observatory, Leiden).

planetary surfaces and atmospheric makeup or density because it does not include light detection capabilities. The method provides most benefits for Doppler shift detection during studies of planets orbiting Sun-like stars because the spectral changes become undetectable in broader spectrums. The mass of exoplanets remains unknown because scientists lack knowledge about their orbital angles. The observed velocity deviation allows scientists to determine planet mass (M_p) information through a relation that depends on the inclination angle.

3.5.2.2 Astrometry

Stellar positions and motions are measured precisely to detect exoplanets using the astrometric method, which comprises some of the earliest methods of planet detection. This technique has historical roots dating back to the late 18th century, when astronomer William Herschel proposed that an unseen companion might be influencing the position of the star 70 Ophiuchi (located in the constellation Ophiuchus). In 1987, Gatewood G. D. briefly discussed the astrometric method [17]. Astronomers employ the radial velocity method to identify exoplanets by evaluating how star velocities oscillate due to orbiting exoplanets similar to astrometry methods. The tracking of exoplanets by astrometric methods has achieved fewer successes than radial velocity measurements did. Scientists use astrometric methods to gather multiple precise sky-based measurements of stellar positions and analyze resulting motion patterns to detect planet gravity-caused wobbles (Figure 3.8). The slight wobble produces detectable shifts in the apparent position of a star. Determining such small position changes in astrometry constitutes a detection challenge, leading to its identification of only three exoplanets.

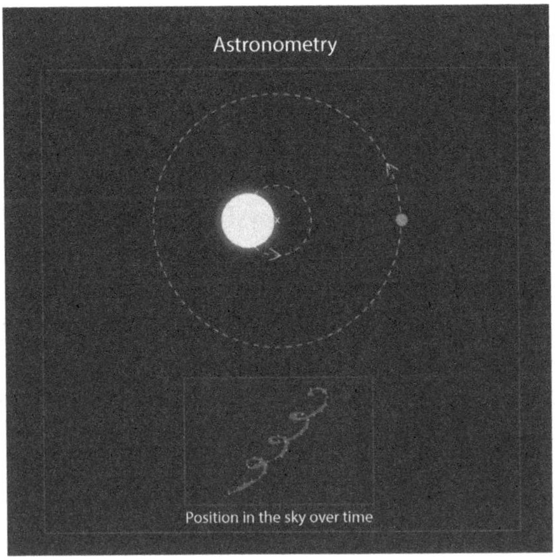

Figure 3.8 Astrometry method. (Image credit: the European Space Agency).

Nevertheless, astrometry is not obsolete. The European Space Agency's (ESA) Gaia mission is currently conducting an extensive survey, making highly accurate measurements of stellar positions across a broad region of the sky. ESA predicts that, in the future, Gaia's astrometric data could uncover tens of thousands of exoplanets, some located as far as 1,600 light-years from the Sun. The characterization of an exoplanet through astrometric methods became successful in 2002 by studying Gliese 876's known planet. As a supplementary technique, astrometry evaluates large orbit planets, yet radial velocity works better to find smaller orbit planets. Astrometry can study a broader variety of star systems through its independent motion observations because it does not need precise Earth-sight alignment, thus it expands exoplanet detection surveys across more stars. Astronomers use this technique to quantify a star's planetary-induced motion as the detection method for obtaining planet mass data while measuring star orbital wobbles. The detection of exoplanets through astrometry faces difficulties because planet-induced star movements remain small enough to be challenging for observations. To measure the star's gravitational wobble properly, both high-precision measurements and advanced optical systems are needed, since Earth's atmosphere causes distortions. The existence of starspots induces distortions that resemble stellar wobble, thus generating incorrect planetary detection signs in astrometric measurements.

3.5.2.3 *Pulsar Timing*

Scientists use the pulsar timing method that can also be called pulsar timing variations or timing anomalies (Figure 3.9) to detect exoplanets when they precisely track the reliable radio pulses generated by pulsars. The spinning pulsars produce predictable radio waves through their rotational motion that resembles an automated signal system identical to a lighthouse. A planet orbiting a pulsar will disturb its motion through gravitational effects that produces minimal variations in the pulsation duration. Measurement inconsistencies reveal that the pulsar shares a central gravitational position with at least one gravitational body that exists as orbiting planets. Through pulsar timing, scientists obtain information about the mass structure of orbiting planets together with their natural orbital patterns. This approach drove scientists to find the very first planets located outside our solar system. Scientists made the historical discovery of two exoplanets orbiting pulsar PSR B1257+12 by using pulsar timing methods in 1992. The gravitational influence of the planets on the pulsar's motion caused the observed timing irregularities, which were essential in identifying the planets and understanding their orbits. Pulsar timing offers remarkable precision, with timing accuracies on the order of microseconds, enabling the detection of small variations even caused by low-mass planets. This method is sensitive

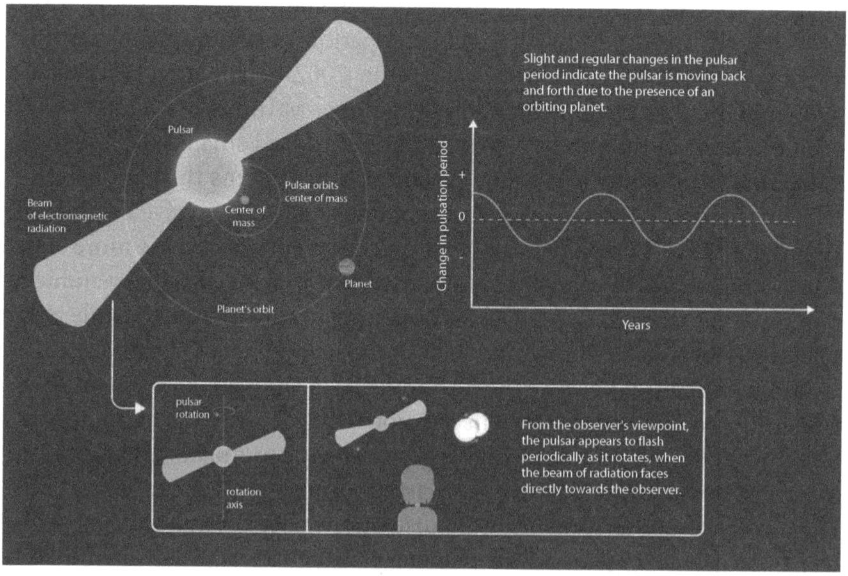

Figure 3.9 Pulsar timing method diagram. (Image credit: Alice Hopkinson, LCO).

enough to detect planets as small as a few Earth masses, especially when they orbit a pulsar. It is particularly effective for identifying planets with large orbits and is largely immune to stellar variability and activity, reducing interference from these factors. Pulsar timing is extremely precise, cost-effective, and able to detect small low-mass planets, making it a powerful tool for finding exoplanets.

However, the pulsar timing method is limited in its applicability, as it can only be used with pulsars, which are relatively rare celestial objects. Furthermore, it requires pulsars that emit strong regular signals and are oriented in a manner that allows for precise timing measurements. While this technique is effective for determining parameters such as planetary mass and orbital period, it has limitations in providing detailed information about the planet's size or composition. In multiplanet systems, the astrometric signal is primarily influenced by the motion of larger planets, which can obscure the contributions of smaller planets. For terrestrial-type planets, the expected astrometric signal is typically too weak to detect, often falling below the theoretical limits of positional accuracy.

3.5.2.4 *Transit Method*

The transit method emerges as one of the best and most commonly used approaches for exoplanet detection (Figure 3.10). Astronomers use the transit method to measure the luminosity drops that happen when a planet blocks the starlight while passing in front of its star. The planet hides some starlight during transits while revealing the original star brightness at the end of each event. Scientists understand essential planetary items including size and orbital information through analyzing light curve depth and duration that result from this method. Furthermore, during planet transits, scientists gain atmospheric composition insights by observing which wavelengths get absorbed. The exoplanet transits and occultation techniques receive a clear breakdown from Joshua Winn during his 2011 publication [18]. Scientists identified exoplanets through the transit method for the first time in 1992. The transit technique continues to gain strength in modern research because of Kepler and TESS space telescopes together with other observational tools. The missions have led to identifying thousands of transiting exoplanets that increased our knowledge about planetary systems. With duplicity of transit photometry and radial velocity observations, astronomers gain planetary size and mass measurements that lead to density calculations. Planetary composition analysis becomes possible through this process that reveals whether a planet possesses a gaseous or rocky structure.

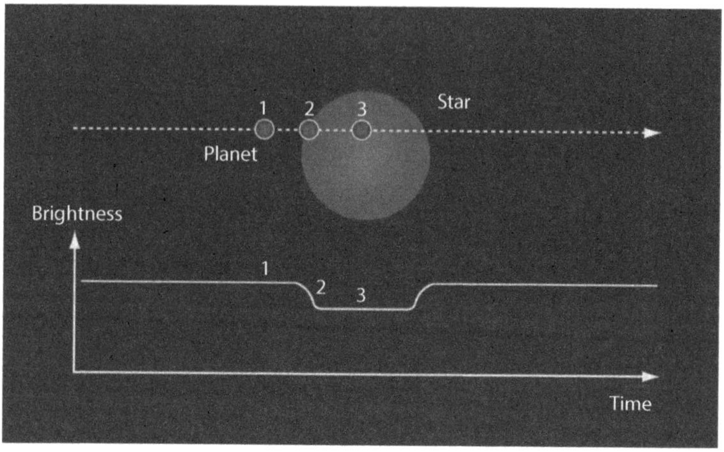

Figure 3.10 Transit method. (Image credit: *NASA, ESA, G. Bacon (STSci)*).

The transit method is particularly useful for calculating the radius of an exoplanet. The ratio of the observed change in flux, ΔF, to that of the stellar flux F is expressed in Equ. (3.1):

$$\frac{\Delta F}{F} = \frac{R_p^2}{R_*^2} \tag{3.1}$$

where R_p and R_* represent the planetary and stellar radii, respectively.

And the total transit duration can be expressed as:

$$t_T = \frac{P\,R_*}{\pi\,a}\sqrt{\left(1+\frac{R_p}{R_*}\right)^2 - \left(\frac{a}{R_*}\cos i\right)^2} \tag{3.2}$$

Here, P is the orbital period of the exoplanet, a is the semimajor axis, and I is the orbital inclination. Now, to determine how close the planet transits to the middle of the star, we can derive the "impact parameter" (b), which is defined as:

$$\mathrm{b} \equiv \frac{acosi}{R*} \tag{3.3}$$

The value of b ranges from 0 to 1.

The transit method has some exciting advantages, especially when it comes to studying the atmospheres of distant planets. When a planet crosses in front of its host star, some of the star's light filters through the planet's atmosphere. This light contains "fingerprints" of the chemicals and molecules present in the atmosphere, offering a unique opportunity to analyze its composition. This can even include the possibility of detecting signs of life, known as biosignatures. However, the method does have its challenges. One big limitation is the need for the planet's orbit to be aligned just right with our line of sight. If the orbit is not perfectly aligned, we would not be able to observe the transit. This means that only a small fraction of planets can be detected this way. For planets with larger orbits, the odds of catching a transit are even slimmer. For instance, the likelihood of detecting a planet orbiting a star like our Sun at a distance of 1 AU is just about 0.47%. Despite these odds, by scanning large areas of the sky, astronomers can still find a decent number of planets using the transit method. Another hurdle is the potential for false alarms. The observed dip in brightness caused by a planet crossing the star's light might not always be due to an exoplanet. It could be the result of other factors like star activity or even instrument noise. To confirm the detection, scientists often need to double-check the signal with additional methods like measuring the star's radial velocity.

In summary, the transit method [19] provides invaluable insights into exoplanetary systems, including the planet's size, orbital properties, mass, density, and atmospheric composition. While highly effective, it is constrained by the need for precise orbital alignment and the potential for false positives, necessitating complementary techniques for verification. The transit method determines a planet's radius by analyzing the amount of starlight blocked during transit and offers insights into the planet's atmosphere by examining absorption patterns in the light. This can aid in the detection of biosignatures. By combining transit data with radial velocity measurements, astronomers can also determine the planet's mass and density, revealing its composition (e.g., gas giant, icy planet, or rocky planet). The transit method shines when it comes to spotting large planets or those that orbit closely around their stars, as these planets cause noticeable dips in starlight that are easier to detect. Space telescopes like Kepler and TESS are particularly suited for this task—they are able to scan vast regions of the sky and have discovered thousands of exoplanets from a variety of star systems, greatly expanding our knowledge of planetary systems. Another benefit of the transit method is that it is relatively inexpensive, since it mainly relies on light curves (photometric data), avoiding the need for expensive and complex instruments. However, the method does have some limitations. The most significant one is that the planet's orbit must be perfectly aligned with our line of sight from Earth. This means only a tiny fraction of planetary systems will have the right alignment for us to detect a transit. For instance, if a planet is orbiting a star like our Sun at 1 AU (the same distance Earth is from the Sun), the chance of detecting its transit is only about 0.47%. The chances get even slimmer for planets with larger orbits or slower movements, so the method works best for planets with close orbits. There is also the issue of false positives. The dip in brightness caused by a transit might not always be due to a planet; it could be from something like stellar activity or noise from the instruments. To confirm that the signal is indeed from an exoplanet, scientists often need to follow up with additional methods, such as radial velocity measurements or extra photometric data. Small planets, which create much smaller dips in brightness, are especially tricky to detect. This is particularly challenging when there are limited data or the instruments are not sensitive enough, making it harder to spot planets like Earth or other smaller exoplanets. For a more in-depth look at the transit method, H. J. Deeg's 2018 handbook on exoplanets [20] offers a thorough explanation.

3.5.2.5 Gravitational Microlensing

Gravitational microlensing (Figure 3.11) is a pretty cool phenomenon that happens when the gravity of a closer object, such as a star or planet, acts like a lens and bends the light from a star farther away. This effect is most

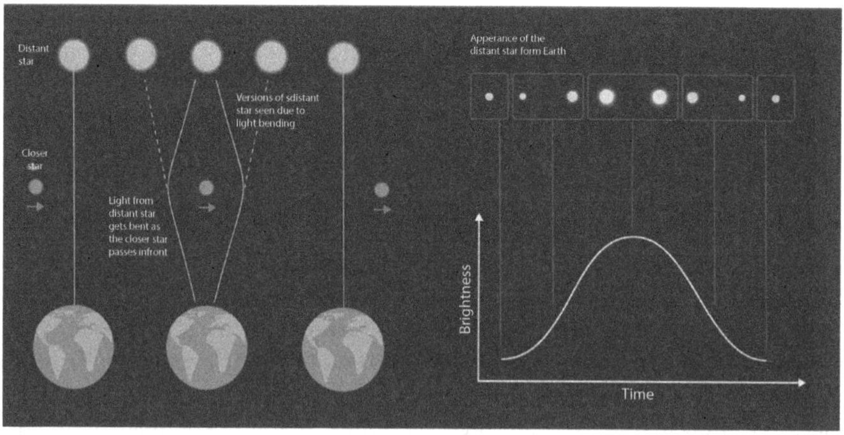

Figure 3.11 Gravitational microlensing method. (Image credit: Alice Hopkinson, LCO).

noticeable when the object and the distant star line up just right. When they do, the light from the background star briefly gets brighter, almost like a flash. It is a temporary effect, but it is something astronomers can use to spot objects that are tough to see otherwise, such as planets or stars that might be hidden in the shadows.

Gravitational microlensing is particularly useful for detecting rogue exoplanets, which are free-floating in space and do not orbit stars. However, microlensing events are pretty rare because they only happen when the alignment between the lensing object and the background star is nearly perfect. Since these events are one-time occurrences, it is hard to catch them again for follow-up observations, which can make studying them tricky. Because of this, the method can sometimes lead to false positives, so scientists need to confirm their findings using other techniques to be sure.

Despite these challenges, microlensing remains an effective method for detecting Earth-like exoplanets around distant stars, especially in regions like the Galactic center, where many background stars are located. The technique involves continuously monitoring a large number of stars because the events are brief, typically lasting only a few days or weeks. In 1991, astronomers Shude Mao and Bohdan Paczyński [21] proposed using gravitational microlensing to detect exoplanets. The method gained popularity through the Optical Gravitational Lensing Experiment (OGLE) project, which led to the detection of several candidate exoplanets in the early 2000s, many of which required further confirmation. Over time, the technique has confirmed the existence of multiple exoplanets, providing a unique means to detect planets that would otherwise be invisible to other detection methods.

A significant limitation of this technique is the inability to replicate the precise alignment between the lensing and background stars, making each detection a one-time event. As a result, studying the planet in detail with microlensing alone is not possible especially because follow-up observations are difficult due to the vast distances involved, which can span several kiloparsecs. In summary, while gravitational microlensing is a valuable tool for detecting exoplanets, particularly rogue planets, its one-time nature and reliance on rare alignments pose significant challenges. The method is most effective when monitoring large portions of the sky, especially in dense star fields like the Galactic center. Gravitational microlensing is particularly useful for detecting small exoplanets located at great distances, often thousands of light-years from Earth. Its sensitivity to subtle gravitational effects enables the detection of planets in distant star systems that might be difficult to observe with other techniques. This method is especially valuable for identifying exoplanets in wide orbits, which are not as easily detected by methods like transit or radial velocity, which favor planets in closer orbits. Beaulieu J. P. demonstrated the effectiveness of microlensing in [22].

Similar to the transit method, gravitational microlensing can detect multiple exoplanets in a single event. If a microlensing event involves a star with multiple planets, each planet may create a distinct feature in the light curve, allowing for the simultaneous detection of several planets. However, unlike the transit or radial velocity methods, which enable continuous monitoring of planets and association with specific stars, planets detected through microlensing can only be observed during the event. Since microlensing events are one-time occurrences, they cannot be revisited, making it impossible to observe the same planetary system multiple times. The distance to a planet detected *via* microlensing, especially for objects at large scales (thousands of light-years), can only be approximated. The technique relies on the gravitational lensing effect, which does not provide precise measurements of the lensing object's distance, introducing uncertainties in determining the planet's actual location, particularly for planets at extreme distances. Additionally, gravitational microlensing depends on the rare and random alignment of two stars along the line of sight from Earth, with one acting as the lens and the other as the background source. This precise alignment is essential for the lensing effect to occur and is inherently unpredictable, making microlensing a stochastic process that relies on chance alignments.

3.6 Missions Launched to Detect Exoplanets

Over the past few decades, several groundbreaking space missions have been launched to observe and study exoplanets, dramatically expanding our understanding of planets beyond our solar system. These missions employ a variety of techniques, from detecting the slight dimming of stars as planets pass in front of them (transits) to measuring the subtle gravitational influences that planets exert on their host stars.

Some of the most significant missions, such as NASA's Kepler and TESS telescopes, have transformed the field by discovering thousands of exoplanets, ranging from Earth-like rocky worlds to massive gas giants. In addition to dedicated exoplanet missions, observatories like **Hubble** and **James Webb Space Telescope (JWST)** have played critical roles in characterizing the atmospheres and compositions of exoplanets, providing deeper insights into their potential habitability and formation. These missions continue to shape the future of exoplanet science, paving the way for more advanced techniques in the search for potentially habitable worlds. We have presented a timeline of all the space-based telescope exoplanet missions in Figure 3.12. We have presented some of the space missions along with their launching year and purpose of launching in Table 3.2.

We have also presented an overview of those missions in detail below:

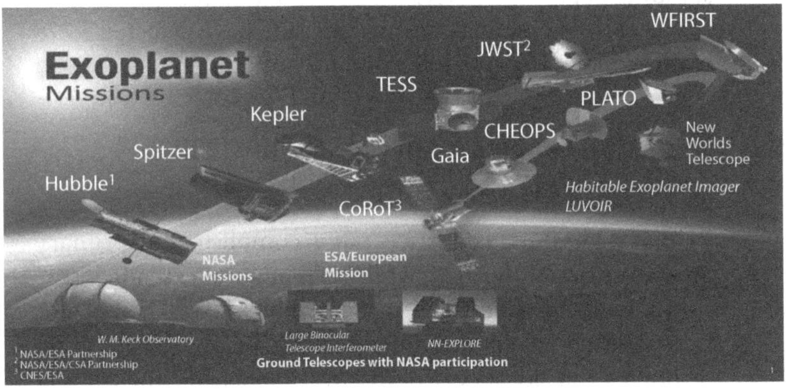

Figure 3.12 Timeline for the launch of all space-based telescope exoplanet missions. (Image credit: NASA/ESA/CSA).

Table 3.2 Details of some major space missions.

Name of the mission	**Launching date**	**End of the mission**
Kepler Mission	6 March 2009	30 October 2018
Hubble Space Telescope	24 April 24 1990	Remains in operation
Transiting Exoplanet Survey Satellite	18 April 2018	Remains in operation
James Webb New Space Telescope	25 December 2021	Remains in operation
CHaracterising ExOPlanet Satellite	18 December 2019	Remains in operation

3.6.1 Kepler Space Telescope (2009–2018)

- **Primary Mission**: The Kepler Space Mission launched from Cape Canaveral, Florida, with the goal of finding exoplanets. It watched the brightness of over 150,000 stars in a single patch of the sky using the transit method [23, 24]. Kepler also aimed to figure out what percentage of stars might have Earth-sized planets that could potentially support life by surveying a small part of the galaxy.
- **Key Achievements**: It significantly expanded our understanding of planetary systems. As of 16 June 2023, the Kepler space telescope and its extended mission have detected 2,778 planets, including super-Earths and hot Jupiters.

3.6.2 Transiting Exoplanet Survey Satellite (TESS) (2018 Present)

- **Primary Mission**: The spacecraft accomplished its primary objectives on 4 July 2020, while its inaugural extended mission ended its operations in September 2022. The observatory scans all stars within 200,000 nearby bright stars through a full sky survey to discover planets during transits [25]. TESS collects information through an observational program that supplements the Kepler/K2 mission NASA has been operating since its beginning.
- **Key Achievements**: Identified thousands of potential exoplanets, including Earth-sized ones located in habitable zones. As of 6 July 2024, NASA's Transiting Exoplanet Survey Satellite (TESS) has detected 7,203 candidate exoplanets, with 482 of these being confirmed.

3.6.3 Hubble Space Telescope (1990–Present)

- **Primary Mission**: The Hubble Space Telescope (HST) was launched in 1990 through a collaboration between NASA and the European Space Agency (ESA). Its main goal was to study exoplanet atmospheres and explore the environments of exoplanets using its ability to observe in ultraviolet, visible, and near-infrared light [26].
- **Key Achievements**: Observed the atmospheres of numerous exoplanets, including the detection of water vapor, methane,

and carbon dioxide. On 4 February 2025, it was reported that NASA's Hubble Space Telescope identified eight visible rings in the massive galaxy LEDA 1313424, surpassing the number previously detected in any galaxy by any telescope. Research teams validated a ninth ring through data obtained from the W. M. Keck Observatory in Hawaii.

3.6.4 James Webb Space Telescope (2021–Present)

- **Primary Mission**: The Webb was launched on an Ariane 5 rocket from Ariane Space Spaceport in Kourou, French Guiana. To use infrared spectroscopy to study exoplanet atmospheres in detail, including identifying biosignatures and analyzing the chemical composition [27, 28], and to better understand how planets, stars, and galaxies are born and evolve over time.
- **Key Achievements**: Will perform in-depth studies of exoplanets, including those in the habitable zone, and detect atmospheric components like water, carbon dioxide, and methane [29]. Surveying new exoplanets is the main goal of JWST, and the telescope has the capability to analyze their elemental makeup as well as environmental elements and potential habitability.

3.6.5 COROT (2006–2013)

- **Primary Mission**: Detect exoplanets using the transit method and study stellar interiors *via* asteroseismology [30].
- **Key Achievements**: Discovered the first multiplanetary system *via* the transit method and provided insights into the sizes of exoplanets.

3.6.6 PLATO (PLAnetary Transits and Oscillations of Stars) (Launch Expected 2026)

- **Primary Mission**: It will survey stars for transiting exoplanets, with a special focus on Earth-sized planets in the habitable zone [31].
- **Key Objectives**: Discover and characterize Earth-like exoplanets orbiting stars similar to our Sun.

3.6.7 CHEOPS (CHaracterising ExOPlanet Satellite) (2019–Present)

- **Primary Mission**: It was launched from Centre Spatial Guyanais, ELS. Its primary mission was to study known exoplanets by precisely measuring their size and refining their orbital parameters [32].
- **Key Achievements**: Provides high-precision photometry of exoplanet transits, improving the determination of exoplanetary radii.

3.6.8 Spitzer Space Telescope (2003–2020)

- **Primary Mission**: Conduct infrared observations of exoplanet atmospheres and outer regions of planetary systems [33].
- **Key Achievements**: Studied the atmospheres of hot Jupiters and made significant contributions to the detection of exoplanetary transits.

3.6.9 Gaia (2013–Present)

- **Primary Mission**: Map the Milky Way in 3D, providing detailed measurements of stars and their planetary systems. Although Gaia primarily focuses on astrometry, it can detect exoplanets by their gravitational effect on their host stars.
- **Key Achievements**: Identified exoplanets through precise measurements of star positions and has contributed to a broader understanding of galactic structure [34].

3.6.10 WISE (Wide-Field Infrared Survey Explorer) (2010–2011)

- **Primary Mission**: Conduct a survey in infrared wavelengths to detect cool objects, including exoplanets [35].
- **Key Achievements**: Discovered rogue exoplanets and brown dwarfs, helping to improve our understanding of how planets are distributed across the galaxy.

These missions collectively advance our understanding of exoplanets, their atmospheres, and their potential for habitability.

3.7 Role of Artificial Intelligence in Exoplanetary Science

Artificial intelligence (AI) plays a crucial role in exoplanetary research to detect exoplanets while determining their characteristics and potential habitability. It accelerates data processing, enabling the identification of exoplanet candidates through methods like transit detection and radial velocity analysis. AI algorithms efficiently analyze light curves and spectra, revealing key planetary characteristics such as size, composition, and orbital parameters. In the search for habitable exoplanets, AI helps model planetary environments, simulate climates, and detect atmospheric biomarkers, such as oxygen or methane, which could indicate potential signs of life. By automating and optimizing detection and characterization processes, AI enhances our ability to identify and study planets that might support life.

3.7.1 Role of AI in Discovering and Studying Exoplanets

AI plays a crucial role in the discovery and characterization of exoplanets, enhancing both the efficiency and accuracy of this complex and data-driven field. AI-based tools facilitate the search for habitable exoplanets by automating data analysis, identifying patterns, and modeling planetary conditions. These tools process large datasets from telescopes, detect exoplanets within habitable zones, and analyze atmospheric signatures for potential biosignatures.

Here are some of the key ways AI is contributing:

3.7.1.1 Data Analysis and Detection

Exoplanet discovery involves analyzing vast amounts of data from telescopes and space missions (like Kepler, TESS, and James Webb). AI algorithms, particularly machine learning (ML) models, are used to sift through enormous datasets and detect potential exoplanet signals. In Figure 3.12, we have represented the diagram of exoplanet detection process using artificial intelligence (AI). Some methods include transit method, radial velocity method, and direct imaging method. Transit method is the most common method of exoplanet detection; it is performed by observing dips in a star's brightness when a planet passes in front of it (transits). AI is employed to analyze light curves, identifying these subtle dips amid noise and distinguishing between real transits and false positives (e.g., caused by instrumental issues). AI can also help detect the small periodic wobbles of stars caused by gravitational interactions with orbiting planets. Machine learning models improve the accuracy of detecting these shifts by recognizing patterns that may be missed by traditional methods. AI aids in processing images captured through direct imaging, helping to filter out noise from other sources, making it easier to detect exoplanets, particularly those far from their stars.

3.7.1.2 Characterization of Exoplanets

AI plays a crucial role in the characterization of exoplanets by enhancing our understanding of their atmospheric composition, climate, and potential habitability. Machine learning algorithms are used to analyze spectral data from exoplanet atmospheres, identifying key chemical signatures, such as water vapor, carbon dioxide, methane, and oxygen, which are critical for assessing habitability [36]. AI models can also help reconstruct atmospheric compositions and predict physical properties, such as temperature and pressure, by analyzing the patterns in light curves and atmospheric spectra. AI models, especially deep learning networks, can analyze the spectra of exoplanet atmospheres obtained *via* transit observations (when a planet passes in front of its star and light filters through its atmosphere). This helps identify chemical signatures such as water vapor, methane, or carbon dioxide. AI-driven climate models play a crucial role in simulating exoplanetary weather systems, surface conditions, and energy balance by integrating various data inputs. These models are essential for helping scientists assess whether a planet's climate could support life. By using AI algorithms, researchers can predict whether exoplanets could be habitable, taking into account factors like how far the planet is from its star, its atmospheric properties, and how stellar radiation affects its surface. Machine learning algorithms classify exoplanets based on their size, density, and other physical properties. For instance, whether an exoplanet is rocky (like Earth), gaseous (like Jupiter), or an icy giant can be inferred using ML algorithms that process the observed data.

3.7.1.3 Pattern Recognition and Anomaly Detection

AI's ability to recognize patterns in large and complex datasets is a critical asset in exoplanet research, especially when dealing with noisy and highly variable data. Traditional methods often struggle to differentiate subtle signals from background noise, but AI excels at identifying these patterns, particularly in time-series data collected from telescopes. This capability is especially useful in distinguishing between normal stellar variability, such as natural star fluctuations, and actual exoplanet transits, where a planet causes a brief dip in its star's light. AI's precision in detecting these faint periodic dimming events significantly improves the accuracy of exoplanet detection and reduces false positives in the data. For example, AI methods like neural networks can help filter out noise and distinguish real signals from false positives in time-series data. Additionally, AI algorithms are adept at identifying anomalies in the data, which could point to unusual behaviors or previously unexplored exoplanetary systems and phenomena. This combination of noise reduction and anomaly detection enables more reliable and efficient identification of exoplanets.

3.7.1.4 Improving Telescope Operations

AI is becoming an essential part of how observatories and telescopes operate, greatly improving data collection and enabling real-time analysis. One major application is automating observations. AI-driven systems can automatically adjust telescope settings to get the best conditions for observing. These systems can also identify and prioritize objects of interest, such as exoplanets that show promising traits like being in the habitable zone of their host star [37]. By recognizing patterns or unusual signals in the data, AI helps focus observation time on the most important targets, increasing the chances of discovering new exoplanets or studying known ones more closely. This automation reduces the need for manual control, allowing astronomers to conduct long-term studies with less human involvement, while also making observations more consistent and accurate.

In addition to automation, **predictive analytics** also plays a vital role in enhancing the efficiency of telescopes. AI models can be trained on vast datasets, such as previous exoplanet discoveries, stellar activity, and orbital data, to predict which regions of space are more likely to host exoplanets. By analyzing historical data and running advanced simulations, these models can guide the telescope to areas that are statistically more likely to yield new exoplanetary systems. This predictive capability allows for a more targeted approach to observation, ensuring that resources are focused on the most promising regions of the sky and reducing the amount of wasted time spent on less likely targets. As a result, AI-driven predictive analytics improves the overall efficiency of exoplanet research by optimizing telescope usage and increasing the probability of discovering new potentially habitable exoplanets.

3.7.1.5 Optimization of Surveys

AI optimizes astronomical surveys by automating data processing, enhancing target selection, and improving pattern recognition. It helps prioritize promising star systems for observation, efficiently classifies data, and detects exoplanet signals while reducing false positives. AI also optimizes survey parameters in real time, predicts the likelihood of exoplanet discoveries, and integrates data from multiple sources, increasing survey efficiency and discovery rates. This leads to faster, more accurate exoplanet detection and characterization. AI can enhance the creation and management of exoplanet catalogs by identifying missing or incorrectly classified entries. Machine learning models help curate and refine datasets to provide more accurate and comprehensive exoplanet databases. AI-driven techniques can help focus on the most promising targets, reducing the time spent on analyzing regions of space that are unlikely to yield new discoveries.

3.7.1.6 Collaborative Efforts with Simulations

AI is used to combine observational data with physical models to simulate exoplanet systems. This enables simulation of planetary formation and integration of multi-model data. AI can help simulate and predict how planets form and evolve in different star systems, offering insights into the types of exoplanets that might exist, and AI can combine data from different telescopes and observational techniques (e.g., radial velocity, transit, direct imaging) to generate a more complete picture of exoplanets.

In summary, AI accelerates the discovery and characterization of exoplanets by automating complex data analysis tasks, improving the precision of measurements, and aiding in simulations of planetary systems. Its application has revolutionized how scientists detect, classify, and understand exoplanets, opening up new avenues for exploration in astronomy and the search for life beyond Earth.

3.7.2 Role of AI-Based Tools in Finding Habitable Planets

AI-based tools accelerate the search for habitable exoplanets by efficiently processing large datasets, detecting planets in habitable zones, and analyzing atmospheric signatures for potential biosignatures. They model long-term habitability and characterize exoplanet properties like size and temperature, streamlining the identification of planets with conditions that could support life.

3.7.2.1 Searching for Exoplanet Habitability Using a Novel Anomaly Detection Method

Indian astronomers have developed a new AI-based algorithm to identify potentially habitable exoplanets with high accuracy. Using this anomaly detection method, they can pinpoint planets that closely resemble Earth among thousands of confirmed and proposed exoplanets.

For centuries, humans have wondered if life exists on other planets. With billions of planets in our galaxy alone, the question remains: are there other worlds that could support life? In this study, astronomers from the Indian Institute of Astrophysics (IIA) and BITS Pilani, Goa, have introduced an innovative approach to address this question [38]. The method is based on the idea that Earth itself is an "anomaly"—a rare planet capable of supporting life. The researchers applied anomaly detection techniques, typically used in industrial systems to find outliers, to the task of identifying similar "anomalous" planets in large exoplanet datasets. The team identified 60 potentially habitable planets from around 5,000 confirmed and 8,000 proposed planets, focusing on those with Earth-like features. The breakthrough AI tool, called the **Multi-Stage Memetic Binary Tree Anomaly Identifier (MSMBTAI)**, uses a multistage memetic algorithm to quickly screen exoplanet data and identify those with potential for habitability. The approach was effective even when excluding surface temperature as a factor, suggesting that the method can simplify future exoplanet analysis. This research opens new avenues for finding habitable planets, making the process more efficient and promising for future space exploration.

3.7.2.2 TOLIMAN Mission to Search for Habitable Worlds in the Alpha Centauri System

A team led by Professor Peter Tuthill from the University of Sydney has partnered with Sydney-based space tech company Spiral Blue for the **TOLIMAN Space Telescope Mission**, which aims to detect potentially habitable planets in the nearby Alpha Centauri system, just 4 light-years away [39]. The mission combines Spiral Blue's cutting-edge **Space Edge Computing** technology, powered by AI, with the University of Sydney's expertise in astronomy. This collaboration will enable real-time data processing directly in space, bypassing the bottleneck of transmitting large data volumes back to Earth. Professor Tuthill highlighted the potential of using AI to process data from Alpha Centauri, opening "a new window onto the nearby Universe" in the search for habitable planets. Spiral

Blue recently achieved a milestone by launching **Space Edge One (SE-1)**, the first Australian edge computer in orbit. This technology has already demonstrated success with **Cloud Detect**, an AI application that processes satellite images at ultrahigh speeds, proving the viability of analyzing data in real-time during space missions.

The **TOLIMAN mission**, named after a star in the Alpha Centauri system, aims to be the first space-based observatory to gather precise data on exoplanets around nearby stars. The mission is set to launch next year with a 12.5-cm space telescope. TOLIMAN follows the success of Professor Tuthill's experiment aboard NASA's **James Webb Space Telescope**. Spiral Blue, founded in 2018, is now advancing the next generation of Earth observation services through AI and edge computing, with applications across defense, city planning, utilities, and more.

3.7.2.3 Machine Learning Techniques to Study the Internal Structure of Rocky Exoplanets

The discovery and characterization of Earth-sized exoplanets, particularly those with a rocky composition similar to terrestrial planets, have been significantly advanced by new observational techniques. A key objective in studying these planets is to understand their internal structure, which is crucial for assessing their habitability. While mass-radius relations are commonly used to infer interior structure, deriving all possible structures for a given planet requires extensive computation.

For this purpose, Y. Zhao and his research team employed mixture density networks through machine learning techniques to forecast how rocky exoplanets are structured [40]. An MDN model processed data about rocky exoplanets within the 0.1 to 10 Earth-mass range utilizing planetary mass, radius, and water content information to develop their database. The model is shown to perform well, accurately predicting layer thicknesses and core properties, with results in good agreement with traditional interior models. Notably, the MDN approach is much faster computationally, offering a valuable tool for rapid predictions of planetary interiors. MDNs form the basis of machine learning technology, which Philipp Baumeister implemented in 2020 to study the interior composition of low-mass exoplanets within a 25 Earth-mass boundary [41].

3.7.2.4 Contribution of Artificial Intelligence in Searching Exoplanets

In the past decade, over a million stars have been observed for transiting exoplanets, but manually interpreting the data is time-consuming and prone to human error. Kyle A. Pearson *et al.* [42] introduced a new method for detecting exoplanet candidates using neural networks, specifically deep learning, which offers clear advantages over traditional methods. Unlike older techniques that depend on hand-coded metrics, the convolutional neural network (CNN) used in this study is trained to automatically spot patterns linked to planetary transits. The detection process for Earth-like exoplanets gets better by utilizing CNN methods instead of least-squares methods while retaining quality performance after interpolating results. Kepler light curves show periodic transits with high detection rates for organizations that utilize deep learning algorithms, since these organizations do not conduct model fitting. Upcoming survey operations in astronomy will gain from machine learning characterizations of exoplanets according to this method. Neural networks alongside artificial intelligence receive assessment from Saima Khan and Md. Abidur Rahman Khan through their published research [43].

AI finds 300+ unknown exoplanets in Kepler data
ExoMiner, an AI algorithm, has discovered over 300 new exoplanets from data collected by the Kepler Space Telescope, which ceased operations in 2018 [44]. Kepler's catalog of potential planets continues to yield new findings, with ExoMiner efficiently analyzing the data and identifying exoplanets more accurately than human experts. By mimicking the process of identifying dimming in stars caused by passing planets, ExoMiner has added 301 new exoplanets to the catalog. Although none of these exoplanets are in habitable zones, ExoMiner's success paves the way for its use in future exoplanet missions such as NASA's TESS and ESA's PLATO.

3.8 Conclusion

In conclusion, artificial intelligence (AI) has rapidly become an indispensable tool in the discovery and characterization of exoplanets, transforming the field by enhancing the efficiency, precision, and scope of astronomical research. AI-driven algorithms have significantly improved the ability to detect exoplanets, especially in large datasets generated by missions like Kepler and TESS, where traditional methods would be computationally infeasible. Machine learning techniques, such as neural networks, have enabled the identification of subtle signals, such as transits and radial velocity variations, with unprecedented accuracy, even in noisy data. Additionally, AI is playing a crucial role in the analysis of exoplanetary atmospheres, helping to model complex chemical compositions and predict habitability potential based on observed features.

Looking forward, AI will continue to play a pivotal role in the next phase of exoplanet research. With the launch of next-generation space telescopes and ground-based observatories, AI will be essential in processing and interpreting the enormous volumes of data generated, helping scientists prioritize the most promising targets for further study. As AI models become more sophisticated, their potential to not only detect exoplanets but also model their climates, predict their habitability, and even identify signs of life will expand, pushing the boundaries of our knowledge about the universe. Ultimately, the integration of AI with traditional astronomical methods will continue to drive innovation in the search for exoplanets, bringing us one step closer to answering the age-old question of whether we are alone in the universe.

Furthermore, AI's capacity for automating repetitive tasks and optimizing data collection is accelerating the pace of discovery, making it possible to analyze vast amounts of data in real time, while improving our understanding of exoplanetary systems. As AI techniques continue to evolve, their integration with traditional methods will lead to more precise characterizations of exoplanets, including the identification of Earth-like planets and the search for biosignatures, thereby advancing the search for life beyond Earth. Overall, AI is reshaping how we explore the universe, offering the potential to uncover new exoplanetary systems, refine our understanding of planetary formation, and guide future missions to distant worlds. As technology keeps advancing, AI will play an even bigger role in unlocking the mysteries of exoplanets and pushing the boundaries of space exploration.

Acknowledgement

I would like to acknowledge the Editor of this book chapter, Yogesh Chandra for his valuable input, which significantly enhanced the quality of this book chapter. I also want to thank the global exoplanet community for their continued contributions to research in this field. I would like to thank the contributors who generously shared their expertise and resources, even in the absence of formal funding, which made the completion of this chapter possible.

References

1. Fontenelle, M.D., Gunning, M., Lalande., J.J.L.F.D., *Conversations on the Plurality of Worlds*, Printed for J. Cundee; sold by T. Hurst., London, 1803.
2. Wolszczan, A. and Frail, D.A., A planetary system around the millisecond pulsar PSR1257 + 12. *Nature*, 355, 145–147, 1992.
3. Mayor, M. and Queloz, D., A Jupiter-mass companion to a solar-type star. *Nature*, 378, 355, 1995.
4. Xin, L., Exoplanets, extraterrestrial life and beyond: an interview with Douglas Lin. *Natl. Sci. Rev.*, 9, 2, 2022.
5. Akeson, R.L., Chen, X., Ciardi, D., *et al.*, The NASA Exoplanet Archive: Data and Tools for Exoplanet Research. *PASP*, 125, 989, 2013.
6. Kane, S.R., Arney, G., *et al.*, Venus as a laboratory for exoplanetary science. *J. Geophys. Res. Planets*, 124, 2015–2028, 2019.
7. Yujuan, L., *et al.*, The Progress of Exoplaneting Extra-Solar Planetary Systems. *Prog. Astron.*, 23, 226–238, 2005.
8. Charbonneau, D., Brown, T.M., *et al.*, Detection of Planetary Transits Across a Sun-like Star. *ApJ*, 529, 45–48, 2000.
9. Fortney, J.J., Dawson, R.I., Komacek, T.D., Hot Jupiters: Origins, structure, atmospheres. *J. Geophys. Res.: Planets*, 126, e2020JE006629, 2021.
10. TRAPPIST-1 JWST Community Initiative, de Wit, J., *et al.*, A roadmap for the atmospheric characterization of terrestrial exoplanets with JWST. *Nat. Astron.*, 8, 810–818, 2024.
11. Inamdar, N.K. and Schlichting, H.E., The formation of super-Earths and mini-Neptunes with giant impacts. *MNRAS*, 448, 1751–1760, 2015.
12. Mazeh, T., Holczer, T., Faigler, S., Dearth of short-period Neptunian exoplanets: A desert in period-mass and period-radius planes. *A&A*, 589, A75, 2016.
13. Scholz, A., Muzic, K., *et al.*, Rogue Planets and Brown Dwarfs: Predicting the Populations Free-floating Planetary Mass Objects Observable with JWST. *Publ. Astron. Soc. Pac.*, 134, 10, 2022.
14. Lagrange, A.-M., Direct imaging of exoplanets. *Philos. Trans. R. Soc. A.*, 372, 20130090, 2014.
15. Chauvin, G., Lagrange, A.M., *et al.*, a giant planet candidate near a young brown dwarf. Direct VLT/NACO observations using IR wavefront sensing. *A&A*, 425, L29–L32, 2004.
16. Dai, Z., *et al.*, Five Methods of Exoplanet Detection. *J. Phys. Conf. Ser.*, 2012, 012135, 2021.

17. Gatewood, G.D., The multichannel astrometric photometer and atmospheric limitations in the measurement of relative positions. *AJ*, 94, 213–224, 1987.
18. Winn, J.N., Exoplanet Transits and Occultations, in: *Exoplanets*, vol. 526, pp. 55–77, University of Arizona Press, Tucson, AZ, 2010.
19. Miguel, Y., The exoplanet revolution. *Europhys. News*, 50, 41–43, 2019.
20. Deeg, H.J., Belmonte, J.A., *et al.*, *Handbook of Exoplanets*, Springer International Publishing AG, part of Springer Nature, Cham, Switzerland, 2018.
21. Mao, S. and Paczynski, B., Gravitational Microlensing by Double Stars and Planetary Systems. *ApJ*, 374, L37–L40, 1991.
22. Beaulieu, J.P., Bennett, D.P., *et al.*, Discovery of a cool planet of 5.5 Earth masses through gravitational microlensing. *Nature*, 439, 437–440, 2006.
23. Borucki, W.J., Koch, D., *et al.*, Kepler Planet-Detection Mission: Introduction and First Results. *Science*, 327, 977, 2010.
24. Howell, S.B., Sobeck, C., *et al.*, The K2 Mission: Characterization and Early Results. *PASP*, 126, 398, 2014.
25. Ricker, G.R., Winn, J.N., *et al.*, Transiting Exoplanet Survey Satellite (TESS). *Proc. SPIE*, 9143, 15, 2014.
26. Meylan, G., Madrid, J.P., *et al.*, Hubble Space Telescope Science Metrics. *PASP*, 116, 790–796, 2004.
27. Beichman, C., *et al.*, Observations of Transiting Exoplanets with the James Webb Space Telescope (JWST). *PASP*, 126, 1134, 2014.
28. Carter, A.L., May, E.M., Espinoza, N., *et al.*, A benchmark JWST near-infrared spectrum for the exoplanet WASP-39 b. *Nat. Astron.*, 8, 1008–1019, 2024.
29. Impey, C. and Henry, H., *Dreams of Other Worlds*, Walter de Gruyter GmbH. Princeton University Press, New Jersey, United States, 2016.
30. Auvergne, M., *et al.*, The CoRoT satellite in flight: description and performance. *A&A*, 506, 411, 2009.
31. Rauer, H., Aerts, C., Cabrera, J., *et al.*, The PLATO mission. *Astron. Nachr.*, 337, 961, 2016.
32. Cessa, V., CHEOPS mission overview. *Proc. SPIE 11180, International Conference on Space Optics — ICSO*, p. 111803L, 2019.
33. Werner, M.W., Roellig, T.L., *et al.*, The Spitzer Space Telescope Mission. *ApJ Suppl. Ser.*, 154, 1–9, 2004.
34. Gaia Collaboration, Prusti, T., de Bruijne, J.H.J., *et al.*, The Gaia mission. *A&A*, 595, A1, 2016.
35. Wright, E.L., Eisenhardt, P.R.M., *et al.*, The Wide-field Infrared Survey Explorer (WISE): Mission Description and Initial On-orbit Performance. *AJ*, 140, 1868–1881, 2010.
36. G, I., Kammari, K.S., Thambi, S.V., Venugopalan, M., Enhanced Classification of Exoplanets Using Machine Learning: Integrating Feature Selection, Ensemble Algorithms, and Explainable AI Insights. *15th International*

Conference on Computing Communication and Networking Technologies (ICCCNT), Kamand, India, pp. 1–7, 2024.
37. Huang, K., Hu, T., Cai, J., Pan, X., Hou, Y., Xu, L., Wang, H., Zhang, Y., Cui, X., Artificial Intelligence in Astronomical Optical Telescopes: Present Status and Future Perspectives. *Universe*, 10, 210, 2024.
38. Sarker, J., Bhatia, K., *et al.*, Postulating exoplanetary habitability *via* a novel anomaly detection method. *MNRAS*, 510, 6022–6032, 2022.
39. Bringing artificial intelligence to the search for habitable planets, news, The University of Sydney, 2 August, 2023.
40. Zhao, Y. and Ni, D., Machine learning techniques in studies of the interior structure of rocky exoplanets. *A&A*, 650, A177, 2021.
41. Baumeister, P., Padovan, S., *et al.*, Machine-learning Inference of the Interior Structure of Low-mass Exoplanets. *ApJ*, 889, 10, 2020.
42. Pearson, K.A., *et al.*, Searching for exoplanets using artificial intelligence. *MNRAS*, 474, 478–491, 2018.
43. Khan, S. and Khan, M.A.R., A Review of Exoplanet Detection Processes using Artificial Intelligence and Neural Networks. *Int. J. Comput. Appl.*, 186, 36, Foundation of Computer Science (FCS), NY, USA, 2024.
44. Valizadegan, H., Martinho, M.J.S., *et al.*, Exominer: A Highly Accurate And Explainable Deep Learning Classifier That Validates 200+ New Exoplanets. *ApJ*, 926, 120, 2022.

4

Cosmology and Dark Matter Research

Arun Kumar Rathore[1]*, B. C. Chanyal[1] and Sirley Marques-Bonham[2]

[1]*Department of Physics, G. B. Pant University of Agriculture and Technology, Pant Nagar, Uttarakhand, India*
[2]*Center for Gravitational Physics, Weinberg Institute, University of Texas, Austin, TX USA*

Abstract

This chapter presents a detailed analysis of cosmological ideas, observations, and models attributed to dark matter and highlights key insights. Approximately 27% of the mass and energy of the universe is made up of dark matter, an enigmatic element that has not yet been directly detected. A fundamental tenet of contemporary cosmology is the connection between invisible matter and the universe. The Big Bang Theory is the cosmological model that is most commonly used to describe how the universe came into being. According to this theory, 13.8 billion years ago, the universe was in a very hot and dense condition. It has been continuously growing ever since, gradually chilling down. The universe experienced an extremely quick and exponential expansion in the first fraction of a second following the Big Bang. Large-scale cosmic structures were formed as a result of this process, which also served to smooth out anomalies. When matter and radiation separated during the Big Bang, the universe cooled enough for electrons to combine with nuclei to form neutral atoms, releasing the cosmic microwave background (CMB) radiation. It should be mentioned that the interaction between the concept of dark matter and the Big Bang sheds light on the early stages of the cosmos and the formation of large-scale structures. In accordance with the Standard Model (SM), the cosmos produces equal amounts of matter and antimatter. This chapter shows that observations also show a significant imbalance, with matter vastly exceeding antimatter. This discrepancy is called the matter-antimatter irregularity. Making the connection between dark matter and invisible energy is extremely difficult. The connection between matter that is dark and physics outside of the Standard Model is examined in this chapter though. The gravitational impact of dark matter

**Corresponding author*: arunkumarrathore706@gmail.com

Yogesh Chandra, Manjuleshwar Panda and Mahesh Chandra Mathpal (eds.) The Intelligent Universe: AI's Role in Astronomy, (129–180) © 2025 Scrivener Publishing LLC

changes the evolution of galaxies and structures, influences cosmic matter distribution, and is essential for understanding the development of the universe. While direct finding of dark matter remains elusive, ongoing research and advancements in observational techniques continue to deepen our understanding of this enigmatic component of cosmology. In addition to providing insight into the structure of the universe, the search for dark matter's characteristics has the potential to revolutionize our understanding of basic physics and the universe. Meanwhile, in modern physics, dark matter is explored through theoretical models and experimental efforts, aiming to identify the particles that constitute this mysterious form of matter. Quantum field theory (QFT) offers a framework for comprehending relativistic particle interactions at the quantum level. As a result, QFT aids in the creation and testing of models of particles of dark matter along with their interactions. Quantum dynamics involves understanding the behavior of dark matter at quantum levels, including its involvement in the evolution and development of cosmic structures. The weakly interacting massive particles (WIMPs), the Lambda Cold Dark Matter (Λ-CDM) Model, the Modified Newtonian Dynamics (MOND) and TeVeS, the Galactic Rotation Curves, the CMB, the Gravitational Lensing, Galaxy Clusters, Direct Detection, and Indirect Detection were also highlighted in this chapter. We have linked the current theories with the experiments that primarily focus on weakly interacting dark matter. On the other hand, artificial intelligence (AI) with astronomical dark matter as well as dark energy shows complex challenges in space science. AI enhances the analysis, interpretation, and exploration of observational data, leading to significant advancements in understanding the secret of dark matter.

Keywords: Dark energy, dark matter, cosmos, quantum field, universe, matter-antimatter symmetry, standard model, fundamental forces

4.1 Introduction

Cosmology is a scientific investigation of the universe's formation, development, destiny, and large-scale structure. It deals with the fundamental questions of existence: How did the universe begin? What does it consist of? How does it evolve? Key empirical findings that have produced strong evidence for the Big Bang Theory, such as the universe's expansion and the existence of the CMB, are the foundation of modern cosmology. Cosmology also deals with structures on a grand scale—galaxies, superclusters, and voids—spread over hundreds of millions of light-years. Understanding how these assemblies form and evolve requires us to consider the role of dark matter, an invisible component that dominates the gravitational behavior of the universe and galaxies itself. At the moment, the Big Bang Theory is the most widely accepted theory to explain how the universe came into being. This hypothesis states that the cosmos has been extending ever since it started as a hot condensed point about 13.8 billion years ago. Edwin Hubble made the first observation of this expansion in 1929 when he realized that galaxies were moving apart at speeds related to their distance from one another. This observation became known as Hubble's Law. A crucial Big Bang Theory prediction was validated in 1965 when Arno Penzias and Robert Wilson discovered CMB. The residual radiation from the early, hot, and dense phase of the cosmos, which was only 380,000 years old, is represented by the CMB, a "fossil" from that period. Cosmologists have shaped our current understanding of the age and composition of the universe with the aid of the CMB's uniformity and minor fluctuations. In contemporary cosmology and astrophysics, dark matter continues to be one of the most profound mysteries. It is predicted to constitute about 27% of the total energy density of the cosmos compared to only 5% of ordinary (baryonic) matter.

However, dark matter does not interact with electromagnetic forces; hence, it neither produces nor absorbs light, making it invisible to all forms of electromagnetic detection. Several lines of reasoning, as discussed below, support the notion that the universe is filled with nonbaryonic matter that remains invisible to our current observational techniques but exerts a gravitational influence on the universe's formation. Fritz Zwicky coined the phrase "dark matter" (or a similar word) in 1933 after researching the Coma cluster and finding that the galaxies there were traveling too fast to be connected by the visible matter's gravitational pull alone [1]. This invisible stuff that appears to be present throughout the universe is called "dark matter." Evidently, one needs compelling evidence—albeit indirect—to presume the existence of a component of the cosmology that is invisible. Dark matter (DM) has been supported by a significant quantity of indirect evidence throughout the last century. Beginning at lower scales, the lack of a "Keplerian fall" and the flatness of spiral galaxies' rotation curves that stretch beyond their visual radius offer compelling evidence that these objects contain hidden mass [2]. Analyzing the value of the velocity dispersion in dwarf galaxies also leads us to the conclusion that, in certain cases, there is 10^2–10^3 times more mass in some of them than can be explained by their luminosity [3]. Another sign of dark matter's existence is the X-ray pattern of emission observed in elliptical galaxies as well as galaxy clusters [4]. Just like in galaxy clusters, where cosmic space is filled with hot gas emitting X-rays through bremsstrahlung at temperatures of millions of degrees, bigger elliptical galaxies also feature expansive atmospheres of hot gas that seem to remain in equilibrium. The cluster's gravitational potential can be calculated by examining the temperature and spreading of the hot gas. This makes it probable to calculate the overall mass of the objects that were quoted. Surprisingly, we find that clusters of galaxies contain five times as much material as we would anticipate from the visible galaxies and hot gas. Gravitational lensing and the combination of lensing with the X-ray method provide additional proof of the presence of DM in clusters and galaxies [5]. Strong proof for the existence of DM and its dissipationless character may originate in the study of baryon placement (by X-ray measurements) and the value of gravitational potential in clusters that experience collisions like the Bullet Cluster [6]. Fritz Zwicky's observations of the Coma galaxy cluster in the 1930s gave rise to the idea of dark matter. The measured velocities of galaxies appeared too great to be explained by the gravitational impact of visible matter alone, according to Zwicky's application of the kind of virial theorem to the motion of galaxies in the cluster area. For the discrepancy, he postulated the presence of unseen, or "dark," matter that provided the additional gravitational pull necessary to grasp the cluster together. Zwicky termed this mysterious material "dunkle Materie." In the 1970s, astronomer Vera Rubin and her partner Kent Ford discovered something revolutionary when researching spiral galaxy rotation curves. The stars in a galaxy's outer edges should orbit more slowly than those close to the center, where the majority of the visible mass—gas and stars—is concentrated, according to Newtonian mechanics. Rubin, however, concluded that galaxies must have more mass than is observable because she noticed that stars in the outer regions of galaxies were traveling at about the same speed as those toward the core. This was another indication of dark matter's presence, influencing galaxy dynamics on large scales. The best evidence for the SM of particle physics' shortcomings comes from cosmology and astroparticle physics. Apart from the mysterious dark energy, which may or may not be simply the cosmological constant, the SM cannot account for DM or matter-antimatter asymmetry, two features of the Universe. A lot of experimental research is being done on the probability that dark matter will be found in the near future. Since there are a lot of theoretical theories regarding the characteristics of the dark matter particles, we focus on a few highly motivated candidates, including sterile neutrinos, axions, and massive particles with weak interactions. We also give examples of more elusive and less motivated candidates, such as fuzzy dark matter. This gives an indication of the variety of possible dark matter possibilities, albeit it is by no means exhaustive. The matter-antimatter asymmetry is next examined, and its possible origin at the 100 GeV–TeV scale is discussed. The earliest period of the cosmos evolution is the last topic we discuss. The latter topic is of significant interest because of its fundamental nature, even though it does not appear related to modern particle physics. We emphasize that the cosmological data, especially in the case of cosmic microwave background anisotropies, clearly show that the commonly recognized hot stage was not the earliest. Even though there are other possibilities, inflation is the better explanation for the earlier stage and is in line with all that is now known. It is now well-accepted that there are close connections between particle physics and astroparticle physics. In fact, the SM of elementary particle physics must be extended in order to account for the existence of dark matter and conventional baryonic matter, two fundamental aspects of the Universe. One intriguing idea is that present or upcoming terrestrial experiments could reveal the physics underlying these events. There is currently much more effort being put into the experimental initiatives in these directions. The earliest period of the Universe's evolution is another facet of cosmology that does not yet appear to have a direct connection to investigations in terrestrial particle physics. On the one hand, it is undeniable that a far less common stage precedes the typical hot epoch. Cosmic microwave background (CMB) anisotropies and the distribution of matter (galaxies, clusters of galaxies, and voids) in the recent and current Universe are used to examine inhomogeneities in the Universe. However, we are certain that the cosmic perturbations were created prior to the hot epoch, even if we only know very general characteristics of them. Because of this, we are unable to pinpoint the exact beginning of the period; inflation is the best estimate, but other results have not yet been considered.

4.2 Role of Dark Matter in the Cosmos

The origin and development of galaxies, clusters' galaxies, and the cosmic web—the vast structure of the cosmos—are depending heavily on dark matter. The early universe's minute density fluctuations would not have gotten big enough to create galaxies if dark matter had not had a gravitational pull. These formations are created by the gravitational pull of dark matter, which unites baryonic matter.

4.3 Future Cosmological Observations

It is amazing that the investigation of the entire cosmos may reveal aspects of the very early epoch, which is characterized by an immense density of energy and rate of evolution. Future cosmological investigations may be able to differentiate between numerous ideas. We will not be able to address cosmology and astroparticle physics to any degree of completeness because they are vast fields of study. We do not even try to mention all of the dark matter object hypotheses that theorists have come up with in the past 30 years because there are so many. Rather, we focus on a few theories that might or might not be connected to reality. We concentrate on possibilities for the development of baryon asymmetry that use physics that can be measured by terrestrial experiments. An example of such a mechanism is electroweak baryogenesis. The earliest cosmology—inflation and its alternatives—is covered in the final section of these lectures. We conclude this introduction by pointing out that most of the subjects we cover have been studied in books in one form or another [7]. Naturally, there are a lot of reviews, some of which will be mentioned where they belong. We will talk about how to investigate the oldest epoch, focusing on future cosmological observations. We will discuss some evidence for DM in the next section.

4.4 Evidence of Dark Matter

From tiny galaxies that span only a few kiloparsecs to almost the entire observable universe, there is evidence for the presence of dark matter across a broad range of astronomical scales. Large-scale formations throughout the universe, galactic clusters, cosmic microwave backgrounds, and individual spiral galaxies are the three primary areas of research for cosmic phenomena [8]. These probes are all reliable and tend to converge on the same fundamental characteristics of DM. All three do not, however, have the same standing. Despite having a simple classical physics foundation and being quite obvious, the first type (galaxies) is arguably the least useful for quantitative calculations nowadays [9].

Among these methods, cosmology offers the most significant evidence and serves as the most precise tool for measuring DM quantities. It is predicated on intricate physics that goes beyond the scope of the SM and is based on the expanding universe framework of general relativity. Notably, all these observational approaches are linked to the gravitational influence of dark matter, as no direct evidence of dark matter exists beyond its gravitational effects. Therefore, changing gravity is a reasonable alternative to describe the behaviors typically ascribed to DM. Later on, a few of the suggestions made in this approach are examined. In the interim, we adhere to the conventional theory that dark matter is composed of unidentified matter particles, whether they are novel macroscopical bodies or new species of elementary particles. With $h = 0.678 \pm 0.009.2$, the present Hubble constant is written as $H_0 = h \times 100\ km/s$ Mpc and density relative to the critical density $\rho_{cr} = 3H^2/8\pi G_N$ is represented by $\Omega_i = \rho_i/\rho_{cr}$. The present cosmological DM density is close to the whole Universe and is represented in terms of a combination of parameters $\Omega_{DM}h^2$. The current best determination reads [10]:

$$\Omega_{DM}h^2 = 0.1186 \pm 0.0020 \tag{4.1}$$

This easily converts to $\Omega_{DM} = 0.258 \pm 0.004$, meaning that DM makes up roughly 26% of the present Universe's entire matter-energy content. Dark matter is thought to make up about 84% of the universe's entire matter content after the density of regular baryonic matter was measured to be $\Omega_b = 0.02226 \pm 0.00023$ or 4.8%. Assuming that dark matter is noninteracting, cold, and stable and has adiabatic inhomogeneities, the astrophysical and cosmological facts are replicated.

- Cold indicates that when structure creation starts, at the critical moment of matter-radiation equality (MReq), DM acts as a nonrelativistic fluid. It follows that since dark matter is made up of particles, at the precise time that corresponds to MReq, the average of their momentum ***(p)*** is much smaller than their mass $(p << M)$, and their velocities ***(v)*** are much smaller than the speed of light ***(c)***, $v << c$. This is true for the Universe's subsequent evolution as well. This suggests that DM must be heavier than a few keV if it is composed of thermalized particles.
- Noninteracting dark matter particles are those whose interactions with other particles or with each other are so small that they may be ignored. What distinguishes dark matter from ordinary matter is its minimal interactions, particularly the absence of electromagnetic interactions. The term "dark" in dark matter originates from its inability to interact with light. This is by no means to say that DM does not interact with traditional objects or with itself. Nevertheless, most theories regarding dark matter production in cosmology, along with nearly all detection methods, are based on the assumption that it possesses some level of interaction.

Interestingly, particles belonging to the "noninteracting" class have SM-weak interactions or interactions of comparable strength. Dissipationless is a corollary of noninteracting matter: DM cannot generate electromagnetic radiation, which means it cannot readily disperse its energy and cool down like conventional matter. This characteristic is fundamental to the ways that dark matter and ordinary matter behave differently in cosmology. Naturally, the ban is not absolute in this instance either; there are intriguing models where DM exhibits a slight degree of loss of energy.

- Continuous indicates that dark matter has existed while the starting of the universe and has not really vanished until recently. If DM consists of particles, either it does not decay or if it does, the cosmic repercussions are minimal since its half-life is significantly longer than the age of the cosmos.
- Adiabatic refers to the idea that dark matter is more concentrated in areas with higher densities of ordinary matter and photons, displaying the same initial density variations as other particles. This phenomenon takes place in an inflationary environment, where quantum oscillations in an inflated field are the source of all anomalies.
- Table 4.1 summarizes key observational evidence for dark matter, including galaxy rotation curves, where stars maintain unexpectedly high velocities at large radii, and gravitational lensing, where light bends due to an unseen mass. The cosmic microwave background (CMB) reveals dark matter's role in structure formation, while the Bullet Cluster provides direct evidence through mass separation in colliding clusters. Dark matter also explains large-scale structure formation by providing necessary gravitational influence, as demonstrated in Zwicky's galaxy cluster mass studies. Big Bang Nucleosynthesis (BBN) confirms that most matter is nonbaryonic, while anomalous galactic motions suggest either dark matter or modified gravity (e.g., MOND). Each phenomenon is supported by equations that quantify dark matter's effects, aiding in its theoretical and observational study.

The presence of DM in the universe is suggested by several lines of evidence:

Table 4.1 Evidence of dark matter and related equations.

Category	Description	Equation
Galaxy Rotation Curves	Observed rotation speeds of galaxies remain constant at large radii instead of decreasing, implying an unseen mass.	Orbital velocity equation: $v = \sqrt{\frac{GM}{r}}$ (Expected Keplerian decline) Observed flat rotation: $v_{obs} = \text{constant}$
Gravitational Lensing	Light from distant galaxies is bent by invisible mass, distorting and magnifying images.	Einstein's Lens Equation: $\theta_E = \sqrt{\frac{4GM}{C^2 D}}$ Einstein ring radius due to dark matter
Cosmic Microwave Background (CMB)	Small fluctuations in the CMB reveal dark matter's role in structure formation.	Density fluctuations (Power Spectrum): Dark matter affects temperature anisotropies
Bullet Cluster (Direct Evidence)	Two colliding galaxy clusters show visible mass separated from gravitational mass (inferred from lensing).	Mass-to-Light Ratio: $\frac{M}{L} \gg \frac{M_{visible}}{L_{visible}}$ (Mass from lensing exceeds luminous matter)
Structure Formation	Dark matter helps form galaxies by providing extra gravity to seed structure.	Jeans Instability Criterion: (Dark matter enhances structure growth)
Galaxy Cluster Mass (Zwicky's Study)	Fritz Zwicky found that galaxy clusters contain far more mass than visible stars suggest.	Virial Theorem for Cluster Mass: $2T + U = 0$ $M_{DM} \approx (Rv^2)/G$ (Mass inferred from velocity dispersion)

(Continued)

Table 4.1 Evidence of dark matter and related equations. (*Continued*)

Category	Description	Equation
Big Bang Nucleosynthesis (BBN)	Dark matter must exist to explain observed element abundances.	Baryon-to-Photon Ratio Constraint: $\eta = n_b/n_\gamma$ (Dark matter does not contribute to nucleosynthesis)
Anomalous Galactic Motions	Galaxy orbits and peculiar velocities suggest extra unseen gravitational effects.	MOND (Alternative to Dark Matter): $F = m\mu(a/a_0)a$ (Modified Newtonian Dynamics, an alternative explanation)

4.4.1 Mini-Galaxies

Dwarf galaxies are small galaxies that have a few billion stars. There are usually only a few billion stars in dwarf galaxies, whereas their larger cousins can have hundreds of billions. Many of these dwarf galaxies orbit more massive galaxies.

4.4.1.1 Rotation Pattern of Spiral Galaxies

Galaxies' vertical axis is where their rotational center is located. Based on the distance they are from the galactic center, the circular motion of stars and tracers (such masers and hydrogen clouds) can be determined by measuring the Doppler shift of atomic lines. One term for this is a rotation curve. The circular velocity $(\boldsymbol{v_c})$ of a test particle, such as a star, with mass $(\boldsymbol{m})$ and its mass $\boldsymbol{M(r)}$ within a radius $(\boldsymbol{r})$ from its center are predicted to have a straightforward relationship by basic Newtonian dynamics:

$$v_c(r) = \sqrt{\frac{G_N M(r)}{r}} \tag{4.2}$$

$$m\frac{v_c^2(r)}{r} = \frac{G_N m M(r)}{r^2} \tag{4.3}$$

The disk's arms, which typically extend O(10) kiloparsec, and a dense central bulge contain the majority of the (visible) mass in spiral galaxies. Gauss' rule allows for the replacement of all apparent mass within the orbit by a constant ($\boldsymbol{M}$) at large enough $\boldsymbol{r}$. The velocity should then follow an "eplerian" fal, $v_c(r) \sim 1/r^2$. The crucial issue is that, in contrast, it has been noted that, in large samples, the rotation curves of galaxies of this type stay flat (i.e., unchanged with ($\boldsymbol{r}$)) to considerable distances. Thus, additional unseen mass is needed to prevent the dissolution of galaxies with the disappearance of peripheral stars. The aforementioned equations can be easily integrated if spherical symmetry is assumed. The extra hidden material ought to be distributed in a thin dense ring of particles $\rho(r) \sim 1/r^2$ out to big $\boldsymbol{r}$ in order for the velocity to stay roughly constant. It is anticipated that the halo will eventually fade and the curve will begin to decline with increasingly bigger ($\boldsymbol{r}$). However, at such great distances, tracers are usually unavailable. In 1970, W. Kent Ford and Vera Rubin traced about 70 hydrogen clouds and made the first precise measurements of the rotation curve of the Andromeda galaxy (M31) [11]. The curve was found to be quite flat out to about 22 kpc [12]. This was ultimately confirmed by additional radio and optical studies in tens of additional galaxies, revealing rotation curves that stay flat up to 50 kpc and beyond. The issue of dark matter gained popularity as soon as the data were interpreted as proof of "missing mass." The flatness of rotation curves is still arguably the most obvious and compelling indication for DM, even after hundreds of spiral galaxies have been discovered. The several luminous components (disk, gases, bar, etc.) of the detected galaxies and eliminate the dark matter involvement are painstakingly modeled in realistic research as opposed to the straightforward proof of principle shown above. It was confirmed by additional measurements in tens of additional galaxies using both optical and radio methods, which revealed rotational curves that remain flat for at least 50 kpc. For a visual representation in Figure 4.1, NASA's educational materials, such as "The Hidden Lives of Galaxies," include detailed figures illustrating these rotation curves. These materials shed light on how dark matter in galaxies is inferred from the observable flat rotation curves. NASA's official website features high-resolution photos of several spiral galaxies, including NGC 4622, taken by the Hubble Space Telescope. While these images primarily showcase the galaxies' structures, they complement studies on galactic rotation by providing detailed views of spiral arms and other features.

The findings were promptly construed as proof of "missing mass," and the dark matter issue gained attention. Even with the observation of hundreds of spiral galaxies, the most obvious and compelling evidence for DM

Figure 4.1 Galaxy rotation curve. (Image credit: NASA, ESA).

is still likely the flatness of rotation curves. The impact of DM is eliminated in realistic investigations, which closely analyze the various luminous properties (bulge, bar, disk, gas, etc.) of the galaxies that have been observed, in contrast to the simplistic proof of the principle illustrated above.

The simplest explanation is that most of the material in galaxies cannot be discussed by the bright stars and objects that live in the disk of galaxies. To account for the observations, galaxies must contain massive dark halos composed of mysterious dark matter. The gravitational attraction that speeds up the circular motion of planets and other objects is produced by the mass inside these halos. In actuality, DM makes up nearly 95% of a galaxy's mass. The galaxy's velocity profile NGC 6503, which is linked to the radial distance from its center, cannot be solely explained with baryonic matter, such as the gas and disk, that is visible. However, a decent fit to the data is made possible by the addition of a dark matter halo. Rotation curves are limited to seeing out to distances of tens of kiloparsecs, or as far as there is light or neutral hydrogen (21 cm). As a result, while DM halos can be seen in its early stages, the majority of the DM cannot be located. These limits are not relevant to the lens experiments detailed in the subsequent section. Researchers now know that the rotation curves of all the galaxies studied, such as our Milky Way, are flat.

4.4.1.2 *Other Evidence At the Galactic Scale*

At the galactic level, more proof of dark matter has been put forth. Early simulations in the 1970s and their analytical interpretation suggested that disk stability in the spiral galaxy field depends on a dark spherical halo. Without it, star distributions would become chaotic as the disk becomes unstable in a matter of rotation cycles. Subsequently, more sophisticated research showed that the dark matter aura contributes to the conversion of a disk becoming a bar structure, ultimately decreasing the spinning of the bar [13]. This phenomenon is likely linked to the widespread presence of spinning disk galaxies. This topic is not now frequently discussed because it appears that there is no straightforward conclusion that can be made regarding the role of DM. The existence of massive DM halos around galaxies can also be shown by galaxy-galaxy lensing. This demonstrates how views of faraway galaxies are distorted by the gravity lensing effect of surrounding galaxies. The investigation of average attributes shows that the total matter in an individual lens galaxy is substantially larger than the observable mass (because the small lensing signal from just one galaxy is too weak to identify). These data can also be used to predict the normal form of dark matter halos that are generally somewhat flattened.

4.4.2 Midi: Cluster of the Galaxies

In 1933, by examining the distribution of velocity in the Coma galaxy formation and coming to the conclusion that more matter was required to hold the galaxies together, scientist Fritz Zwicky was the initial researcher to propose the presence of dark matter [1]. Galaxy clusters, which span several megaparsecs and comprise 100 to 1,000 galaxies, are the main gravitationally bound formations in the cosmos. Galaxy clusters are a great way to analyze the "average" cosmos because of their size [14]. Gravitational interactions enable us to calculate the total mass $\boldsymbol{mN}$ of a simple system with many particles $N >> 1$, each having mass m and spaced apart by r. This is done by using the velocity (v) and the size of the cluster r:

$$mN = \frac{2rv^2}{G_N} \tag{4.4}$$

Although the most accurate measurements of the average dark matter density today do not originate from galaxy clusters, they still suggest $\Omega_{DM} \approx 0.2$. This value differs from the circulation of whole mass inferred through weak lensing, as represented by the green contours. Zwicky's estimation was derived using the virial theorem, and due to its historical significance, we will briefly summarize the key idea. The virial theorem relates the average kinetic energy (KE) of a system to its average potential energy (PE).

$$\langle K \rangle = -\frac{1}{2}\langle V \rangle \tag{4.5}$$

The hot baryonic gas's X-ray picture is depicted on the colored map. The length of 200 kpc is represented by the white bar. When this kind of study is used to galaxy clusters, Zwicky claimed that more dark matter is needed because the overall mass of the cluster is more than the visible mass. As a result, the visible and components of dark matter are now separated, as shown in Figure 4.2. Although it was not widely believed, the DM idea was not ignored. It was believed that more data would be required to comprehend these systems. A galaxy cluster, also known as a cluster of galaxies,

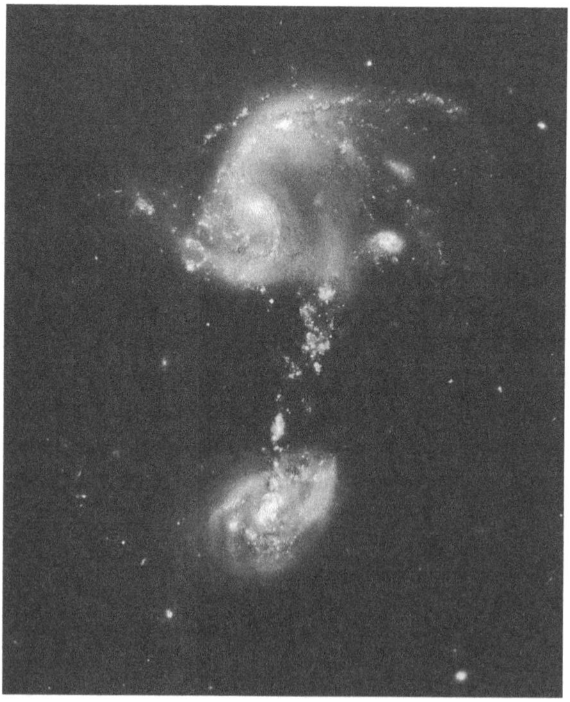

Figure 4.2 Collision between two galaxy clusters. (Image credit: NASA).

is an object made up of hundreds to thousands of gravitationally connected galaxies, with masses typically between 10^{14} and 10^{15} solar masses [6]. After some superclusters, they are the second-biggest known bound-by-gravity structures in the cosmos. Superclusters can form when galaxy clusters and individual galaxies group together under the influence of gravity.

4.4.2.1 *Effect of Gravitational Lensing*

One of the most compelling pieces of evidence is the phenomenon known as gravitational lensing, which had been anticipated by Einstein in his General Theory of Relativity. Light from a far-off object, like a galaxy, bends and creates several pictures or arcs when it approaches a big object, such as a galaxy cluster. The quantity of bending reveals the mass of a lensing object. The lensing mass often greatly exceeds the apparent mass, suggesting the presence of DM. Einstein's General Relativity Theory predicted notions such as light and mass bends, or lenses. This phenomenon allows for the gravitational determination of mass even in the absence of light. Lensing observations reveal that clusters of galaxies and galaxies contain huge masses of dark matter. This makes cosmic shear more important than just providing additional evidence for the DM as shown in Figure 4.3. Faraway luminous objects like galaxies or quasars are observed. The light from distant objects is bent toward the areas of great mass due to intervening materials. As a result, if the distant objects cannot be resolved separately, there may be many images of them or the surrounding item may appear brighter. Some of these pictures might be sheared or warped. Based on weak lensing, the Sloan Digital Sky Survey has found that galaxies, including the Milky Way, are far larger and more massive than previously thought, and that they need even more dark matter out to enormous distances. It is again pointed out that galaxies are dominated by dark matter. The panel on the right shows a computer construction of a foreground cluster based on lensing data created by Tyson *et al.* using the Hubble Space Telescope's camera. The number of galaxies in this incredibly rich cluster is indicated by the peaks in the image. Aside from these galaxies, a smooth component is evidently present: the dark matter found in clusters between galaxies. The fact that DM is detected at considerably more distant sites than rotating curves would probe—in galaxies up to 200 kpc far from the centers, in accordance with N-body calculations—is the main reason why lensing of DM has been successful thus far. There is evidence for DM in the filament (the cosmos web) on even bigger Mpc scales. The investigations of two clusters colliding called the "bullet cluster" (3.7 Gyr away, cataloged

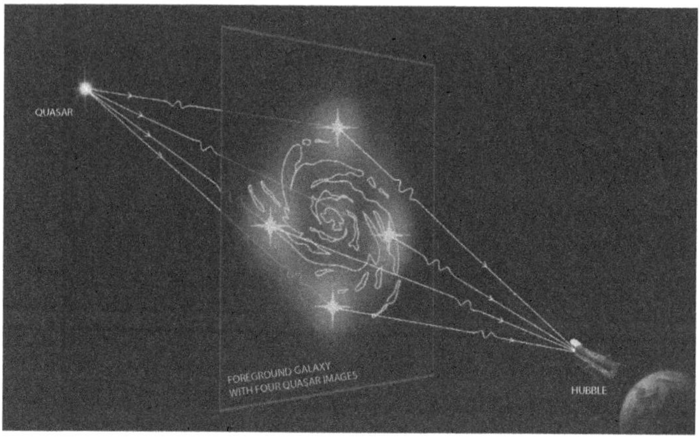

Figure 4.3 Weak gravitational lensing. (Image credit: NASA/JPL).

as 1E0657-558 and first observed in 2006) and the strongest proof of dark matter's existence to date for galaxy cluster measurements come from similar systems [15].

The bullet cluster's X-ray emissions reveal the dispersion of heated gas that constitutes the majority of its baryonic matter mass. The visible and dark components of the total mass distribution were examined independently using weak lensing. The physical distinction between visible and invisible matter is a distinctive feature of the system. According to the explanation, the two objects were once both regular systems with a mixture of dark and visible matter. A collision between the two objects occurred 150 million years ago. The heated gas obtained from the pair of clusters suffered a collisional tremor wave because visible matter interacts with itself in a substantial way. The clouds in terms of DM between the two systems, however, merely go through one another because DM had very few collisions with regular matter and with itself. Many equivalent processes have been observed since the bullet cluster was observed. The outcomes of 72 of them are reported by Harvey *et al.* (2015), who conclude that a significance of greater than 7σ may be used to show the DM [16]. Such data severely constrain any other possible interpretations in which DM is substituted with a gravity modification that is spatially indistinguishable from regular matter, causing the lensing to follow visible matter.

$$\frac{\sigma}{M} \leq 1\frac{cm^2}{g} = 1.8\frac{mb}{GeV} = \frac{4580}{GeV^3} \tag{4.6}$$

Remember that an average QCD cross section is 50 mb, for instance, in pp scattering. The finding of cosmic shear also supports DM at length gauges that are a little in between the galaxy, cluster, and cosmic ones [17]. When light from far-off galaxies is bent by the gravitational pull of foreground mass distributions, this phenomenon is known as cosmic shear. Unlike galaxy collide with one another galaxy lensing or cluster collisions, this effect arises from much larger and more diffuse structures, such as massive filaments and loosely bound matter clumps, rather than the dark matter halos of individual galaxies or clusters. To statistically analyze this phenomenon, multiwavelength investigations are conducted, covering both vast areas of the sky and reaching great depths to detect extremely distant galaxies. These surveys enable measurements across millions of background galaxies. Furthermore, these processes suggest valuable understandings of the high-energy physics characteristics of DM: the halos of DM did not slow down in comparison to the collisionless paths, which suggests a maximum value for the DM self-interaction cross section under the assumption of point interactions. They make it possible to calculate $\Omega_{DM} \approx 0.25$. By reproducing the matter distribution across the line of sight, one is essentially looking backward at the evolution of the huge DM frameworks [18] that form the scaffolding of the Universe (discussed in the next section).

4.4.3 The Universe: Maxi

At the moment, the universe as a whole provides the most convincing and precise proof for the possibility of dark matter. One of the main claims is that the cosmos would not have developed into what it is today if dark matter had not existed. Galaxies would be distributed very differently, and the cosmic microwave background's (CMB's) temperature variations would not show up as they do. The creation of cosmic structures is mostly driven by dark matter. It transforms the Universe into a state rich in structures of various sizes, from its starting state of nearly flawless smoothness with only minor inhomogeneities. Ordinary matter is unable to perform this role, mostly due to its coupling to radiation.

4.4.3.1 Large-Scale Structure Formation

According to computer models of the development of large-scale structures like clusters and galaxies, dark matter is necessary to explain how the minuscule density variations seen in the early cosmos gave rise to such formations. Today, the universe looks incredibly clumpy. Consider the findings of galaxy surveys, which enable the appreciation of a wide range of structures at various scales, including lumps, filaments, walls, voids, and more. The picture is equivalent to a three-dimensional map of nearly all of the observable universe's galaxies. The idea of a chunky universe is supported by weak lensing evaluations, cluster counts, the Lyman-α forest, and other astronomical evidence. In contrast, primordial inhomogeneities with amplitudes of $\delta \approx 10^{-5}$ were produced by inflation in the conventional cosmological model. Analysis of temperature anisotropies in the CMB reveals a distinctive spectrum with several auditory peaks. These peaks offer crucial information on cosmological parameters and are proof of density oscillations in the early universe. These anisotropies have been measured in detail by NASA's Wilkinson Microwave Anisotropy Probe (WMAP). The WMAP-derived angular power spectrum shows the magnitude of temperature variations at various angular scales, emphasizing the strong acoustic peaks.

The distribution of regular matter (red) and dark matter (blue) throughout the universe is contrasted in Figure 4.4. By showing how regular matter, such as stars and galaxies, is constructed upon an underneath substructure of dark matter, the clusters' brightness correlates their bulk density.

In reality, this is supported by the CMB's almost perfect homogeneity. Consequently, how could these small primordial lumps have grown from such small amplitudes to the massive contrasts seen today? As anticipated above, the response is that the increase in density perturbations is largely driven by the accessible supply of dark matter. We must outline their evolution in the early Universe to statistically comprehend how that occurred [19]. Let a universe that contains a fluid of generic matter, the precise properties of which are not yet specified. By substituting the Newtonian limit for the entire general-relativistic calculation, the key characteristics of the

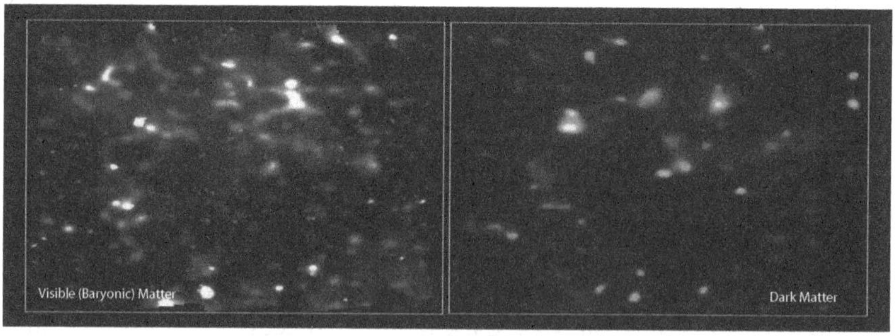

Figure 4.4 Large-scale structure formation. (Image credit: NASA and R. Massey, California institute of technology).

development of density agitations in such type of a medium may already be recognized. The matter (nonrelativistic) with length measurement scales far less than the distance can be explained in this way. Complete properties of a nonrelativistic fluids are provided by the density $\rho(\vec{r}, t)$, the velocity field $\vec{v}(\vec{r}, t)$, and the equation of state for pressure $P(\rho)$. The Newton potential Φ describes its gravitational interactions. The evolution (or "Euler") equations control these quantities. The continuity equation, conservation of momentum, and Poisson equation are given in Equ. (4.7), (4.8), and (4.9), respectively:

$$\frac{\partial \rho}{\partial t} + \vec{\nabla}.(\rho\vec{r}) = 0 \tag{4.7}$$

which create a set of differential equations that are not linear.

$$\frac{\partial \vec{v}}{\partial t} + (\vec{v}.\vec{\nabla})\vec{v} = -\frac{\vec{\nabla}P}{\rho} - \vec{\nabla}\$ \tag{4.8}$$

$$\nabla^2\Phi = 4\pi G\rho \tag{4.9}$$

4.4.3.2 *Cosmic Microwave Background*

The little temperature changes in CMB radiation include information about the universe's makeup. Dark matter must make up a sizable amount of the mass of the cosmos in order to fit into the observed trends, according to observations from satellites like WMAP and Planck. Cosmological-scale studies of CMB anisotropies provide additional evidence for dark matter (WMAP Working together 2003, 2008). The remaining radiation from the early hotter periods of the universe makes up the CMB. The oscillations that the photons underwent at a redshift of 1,000 froze just before they decoupled from the baryonic matter is appropriate sentence. These oscillations' angular scale and peak heights are good predictors of cosmological quantities like the baryonic proportion, the DM component, and the whole energy density. The sound boundary at final scattering refers to the geometry of the universe: if light is traveling in a one-dimensional straight line, as it might in a flat geometry, the angular scale of the initial Doppler peak should be at one degree, which is really the case. The energy concentration of the cosmos is approximately $\sim 10^{-29}\ gm/cm^3$ due to its

Figure 4.5 CMB (Image credit: planck Collaboration/ESA).

flat geometry. The magnitude of the second peak indicates that 4% of the entire amount is composed of regular atoms, whereas comparing all the peaks indicates that 23% of the entirety is DM. A Fourier transformation for the photonic field temperature is comparable to the cosmic microwave background's power spectrum. It provides a plethora of information on the contents of the cosmos, particularly DM, and is precisely measured. The DM density dictates the acoustic peaks' position, and their amplitude is extremely sensitive to the baryon-to-DM density ratio. The minimum Λ_{CDM} model makes it possible to derive the exact results of the cosmological quantities from a global fit and reproduce the specific characteristics of these disparities (as long as the primordial inhomogeneities are adiabatic). The European Space Agency's (ESA) Planck satellite has produced one of the most detailed maps of the CMB.

The map exposes minute temperatures corresponding to regions of varying densities, which are the seeds of all current cosmic structures. The image below is based on data from the Planck Legacy released in July 2018 as shown in Figure 4.5. The exact values for Ω_{DM} and Ω_b that are provided above are calculated in this manner. However, by significantly simplifying the study and concentrating solely on the key characteristics, which are listed below, it is also feasible to gain a qualitative understanding of how DM influences the spectrum's structure.

4.4.3.3 *Big Bang Nucleosynthesis*

A worldwide fit to the CMB data yielded a density of normal matter that nearly matches another exact determination from the Big Bang Nucleosynthesis (BBN). At a significantly higher temperature, T ~ MeV, BBN investigates an older stage of the cosmos. The synthesis of light elements (D, ^{3}He, ^{4}He, and ^{7}Li) is described by this theory, beginning with the free particles of protons (p) and neutrons (n) found in the state of plasma.

The baryons to photons ratio $\eta \equiv n_b/n_\gamma$ is complicated in the Big Bang Nucleosynthesis and can be converted into:

$$\Omega_b \equiv \frac{\rho_b}{\rho_{crit}} = \frac{n_b m_p}{\rho_{crit}} = \eta \frac{n_\gamma m_p}{\rho_{crit}} \tag{4.10}$$

In this case, $(\boldsymbol{m_p})$ represents the proton's mass and the photon's density, $n_\gamma = 410.5\ \text{cm}^{-3}$, which is determined using the CMB's temperature. Because the two findings are compatible, it is certain that a firm understanding of the evolution of the early Universe exists. Additional quantitative evidence supporting the need for an extra nonbaryonic component essential for cosmic structure formation arises from the fact that $\Omega_b < \Omega_M$. The total matter density, represented by ΩM, accounts for all forms of matter and is approximately 0.30. Before the ability of CMB and LSS observations to independently determine the relative proportions of DM and baryons, this was significant.

4.5 Theoretical Models of Dark Matter

Even though there are many potential dark matter particles, no direct discovery has been achieved as of yet. The dark matter candidates are numerous. Conventional matter, such as faint stars or stellar debris, makes up MAssive Compact Halo Objects; they can also be mirror matter or primitive black holes [20]. These are insufficient to fully address the question. Among the most widely considered nonbaryonic candidates for dark matter are axions and WIMPs, each proposed for distinct reasons within particle physics. While sterile neutrinos remain a potential option, standard large neutrinos surpass the required mass threshold to be cosmologically significant. Other possible candidates include primordial black holes, Kaluza-Klein particles, and nonthermal WIMPzillas, which emerge from higher-dimensional theories.

4.5.1 MAssive Compact Halo Objects

MACHO candidates include neutron stars, white dwarfs, black holes, planetary objects, and faded stars. Microlensing tests (MACHO and EROS experiments) and a mixture of other observational and theoretical conclusions [21] have demonstrated that MACHOs with masses $\sim 0.5 M_\odot$ provide little or nothing to the Galaxy's energy density. Nevertheless, a detection of a 20% halo fraction composed of $\sim 0.5M_\odot$ objects may be composed of planetary objects like white dwarfs. In conclusion, stellar objects can account for no more than 20% of the halo due to several constraints. Their progenitors must have a significant mass budget, generate detectable abundances of elements such as carbon, nitrogen, and helium, exhibit a highly pronounced peak in the initial mass function, and, most critically, emit observable infrared radiation [22].

4.5.2 WIMPs

WIMPs refer to a prominent DM applicant. It is thought that the weak nuclear force and gravity are the only ways in which these particles interact, making them hard to detect. Many extensions of the SM, including supersymmetry, anticipate WIMPs. In particle physics, WIMPs are also a kind of dark matter candidates. If there were thermal abundances of these types of particles in the early cosmos, they destroy one another, leaving a predictable quantity of them in the present. The relic density of these particles is found to be the appropriate value:

$$\Omega_\chi h^2 = (3\times 10^{-26} cm^3/sec)/\langle\sigma\upsilon\rangle_{ann} \tag{4.11}$$

where the value of annihilation cross section automatically provides the correct answer $\langle\sigma\upsilon\rangle_{ann}$ of the weak interaction strength. WIMPs are given a lot of weight as DM particles because of this synchronicity, which is referred to as "the WIMP miracle." Supersymmetry (SUSY), the smallest neutralino in the Minimal Supersymmetric SM, is the driving force behind the WIMP candidate. The supersymmetry in the study of particle theory aims to maintain the proper value of particle masses. Consequently, every known particle has a partner: the electron's partner is the selectron, the partner of the quark is the squark, and the photon's partner is the photino. Direct detection tests, in which the WIMP shoots off a nucleus in the detector, provide a variety of experimental characteristics of the interaction [23]. The lightest supersymmetric partner is an encouraging dark matter contender. Although it is unknown if SUSY particles are long enough to be dark matter, it is known that they are long enough to evade detection if they are found at the LHC as an elimination of energy in an event. Consequently, comparable astrophysical observation experiments are required. Several experimental traces of the interaction are detected in direct detection studies, where the WIMP fragments bounce off inside the nucleus in the detector [23]. As annihilate products of trapped WIMPs, neutrinos are released from the Earth or Sun in indirect detection studies. Silk *et al.* [48] first put forth this idea in regard to the Sun, while Freese [22] and Krauss *et al.* [49] subsequently developed it with regard to the Earth. Searching for anomalous cosmic radiation coming through a galaxy's halo is another way to find WIMPs. In the halo, WIMPs can destroy each other to produce neutrinos, positrons, or antiprotons. Additionally, it is possible to identify radio waves, gamma radiation, and neutrinos as products of the Galactic Center WIMP annihilation [24]. There will be a lot of discussions about planned and existing DM searches.

4.5.3 Lambda Cold Dark Matter

The Big Bang Theory uses the ΛCDM model as a mathematical framework. It is composed of three main parts: ordinary matter, cold dark matter (CDM), which is thought to exist, and the cosmic constant (Λ), which stands for dark energy. The simplest model that plausibly describes a variety of cosmic events is the SM of Big Bang astrophysics.

- the presence and construction of CMB
- the large-scale organization in galaxy spreading
- the measured levels of lithium, hydrogen, and helium
- The light from galaxies far away and supernovae is proof of the universe's observed accelerating expansion.
- According to this concept, the proper theory of gravity on cosmological scales is general relativity. After a time of dispute on the energy density of the universe and a number of observed elements of the cosmos that seemed to contradict one another, the result was the accepted consensus of astrophysics in the late 1990s. In the 1980s, cold dark matter was the main focus of research, which was found to have a critical density consisting of approximately 95% cold dark matter and only 5% hadrons (baryonic matter). These were promising for the formation of both clusters and galaxies of galaxies, but problems persisted. For example, observations made between 1988 and 1990 showed more large-scale galaxy grouping than was first thought. Furthermore, a smaller Hubble constant was needed by the model than what was shown by observational evidence. After the Cosmic Background Explorer (COBE) discovered CMB anisotropies in 1992, these difficulties became much more apparent.

Numerous modified CDM theories, such as mixing cold and hot DM and ΛCDM, were being intensively investigated until the mid-1990s. Later, the ΛCDM model became the mainstream paradigm following the observations of accelerated development in the 1998s.

- The core presumptions of the ΛCDM concept are contested by a number of other models. These include hypotheses that suggest disintegrating dark matter, bimetric gravity, entropic gravity, modified gravity, modified Newtonian dynamics (MOND), scale-invariant space, other variations of gravitational and dark matter concepts, and large-scale changes in cosmic matter density. The ΛCDM model describes widely known instances of metric space expansion, including the time dilation found in the fading brightness curves of supernovae and the redshift detected in important spectral lines in the light from distant galaxies, whether from absorption or emission. The Doppler shift in the electromagnetic spectrum as it passes through the expanding universe causes both of these phenomena. That expansion does not increase the size of the galaxies in space, but it does increase the distance between things that are not subject to the same gravitational pull. The modern rapid expansion of space against the pulling effects of gravity is explained by the cosmic constant, represented by the letter Λ (lambda), which is linked to the vacuum energy or dark energy that exists in empty space. Additionally, the concept allows galaxies that are far apart to move apart at speeds faster than light; although local expansion stays below this limit, the total effect across enormous cosmic distances may exceed it. Furthermore, the gravitational effects seen in large-scale cosmic structures that cannot be described by the visible matter alone are explained by the presence of dark matter [8]. The ΛCDM model specifically hypothesizes that the cold dark matter is:
- Nonbaryonic: Consists of materials that are not baryons but are instead composed of protons, neutrons, and, conventionally, electrons.
- Cold: Neutrinos, which are nonbaryonic but not cold, are excluded since their velocity is much less compared to c (speed of light) during the period of radiation-matter equivalence epoch.

- Dissipationless: Photons cannot cool
- Collisionless: Only gravity and perhaps the weak force allow DM particles to interact with other particles and with one another.

4.5.3.1 *Extending the ΛCDM Model*

The ΛCDM model, originally formulated for large-scale cosmological applications, has been adapted to various domains, including the mesoscale structure of individual galaxies. In these cases, both the slow-motion approximation and the weak-field approximation—considered valid for galaxies and galaxy clusters—allow the field equations to be simplified into Newtonian dynamics for describing mesoscale structures. This enables the ΛCDM model to be applied in describing the rotational dynamics for the study galaxies. The existence of dark matter within the ΛCDM framework was historically supported by Vera Rubin's study, which dates back to the 1930s, as well as studies of individual galaxy rotation curves. The link between the radial distance of visual stars or gas from the galactic center and their orbital velocity is depicted by a galaxy rotation curve. The evidence from early galaxies, however, did not match the expected curve. The masses of galaxies must be much greater than what we can observe if the galaxies are to spin as the data suggested. According to a model-centric interpretation, dark matter was introduced into the model and hypothesized as a characteristic of the target systems.

4.5.4 MOdified Newtonian Dynamics

MOND suggests altering Newton's second rule to explain the observable characteristics of galaxies. One of the most well-known ideas of this class, its main goal is to describe rotation curves of galaxies without referring to dark matter. Most astrophysicists, however, believe that the ΛCDM model fits observations better; hence, it has not been usually known. Israeli physicist Mordehai Milgrom created MOND in 1982 and gave its presentation in 1983 [25]. Milgrom proposed that the observed galactic rotation curves, which appeared to suggest the presence of greater amounts of matter than is observable, might be explained unless the force of gravity that a star encounters in a galaxy's outer regions dissipates more gradually than Newton's law predicts. By modifying Newton's rules to account for extremely low accelerations, which are common in galaxies' outer regions or in the intergalactic forces within galaxy clusters, the MOND hypothesis analyzes these rotation curves. According to the idea, the orbits of objects in the Kuiper Belt should also be impacted by the masses of the center of Galactic [26]. MOND's basic premise is that while the equations developed by Newton have been extensively studied in environments with significant acceleration (such as Earth's surface and the Sun's Solar System), these have yet to be verified for objects with comparatively modest accelerated motions, such as galaxies' stars. As a result, Milgrom proposed a novel effective gravitational force law (i.e., Milgrom's law), which connects an object's actual acceleration to the acceleration predicted by Newtonian physics [25]. The foundation of MOND, this law is selected to replicate the Newtonian outcome at higher acceleration, but at low acceleration, it produces distinct ("deep-MOND") behavior:

$$F_N = m\mu\left(\frac{a}{a_0}\right)a \tag{4.12}$$

Newtonian force (***F_N***), object gravitational mass ***(m)***, acceleration ***(a)***, an unknown function ***($\mu(x)$)*** called the interpolating function, and a new fundamental constant (***a_0***) that characterizes the boundary between both the Newtonian and deep-MOND domains are all used in this context. According to MOND, the gravitational force diverges substantially from the classic Newtonian connection $M \cdot G / r^2$ at very low accelerations. In this relationship, the gravitational strength is typically proportionate to mass and inversely proportionate to the square of the distance. In particular, MOND suggests that gravity's rate of change—along with spacetime's curvature—scales with the squared root of mass when gravity is far less than the value of a0, rather than the linear relationship observed in Newtonian physics. Additionally, the gravitational force weakens linearly with distance, rather than following the inverse square law. When a tiny mass, m, is close to a much bigger mass, ***(M)***, as a body on Earth or a star that lies near the Galactic center, MOND produces dynamics that are almost the same as those of Newtonian gravity. For accelerations up to roughly 1.2×10^{-10} m/s^2, there is a direct relationship between MOND and Newtonian dynamics. However, as accelerations fall below the a0 threshold, the forces predicted by MOND gradually deviate from the Newtonian model of gravity. In a galaxy, the gravitational acceleration reaches a_0 at a specific distance from its center. According to Newtonian gravity, at a distance 10 times greater, the gravitational force should decrease by a factor of 100. MOND, however, only anticipates a 10-fold decrease. MOND effectively explains important galactic-scale rotating effects without depending on the existence of as-yet-undiscovered dark matter particles outside of the

Standard Model, despite its negligible and undetected effects on objects on Earth, inside the Solar System, or even close to other planetary systems. The main reason for this is that MOND suggests that the incredibly weak gravitational pull that holds galaxies together on a large scale decreases at a considerably slower, almost linear pace with range from a galaxy's nucleus rather than according to the inverse-square equation.

4.5.4.1 Observational Evidence of MOND

These do not support the theory because MOND was created expressly to generate flat rotation curves, but each corresponding observation strengthens the empirical law. However, supporters assert that the MOND framework cleanly accounts for a wide variety of galactic-scale astrophysical phenomena [27]. It is challenging to use the concept of the dark matter hypothesis to explain many of these, which were discovered after Milgrom's original publications were published. The following are the most notable:

- Rotating curves in the computational model of MOND have been found to be flat and to give a particular link between a galaxy's asymptotic rotation velocity and the entire mass of baryonic substance, which is the sum of its mass in stars and gas. Milgrom dubbed this expected relationship the mass-asymptotic speed relation (MASSR); it is shown to closely match the MOND prediction in its observable manifestation, the baryonic Tully-Fisher relation (BTFR) [28].
- Given just the distribution of a galaxy's baryonic mass, Milgrom's law completely describes the galaxy's rotation curve. MOND specifically expects an even stronger link between aspects in its rotation curve and properties in the baryonic distribution of mass than the dark matter theory does (because DM triumphs over its mass budget and is anticipated to not nearly mirror the composition of baryons). Many spiral galaxies are said to exhibit this association, which is dubbed "Renzo's rule" [27].
- Recent studies have revealed that many of the tiny galaxies in the Milky Way and Andromeda are preferentially located in only one direction and display related movements. Given that they formed during a close collision with another galaxy, this suggests that they might be tidal miniature galaxies. If true, MOND is supported by the existence of mass disparities in these systems. Moreover, it has been underlined that

galaxies require a gravitational force stronger than Newton's, such as Milgrom's, in order to sustain their orbits over time [29].

- In the context of open clusters of stars published in 2022, Kroupa *et al.* [50] made the case that the imbalance in the population of resulting leading tidal tails and clusters' measured lifetimes are reliable with MOND but incompatible with Newtonian dynamics.
- The finding of particles that could make up cosmic DM would significantly imply that ΛCDM is accurate and that the laws of Newton do not need to be changed.
- If MOND is adopted as an idea of enhanced inertia, it predicts that unusual accelerations would occur on Earth at specific times and places throughout the year. These might be discovered in an exact experiment. The computation would be incorrect if MOND is understood as a concept of modified gravity since the Earth's exterior field effect would cancel out MONDian impacts at the Earth's surface.
- Tobias Mistele and Sabine Hossenfelder suggest that better acceleration evaluations could aid in distinguishing between different MOND systems and DM particle by using the redshift dependency of the radial accelerated motion to evaluate MOND inside their parameter-free MOND framework, called covariant emergent gravity.
- Their analysis suggests that MOND predicts a significantly weaker redshift dependence compared to particle DM models, offering a potential way to distinguish between these competing explanations as observational data become more precise [30].

4.6 ΛCDM and MOND

The ΛCDM model is the accepted position in both astrophysics and community and is considered the SM of cosmology. DM, a hitherto unidentified type of substance that can only be detected through its gravitational interactions and cannot be directly measured at any EM wavelength, is one of the greatest prominent aspects of the ΛCDM model. Based on the ΛCDM model, DM makes up 24% of the universe's total mass and energy. While the majority of astrophysicists accept the existence of dark matter, other astrophysicists contend that the incorporation of DM within the

ΛCDM model is a prime example of the model's overly flexible ability to include a wide variety of data into its parameters. To make sure the model fits the empirical facts, some astrophysicists believe that DM is an unnecessary *ad hoc* addition. To describe the same empirical facts, some of these opponents have put out alternative models that modify gravity to a MOND, eliminating the need to suggest a new type of matter. However, some people also think that the MOND approach is not acceptable. The modern understanding of the cosmos is based on general relativity. General relativity would need compelling arguments to be abandoned. As a result, the majority of astrophysicists believe that MOND is unnecessary. Although proponents of MOND assert that its models describe structural processes in the cosmos on par with ΛCDM, or perhaps better in some ways [31]. These two methods of simulating cosmic structures have various components and are essentially different in terms of the theories they are predicated on. Both models assert that they adequately anticipate, explain, and fit the empirical observations. It is not always troublesome for a phenomenon to have more than one incompatible model; in these situations, there is an expectation that the incompatible theories will eventually be decided [32]. Despite fitting the pattern of many incompatible models, the example of ΛCDM and MOND is surprising because the discrepancies do not stem from various representational decisions but rather from disagreements over theory. One way to read the ΛCDM versus MOND argument from a theory-focused perspective is as a straightforward fundamental disagreement: since MOND models are not founded on the best theory of gravity at the moment, it is incorrect to consider them successful. On the other hand, it might be considered a situation of incommensurability between two paradigms or a subjective decision between two possibilities. A third possibility is to interpret the problem as a sign of potential underdetermination of the theory since both claim to describe the empirical evidence [33]. These calculations are based on a particular wording of the question: is there solid evidence to support the belief in the entities that ΛCDM posits and to believe that it is the best model at the moment? By emphasizing how much evidence supports each theory, this formulation appears to lead to an argumentative dead end that might only be broken by acquiring more observational data or by the discovery of a DM particle. In 2018, Michela Massimi [51] offered a different approach to understanding the success claims of MOND and ΛCDM, suggesting that the argument be analyzed using multiscale modeling. The ΛCDM versus MOND debate, she claims, raises philosophical questions that represent Robert Batterman's 2013 "the oppression of scales," and the argument emphasizes the challenges of multiscale modeling when taking cosmological scales into account.

The following are the first three of her five assertions regarding the context of the ΛCDM–MOND dispute:

- Both MOND and ΛCDM function best at a particular scale (the galactic or galactic scale for MOND and the vast scale for ΛCDM).
- When modeling on several scales, each model has difficulties. The "scale down problem" that the ΛCDM faces as it transitions beyond large-scale creation of structures to the mesoscale of particular galaxies is what she is referring to. From the microscopic level of isolated galaxies to the vast scale of clustering and pattern formation, MOND faces what she calls the "upscaling problem."
- These problems have come up from a variety of philosophical perspectives, and several physical solutions have been proposed. At the mesoscale of individual galaxies, the ΛCDM model faces a challenge known as the "reduction in scale problem," which questions its ability to fully explain certain persistent anomalies. While ΛCDM effectively describes large-scale cosmic phenomena and observational techniques such as gravitational lensing—both of which are based on the framework of general relativity (GR)—it struggles with specific small-scale dynamics. Conversely, MOND encounters an "upscaling problem," which raises concerns about its consistency when extended to large scales where GR remains the dominant and well-tested theory. The difficulty lies in reconciling MOND's modified dynamics at galactic scales with the broader relativistic framework governing the cosmos.

4.6.1 Interpretation of the ΛCDM–MOND Debate

The ΛCDM model is based on the general theory of relativity, which is currently the best theory of gravity. In general relativity, Einstein's field equations (EFEs) specify the key relations. On long-length scales, general relativity is the foundation for models of the universe's evolution because gravity is predicted to be the dominant factor. The structure of the universe can be modeled using EFE by supplying assumptions, or conditions describing the global features of the space-time with the help of general relativity. The ΛCDM model provides a precise solution to EFE.10. Good estimates of the mass ***(m)***, energy content *(Ωm)*, general curvature *(Ωk)* of the universe, and other parameter values are incorporated into the parameterization of a model (ΛCDM) on the basis of empirical data. He briefly discussed the approximations and idealizations used to derive this model from EFE.11 in what follows. These events are vital because they shed light on the presumptions, idealizations, and approximations involved in creating a cosmological model. The target system, as defined in the modeling literature, is the whole universe; yet, the domain that this model is intended to use consists of only very vast distance scales. Such a model is intended to provide a mathematical explanation of the evolution of the cosmos and the formation of massive structures over extraordinarily long periods. More specifically, the model primarily performs a predictive role because its objective is to make claims about the possible theory of growth, past, take, and future configurations of the universe. The presence of smaller characteristics, including stellar populations within galaxies, is not a cause for concern; instead, the model is limited to regularities that are above the Hubble radius but below the galaxy length scale. After beginning with the General Relativity Theory, the model has attempted to describe and explain the configuration of the universe by assuming a certain explanation for the system's dynamics.

4.6.2 Justified Extension Explanation of Models

It appears that the reasoning for both MOND and ΛCDM has been a little "top-down" from gravitational theory. Given the widely accepted fundamental physical theories (such as general relativity) in the setting of the ΛCDM model, a model that significantly depends on a hypothesis is likely to be universally true on all boundaries, which is a sensible assumption. Similarly, a model is seen to be explaining anything when it is built with a significant theoretical component, such as a gravitational theory that explains space-time's dynamic character. The desire to expand the model to additional domains is understandable given these factors. The entities that are proposed follow a similar logic. Regarding κCDM, this entails acknowledging DM as a characteristic or feature of the actual target system, which necessitates its representation in that model. The model ought to possess descriptive extendibility to other domains if it includes explanatory processes that are thought to be both essential to the model's design and an actual component of the target system. These models are thought to have justified extensions beyond their original realm of applicability and construction, in addition to being regarded as explanatory in that regard.

This model-focused perspective, however, highlights a second point: a model can be enlarged in two different ways. The first is an extension to other domains or scales, such as the meso or cosmic scale, from the big scale. The second is an expansion of the model's intended use or claim of success, i.e., effectively taking into consideration phenomena as opposed to providing accurate forecasts or an explanation. Regarding ΛCDM and MOND, both models are effective in their initial specified domain of application (large-scale for ΛCDM, meso for MOND) in terms of their descriptive and predictive goals. This analysis can clarify this complexity by focusing on the model's objectives. When building a model, consideration is given to a specific domain and goal. It is therefore possible to evaluate a model's success or excellent fit concerning certain goals. However, a model that is thought to be appropriate for result prediction might not be a good fit for other applications, including explanation. Moreover, these two approaches to expanding a model can be integrated. A model is initially designed with a specific purpose and application domain in mind. For instance, the MOND framework was originally formulated to explain and predict galactic rotation curves at the scale of individual galaxies. By focusing on this particular scale, MOND aimed to address discrepancies in observed rotational velocities without invoking unseen dark matter. However, attempts to extend such models beyond their original scope—either by scaling down ΛCDM to explain small-scale structures or scaling up MOND to remain consistent with general relativity—pose significant theoretical challenges. It is not always the case that a model that is deemed successful for one domain will also be considered a good fit when applied to another; this could entail either moving from mesoscale prediction to large-scale prediction or transitioning from mesoscale prediction to mesoscale explanation.

4.6.3 Axions

Another dark matter candidate is the axion, an imaginary particle that was first proposed to solve a powerful CP issue associated with the concept of quantum chromodynamics.

Although axions are far lighter than WIMPs, if they are present in sufficient quantities, they may constitute dark matter. Fortunately, there is no need for cosmologists to "invent" new particles. For other reasons, particle physics already has two candidates: WIMPs and axions. Axions with masses in the $10^{-(3\text{-}6)}$eV range arise from the Peccei-Quinn solution to the stronger-CP issue associated with the concept of strong couplings. Axial bounds determined by the ADMX cavity investigation are approaching the end of the parameter-to-be-used range [34].

4.7 Sterile Neutrinos

Sterile neutrinos, also known as inactive neutrinos, are hypothetical particles that interact only through gravity rather than one of the basic interactions. Examples of these particles are neutral leptons-neutrons. To distinguish right-handed neutrinos—which can be incorporated into the Standard Model—from the active neutrinos that are already a part of the SM, the phrase "sterile neutrino" is employed. Sterile neutrinos are different in their behavior and possible involvement in particle physics since they do not participate in weak interactions, whereas active neutrino particles have an isospin charge that is $\pm 1/2$. Particles known as heavy neutral leptons or neutral heavy leptons have masses high enough to not interfere with the existing Big Bang Nucleosynthesis scenario and sterile neutrino-like quantum numbers [35]. Since sterile neutrinos only interact with ordinary matter through gravity rather than the weak nuclear interaction that genuine neutrinos do, they are a potential DM candidate. All other known fermions have both left- and right-handed chirality, which leads to the theoretical discussion of right-handed neutrinos. Since the observed active neutrinos are exclusively left-handed, the existence of their right-handed counterparts is a natural extension within the framework of particle physics [36]. Right-handed neutrinos could provide a natural explanation for the small masses of active neutrinos, as suggested by neutrino oscillation experiments. However, their own masses remain uncertain and could range from as low as 1 eV to as high as 10^{15} GeV [37]. According to the DM and leptogenesis theories, there must be a minimum of three different kinds of sterile neutrinos [38]. To maintain an anomaly-free electroweak interaction, exactly three active neutrino types are required—corresponding to the generations of charged quarks and leptons. The quest for sterile neutrinos remains ongoing in particle physics. If their mass is lower than the energy scale of high-energy particle collisions, they could be generated in experiments through interactions between active and sterile neutrinos. Conversely, if they are significantly heavier, their immediate impact would primarily manifest in the observed masses of active neutrinos. However, they might be important in DM, baryogenesis, or imaginary dark electromagnetic radiation, answering open concerns in cosmology and astrophysics [38]. A stronger-than-expected neutrino fluctuation output was reported in 2018 by the MiniBooNE experiment, which may have revealed the presence of sterile neutrinos. Yet, information from the MicroBooNE investigation in October 2021 revealed no proof of their existence. A particle needs to be nonzero mass and electromagnetically charge-free in order to be a good dark matter contender. Neutrinos and related particles continue to be a major area of study in dark matter because they satisfy these requirements. Additionally, observational evidence indicates that most dark matter is not relativistic (hot dark matter) but rather nonrelativistic (cold dark matter). It remains uncertain whether the Standard Model's active neutrinos, which have extremely low mass and move at high velocities, can fully account for dark matter. Unlike active neutrinos, sterile neutrinos have no strict constraints on their mass, meaning the possibility of their role in DM has not been ruled out. If sterile neutrinos do make up dark matter, their properties must meet certain conditions. Specifically, their mass would need to be sufficiently large (on the keV scale) to suit the current framework development of the cosmos based on the parameter space of extant supersymmetric theories that experimental data have not yet disproved.

Second, even though DM stability is not required, the particle lifetime should be longer than the age of the universe today. As a consequence, the seesaw mechanism's sterile-active neutrino mixing strength has an upper limit [36]. The sterile neutrinos are possible dark matter candidates based on current knowledge, but like all other postulated dark matter particles, their presence has not yet been verified.

4.7.1 Creation and Decay of Sterile Neutrinos

Sterile neutrinos may be produced and decayed as a result of mixing with virtual neutrinos. Heavy neutral leptons (NHLs) have been detected or observed in a number of experiments, including the NuTeV (E815) research at Fermilab Laboratory and the LEP-L3 operation at CERN. However, instead of confirming their existence, these experiments only helped establish observational limits. Numerous experiments were carried out, including the LEP-L3 experiment at CERN and the NuTeV (E815) research at Fermilab. To monitor the X-rays released from their demise, extremely sensitive X-ray detection equipment would be required if NHLs are a component of dark matter. Once the electroweak symmetry is violated, sterile neutrinos can combine with conventional neutrinos through a Dirac mass, just as quarks and charged leptons [39]. Both sterile and ordinary neutrinos can have Majorana masses, especially in more intricate theoretical models. Sterile neutrinos in the type-1 seesaw mechanism are often much heavier than the interacting neutrinos in the Standard Model since both the Dirac and Majorana mass components help to lower the masses of standard neutrinos and observe NHLs or heavy neutral leptons.

In seesaw models aligned with the Grand Unified Theory (GUT) scale, heavy neutrinos may reach masses close to the GUT scale (~10^{15} GeV). However, in alternative models such as the νMSM, sterile neutrino masses can range from keV to GeV, potentially lower than weak gauge bosons Z and W. A sterile neutrino with a light mass (~1 eV) was initially proposed to explain anomalies observed in the liquid scintillator neutrino detector (LSND) experiment. On April 11, 2007, however, researchers from Fermilab's MiniBooNE investigation declared that they had not discovered any evidence for this kind of sterile particle like neutrino. Despite this, subsequent analyses and experiments have provided some indications of its possible existence. The existence of a fourth type of neutrino with an estimated mass equal to 1.2 eV, for example, could account for the 3% discrepancy in detected antineutrinos reported by two separate detectors near a French nuclear reactor. Furthermore, the Daya Bay experiment successfully ruled out specific mass ranges by examining the potential of low-mass sterile neutrinos [40]. The Daya Bay collaboration's measurements of the antineutrino energy spectrum revealed an excess of antineutrinos around 5 MeV compared to theoretical predictions. Additionally, the overall antineutrino flux was observed to be approximately 6% lower than expected. This disparity may suggest the presence of sterile neutrinos or highlight knowledge gaps on neutrino interactions and characteristics. The CMB's appearance can be significantly influenced by the particle masses and neutrino counts. For example, in the early cosmos, the number of neutrino species had a direct effect on the expansion rate: the more neutrinos, the faster the expansion. The 2013 release of information from the Planck Satellite is consistent with the presence of sterile neutrino. Between 0 and 3 eV is the assumed mass. In 2016, scientists from the IceCube Neutrino Observatory discovered no proof that sterile neutrinos exist. However, a stronger-than-expected neutrino oscillation signal was observed by the MiniBooNE experiment in May 2018, raising the potential of sterile neutrinos. Despite this, the first findings from the October 2021 MicroBooNE experiment were consistent regarding the 3-neutrino model anticipated by the Standard Model and found no signs of sterile neutrinos. However, this discovery does not explain the unexpected results of MiniBooNE.

4.8 Method of Direct Detection

Cosmological and astrophysical data can be used to indirectly infer data concerning the invisible (DM) particle. The strength of its self-interaction is constrained by the Universe's ionization history, large-scale structure restrictions, and collision galaxy clusters, such as the Bullet clusters [6–16]. Three main strategies are being investigated at the moment to look at the mass, the coupling, and interactions of the cross section of dark matter particles with baryonic substances. These techniques entail looking for the decay products of DM annihilation in areas with a high density of dark matter, like the Sun, dwarf galaxies, and the galactic core. Researchers are also working on creating DM particles in hadron colliders like the LHC and directly identifying WIMP scattering events with highly sensitive low-background studies [41]. The physics of direct detection will be reviewed in this section, along with the features of the detectors that are being used.

4.8.1 Rates, Spectra, and Interactions

The possibility of directly detecting DM particles in the way of weakly interacting massive particles was first discussed by Goodman and Witten [23]. For most cases, WIMP will scatter off the nucleus rather than interact with the atomic electrons since it is electrically charged. A potentially detectable nuclear recoil results from the momentum transfer. The nuclear recoil (NR) energy is represented by ($\boldsymbol{E_{nr}}$), the WIMP mass by ($\boldsymbol{m_\chi}$), and the scattering cross section by (σ). ($\boldsymbol{M}$) is the detector's target mass. When WIMPs scatter off a target nucleus of masses, the anticipated rate is determined by:

$$\frac{dR}{dE_{nr}} = \frac{\rho_0 M}{m_N m_\chi} \int_{v_{min}}^{v_{sec}} v f(v) \frac{d\sigma}{dE_{nr}} dv \tag{4.13}$$

The normalized WIMP velocity distribution is represented by ($\boldsymbol{f(v)}$), and the local density of DM in the Milky Way galaxy is $\rho_0 = 0.3\ \mathrm{GeV/c^2/cm^3}$. All velocities are defined by the detector's reference frame. The minimal velocity ($\boldsymbol{v_{min}}$) required for a WIMP to produce a nuclear recoil of energy ($\boldsymbol{E_{nr}}$) is:

$$v_{min} = \sqrt{\frac{E_{nr} m_N}{2\mu^2}} \tag{4.14}$$

WIMPs will not be confined to the Milky Way's potential well if their velocity is higher than the escape velocity $v_{sec} = 544\ \mathrm{km/h}$ [42].

The nucleon-WIMP system's reduced mass is indicated by (μ). Where $m_\chi = 100\,\mathrm{GeV/c^2}$ WIMP-induced nuclear recoil spectra in a range of typical target particles, considering a spin-independent Weak Interacting Massive Particles-nucleon cross section $\sigma_{SI} = 1 \times 10^{47}\,cm^2$. A^2 preferentially interacts with heavier target materials by increasing the amount of spin-independent interactions. However, the form factor suppression causes the rate of reaction for big nuclei (Xe, I, and W) to drop at higher energies. Given that the incident WIMP's most likely kinetic energy was E0, which depicts a WIMP colliding with an argon atom and producing ionization and scintillation signals, it is likely developed by researchers involved in dark matter detection experiments such as the WARP (WIMP Argon Programme), XENON, LUX-ZEPLIN (LZ), or DARWIN collaborations. These experiments focus on detecting dark matter interactions using noble liquid detectors (such as argon or xenon) (Figure 4.6).

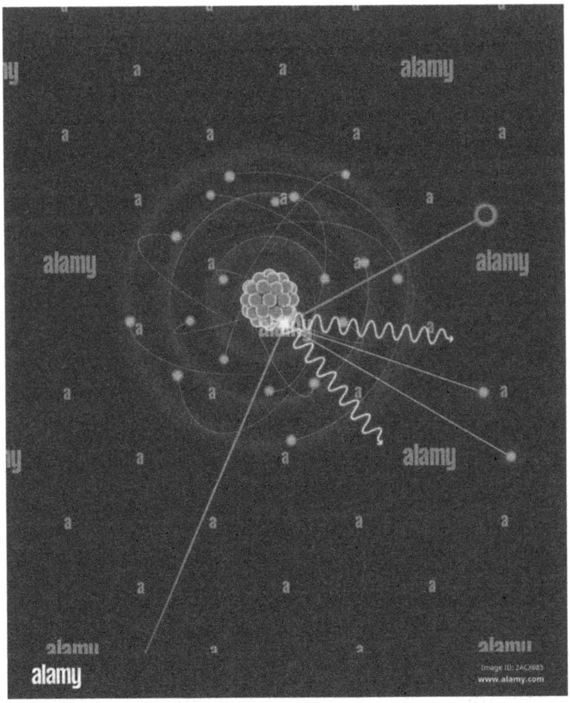

Figure 4.6 WIMP conceptual illustration and diagram. (Image credit: INFN Italian National Institute for Nuclear Physics).

4.9 Indirect Detection

Pairs of DM particles that decay into SM particles or DM particles that annihilate into SM particles are the signals that indirect detection investigations look for. Finding a great deal in cosmic rays gathered on Earth relative to the anticipated astrophysical input is one method of detecting dark matter. Targeting the areas with the highest predicted density of dark matter—the center of the Earth, the core of the Sun, our galaxy's inner halo, neighboring galaxies dominated by dark matter, and the Galactic Center—is the most promising. However, these regions, particularly the Galactic Center, are astrophysically complex, which could complicate detection. To improve the chances of successful detection, it may be more effective to focus on areas that offer the best signal-to-background ratio, rather than those with the highest dark matter concentrations. With the aim of detecting particles like photons as well as neutrinos, positrons, antiprotons, anti-deuterium, and possibly rarer anti-nuclei like antihelium, the kind of cosmic ray being targeted is also very important. Every one of these messengers has benefits and drawbacks.

- The galactic environment allows high-energy photons (γ-rays) to easily circulate, meaning that information can be found in both their energy spectra and angular spectra. Since dark matter is electrically neutral, photons are only produced by secondary radiation or some other subdominant mechanism, such as loops involving charged particles. As a result, a suppressed and highly model-dependent photon spectrum is anticipated.
- Low-energy photons, such as radio waves and X-rays. In the case of heavy DM, which has mass in the ~ GeV-TeV range, electrons and positrons from DM annihilation or decay often produce comparatively weak photons as secondary electromagnetic radiation. After that, bremsstrahlung or synchrotron radiation is used to create the radio waves and X-rays. Despite being "doubly indirect" and reliant on the cosmic environment that reprocesses the $e^{\pm}$, these photons make up a DM signal. There are uncertainties involved due to factors such as gas density and magnetic field strength. For example, in sterile neutrino theories, X-rays and other low-energy radiation could be directly emitted by the disintegration of light-dark material particles with masses between ~keV and MeV.

- The directions of photons are randomized due to their diffusion in the galactic magnetic fields. They lose energy by bremsstrahlung, ionization, inverse Compton (IC) scattering, synchrotron emission, and Coulomb scattering. The nearest parts of the galaxy dominate the possible DM contribution to positron flux. The energy spectrum contains information about DM. Furthermore, below a few GeV, the e^+ energy spectra are distorted by solar activity.
- **Electrons**. The electrons are subject to similar remarks as the positrons, with the benefit that it is simpler to estimate the total $e^- + e^+$ flux at higher energy than the positron flux alone, and the drawback of having a higher astrophysical background.
- **Antiprotons**. The directions of antiprotons are randomized due to their diffusion in the galactic magnetic fields. They have minimal energy losses up to modest scatterings at the galactic plane matter, in contrast to $e^{\pm}$. As a result, the flux gathered on Earth is influenced by even distant parts of the Galaxy, and as a result, there are substantial astrophysical errors in its normalization. Below are a few GeVs, but solar activity distorts the energy spectrum, which contains information about DM.
- **Antideuterons**. An antineutron and an antiproton created during DM annihilation or decay can merge to create anti-deuterium nuclei. The anticipated yield is minuscule. However, the astrophysical process is also expected to be modest and, more importantly, to reach its highest point in an amount of possible energies that deviate from the one of a typical dark matter signal due to the different dynamics of the two-generation mechanisms. The antiproton scenario is comparable to the propagation in the cosmic environment. Because more antinucleons must coalesce, heavier antinuclei, such as antihelium, can be generated identically but with the significant drawback of a significantly suppressed flow.

4.9.1 Energy Spectrum at Production

In theory, DM can decay (or annihilate) into any type of pair of particles from the SM. The most pertinent channels and/or branching ratios will be determined by the particle physics model that underlies DM.

Here, we discuss dark matter annihilations and decay into all of the following major channels to be as model-independent and instructive as possible. The subscripts (L) and (T) represent the longitudinal and transverse polarizations of the vector boson channels, whereas R and L represent the right and left polarizations of the leptonic channels. For electroweak radiation, which we will address in a moment, it is crucial to distinguish between such polarizations.

$$\begin{aligned} e_L^+e_L^-, e_R^+e_R^-, \mu_L^+\mu_L^-, \mu_R^+\mu_R^-, \tau_L^+\tau_L^-, \tau_R^+\tau_R^-, \nu_e\bar{\nu}_e, \nu_\mu\bar{\nu}_\mu, \nu_\tau\bar{\nu}_\tau, \\ q\bar{q}, c\bar{c}, b\bar{b}, t\bar{t}, \Upsilon\Upsilon, gg, \\ W_L^+W_L^-, W_T^+W_T^-, Z_LZ_L, Z_TZ_T, hh, \end{aligned} \tag{4.15}$$

where $q = u, d, s$ denotes a light quark and ***(h)*** stands for the SM Higgs boson, which has a mass of 125 GeV.

These fundamental particles experience hadronization and Parton shower, resulting in fluxes of $e^{\pm}, \bar{p}, \bar{d}, \gamma, \nu_{e,\mu,\tau}$. Most DM research uses Monte Carlo simulation programs, like Pythia, which are primarily made for collider physics, to carry out this process. From a model-independent perspective, a few channels (such as $\gamma\gamma, \nu\bar{\nu}, gg$) are just as viable as any others, despite being "unusual" because they are frequently suppressed in many models. The following averages can be used to retrieve the corresponding unpolarized channels:

$$W^+W^- = \frac{2W_T^+W_T^- + W_L^+W_L^-}{3}. \tag{4.16}$$

$$e^+e^- = \frac{e_L^+e_L^- + e_R^+e_R^-}{2}. \tag{4.17}$$

Only lately has electroweak radiation impact been acknowledged as important for the indirect detection of DM. Such bremsstrahlung corrections are compensated for the suppression caused by the increased weak coupling at large DM masses by more than on powers of $ln(M_{DM}/M_W)$ logarithms, which grow larger for $M_{DM} \ggg M_W$. Currently, some but not all of the necessary EW effects are included in MC programs like Pythia. However, by "postprocessing" the codes' output, electroweak bremsstrahlung can be included. For a focused discussion or the thorough explanation of the computational methods. The boosted components are independent of the model; however, in the numerical findings shown here, they are suddenly activated when $M_{DM} \ggg M_W$. In a comprehensive DM model, these impacts would actually show up in a straightforward model-dependent way as the DM mass increases. The finite non-logarithmic components that cannot be computed in a model-independent way have been disregarded instead.

4.9.2 Secondary Photons

In the diffusion volume, galactic $e^{\pm}$ produced by DM lose nearly all of their energy as photons through two processes: synchrotron radiation and inverse Compton. Bremsstrahlung radiation is also significant for low-energy electrons, particularly in areas near the galactic disk, where gas is plentiful. Thus, the resulting fluxes of microwave synchrotron radiation and ICS (and bremsstrahlung) γ rays could be indicators of DM. In particular, the ICS flow shows promise. One of its characteristics is that it comes from "anywhere" in the galactic halo's diffusion volume, including places with lower astrophysical backgrounds (such as high latitudes). Furthermore, while synchrotron losses of energy are subdominant with respect to reverse Compton energy losses practically everywhere, the consequent ICS γ flux has just slight astrophysical uncertainties because of energy conservation. Since the region near the Galactic Center is where the magnetic field intensity and DM density are highest, it generates a substantial quantity of microwave synchrotron radiation, which is hampered by more background and uncertainty. But it can also originate in far-flung latitudes. Lastly, under the previously stated circumstances (high gas density and $\lesssim 10\,\text{GeV}\ e^{\pm}$ energy), the bremsstrahlung emission can be a significant indication of dark matter. In any event, if one wishes to mimic the other emissions with which it competes consistently, it must be considered.

4.10 Role of Artificial Intelligence in Dark Matter and Cosmology

Artificial intelligence (AI) is becoming a powerful tool in the search to understand dark matter, aiding researchers in ways that were unimaginable just a few decades ago [43]. Massive datasets from telescopes, satellites, and simulations are essential to cosmology, the study of the universe's origin, and large-scale structures. Galaxy investigations, CMB observations, and computer models of the cosmos are some examples of these datasets. Because of their volume and complexity, these data require sophisticated computing methods in order to yield insightful information. AI provides strong techniques for examining these enormous datasets, especially machine learning in general (ML) as well as deep learning (DL). By identifying patterns, making predictions, and even revealing hidden correlations, AI accelerates the discovery process in cosmology.

Here are some key areas where AI plays a crucial role in dark matter research:

4.10.1 Data Analysis and Pattern Recognition

Modern telescopes and detectors generate vast amounts of data, including images, particle interaction records, and cosmic background data. AI algorithms, particularly machine learning models, can process and analyze these huge datasets far more efficiently than traditional methods. AI can be trained to detect patterns associated with dark matter's gravitational effects like subtle distortions in light (gravitational lensing) or unexpected galaxy rotation curves. Deep learning models are especially useful in identifying these complex patterns and anomalies within noisy data.

4.10.2 Simulating Cosmological Models

Researchers can model the impact of dark matter on galaxy evolution and large-scale cosmic structures with the use of AI-driven simulations [44]. Scientists can test several ideas, such as Λ-CDM hypothesis, using these simulations to evaluate how well they suit observable data. In order to see how various dark matter particle candidates—such as WIMPs or axions—would affect cosmic development, algorithms using machine learning can also model scenarios based on these particles.

4.10.3 Enhancing Observational Techniques

One of the main techniques for identifying the concept of dark matter is gravitational lensing analysis, which AI has transformed. Machine learning models may analyze lensing effects across large datasets, identifying dark matter's distribution by mapping the way it bends light from distant galaxies. Researchers can increase sensitivity and precision while looking for indirect indicators of mysterious matter, which include gamma radiation or neutrinos generated through dark matter annihilations, by using AI algorithms to optimize telescope and detector settings.

4.10.4 Deep Learning in Direct and Indirect Detection

In direct detection investigation (e.g., LUX-ZEPLIN or XENON1T), AI can help filter out background noise and isolate potential signals of DM particles interacting with detector material. AI is also used to examine cosmic rays and gamma rays in indirect detection experiments. For illustration, machine learning tools can differentiate between signals potentially produced by dark matter and those from known astrophysical sources, helping refine data from space observatories like Fermi-LAT.

4.10.5 Accelerating Theoretical Model Testing

AI can quickly sift through complex data and simulations to help identify which theoretical models align best with observations. Machine learning models such as neural networks and genetic algorithms can evaluate and optimize theoretical parameters in dark matter models far faster than traditional methods. AI can predict the results of upcoming experiments based on existing data, helping researchers refine their approach, saving time and resources.

4.10.6 Enhancing Survey Efficiency

AI makes it possible to quickly analyze sky survey data from initiatives like the Legacy Survey of Space and Time (LSST) at the Vera C. Rubin Observatory, which intends to plan billions of galaxies. This mapping is crucial in understanding dark matter's influence on galaxy clustering and distribution. Machine learning helps classify celestial objects in survey data, such as separating galaxies, stars, and quasars. This is essential in identifying regions influenced by dark matter without manually sorting through millions of images.

4.10.7 Accelerating Discoveries with AI-Driven Hypotheses

AI can analyze data to suggest new connections or hypotheses about dark matter behavior, leading to potential discoveries. For example, AI may identify unexpected correlations in data that could hint at new dark matter properties. Based on analyzed data and generated hypotheses, AI can recommend modifications to experimental setups or observational priorities to maximize the chances of detecting dark matter.

4.11 AI's Role in Quantum Simulations of Dark Matter

Since DM is believed to interact feebly with normal matter, AI aids in quantum simulations that model these rare interactions. Researchers may investigate how dark matter might interact with observable particles in high-energy situations using machine learning-supported simulations of quantum field theory. AI algorithms streamline the computation of quantum dynamics involving dark matter, helping to model its potential role in the universe and particle detectors. Although distinct from dark matter, dark energy research overlaps with dark matter research in cosmology. AI is instrumental in analyzing cosmic expansion data, which provides an understanding of both dark energy and DM through their combined influence on cosmic structure.

4.12 Challenges and Future Prospects

One challenge is interpreting AI's predictions accurately in the context of physical theories. AI's "black box" nature means that insights are not always immediately explainable, requiring careful validation. AI models need continuous training on high-quality labeled data, and as our understanding of DM improves, these models will need frequent refinement to stay accurate.

4.12.1 Toward a Dark Matter Breakthrough

As AI techniques continue to advance, they promise to push the boundaries of dark matter research, providing a clearer picture of this elusive component of the universe [45].

In space research, AI is revolutionizing how we tackle difficult problems, particularly the enigmas surrounding DM and dark energy. Here are some ways AI is advancing the area and assisting scientists in overcoming these obstacles:

4.12.2 Handling Vast and Complex Datasets

Large datasets are produced by space science research, especially in cosmology. Examples of these include the European Space Agency's Gaia project, the Vera C. Rubin Observatory's LSST, and other extensive surveys.

AI-driven data-processing algorithms can sift through these vast amounts of information far faster and more accurately than manual or traditional methods. AI can help eliminate sound from observational data, such as background signals or irrelevant artifacts that might obscure critical information about dark matter or dark energy effects.

4.13 Enhancing Analysis and Interpretation of Astronomical Data

AI, particularly deep learning and neural networks is used to detect subtle patterns in data that might signify dark matter's gravitational effects, such as gravitational lensing or cosmic background fluctuations. This helps map out where DM is distributed throughout the universe. AI systems can identify unusual signals that may indicate new insights about dark matter or dark energy, often highlighting data points that human researchers might overlook. For example, certain galaxy rotation anomalies or unusual gravitational lensing patterns can provide hints about dark matter presence.

4.13.1 Advanced Cosmological Simulations

Scientists can study the effects of DM and dark energy on the distribution and creation of galaxies, galaxy clusters, and cosmic voids using AI-powered simulations. These simulations help refine cosmological models like ΛCDM model [46]. Machine learning can efficiently simulate different scenarios, such as how hypothetical dark matter particles might interact, which speeds up the testing of several types of theoretical models for DM and dark energy.

4.13.2 Direct and Indirect Detection Support

AI is used in experiments that aim to detect particles of DM by analyzing the outcomes of high-energy particle collisions or detecting faint interaction signals. AI algorithms help filter through background noise to highlight potential dark matter events [47]. In space-based experiments, AI assists in analyzing gamma-ray and cosmic-ray data, which may indicate dark matter interactions. AI can distinguish between signals from dark matter and those from other astrophysical sources like pulsars or black holes.

4.14 AI in Theory Development and Hypothesis Generation

AI algorithms are capable of analyzing vast datasets to suggest new connections or hypotheses regarding the behavior of DM and dark energy. For example, AI might identify correlations in observational data that suggest new properties of dark matter or point toward unknown forms of interaction. AI models help identify which cosmological theories best align with observed data, allowing researchers to refine theories and focus on the most promising avenues in their studies.

4.15 Challenges and Future Prospects

AI can produce remarkably accurate predictions, but understanding the "why" behind those predictions remains challenging. Developing explainable AI models is essential for interpreting results in the context of known physics. As more data become available, AI models require continuous updates to stay accurate. Advances in observational technology mean that AI models must adapt to increasingly high-quality data, pushing the limits of their predictive power.

4.16 Conclusion

An unidentified kind of DM makes up 95% of the mass of clusters and galaxies. Rotation curves, gravitational lensing, and hot gas in clusters all provide evidence for this. This estimate of DM's impact to the universe's global energy density is consistent with calculations of the CMB, nucleosynthesis, and the value of large-scale structures and supernova. According to a widely accepted theory, the DM accounts for 23% of the universe's total energy density. White dwarfs or other MACHO candidates may make up as little as one fifth of a galaxy's DM, but exotic particle candidates are more likely to make up the majority. Research into dark matter continues, particularly focusing on promising candidates like axions and WIMPs. The annihilation cross section of WIMP applicants is a crucial characteristic that establishes the current level of relic density. This annihilation process could also help test the dark matter hypothesis by providing a potential source of energy for the first stars in the universe, which may be detectable with future telescopes. We have explored how both the MOND and ΛCDM models are highly successful and how interpreting their achievements through this model-centric approach offers an alternative perspective on their success. ΛCDM provides a good description and prediction of the universe's large-scale structure. However, to maintain the theory of general relativity as its main explanatory basis, it must specify that DM particles are a feature of the particle target system. MOND is effective at describing and predicting the mesoscale structure of rotations of solitary galaxies. To generate this accurate description, however, modified Newtonian dynamics is employed as the underlying theory rather than general relativity. Both can be seen as successful even if they were founded on very different ideas. The models are effective at producing accurate descriptions and forecasts that match the data in their respective fields. Finding the third dimension—explanation—at which to evaluate a model's suitability is crucial to understanding the dispute. The inclusion of potentially problematic explanatory structures in each model is a concern that can be separated using this third evaluation component. As a consequence, this model-based investigation of ΛCDM–MOND conversation debate exposes a few commitments that could help clarify other aspects of the argument. Since the dynamics in the mathematical framework for gravitational forces originate from current good theory, ΛCDM has the stronger representational link as far as our understanding of our particle target system is concerned. According to general relativity, its dubious components—dark matter and dark energy—do not pose a theoretical challenge. Evidence for specific characteristics, traits, or mechanisms of the model as they relate to the particle target system may be required, rather than evidence for the overarching theory itself. Instead, we wish to highlight the necessity of considering the expansion of models outside of their initial areas with more careful thought and examination. When a model is extended to a new domain, it involves specific commitments, and the justification for such extensions often includes underlying assumptions about the model's explanatory strengths. According to analysis, this discrepancy results from the various rationales used in the model's development, including approximations and idealizations, as well as doubts regarding the model's applicability to the actual system it depicts. We can have a better understanding regarding the way two approaches with essentially different physics can both accomplish their goals in their respective fields by looking at this. A model's extension beyond its original scope leads to conflicts, and such extensions are only justified if the model is considered a reliable informative tool. If a model is regarded as a strong explanatory tool, one is committed to its properties having a meaningful relationship with the real system it aims to represent. In a broader sense, the crucial issue that needs to be recognized is what these models bind us to when assertions transcend the model. It must consider the commitments made by the model when related to the real world. There is a lot of strong evidence from cosmology and astrophysics that the SM of particle physics is not complete. It must be expanded to account for DM, baryogenesis, and neutrino masses. Although Einstein's concept can explain dark energy, its existence is unknown, and there appear to be too many coincidences between the numerical values of cosmological parameters. However, cosmological evidence is increasingly supporting an inflationary paradigm that uses a type of dark energy to explain the Universe's formation. Despite only recently becoming a precision science, cosmology is already full of surprises and is contributing to the development of a realistic model of microphysics. The task of identifying this new physics falls to high-energy physicists. AI has improved data processing, improved observational methods, and expedited the testing of theoretical models, all of which have transformed dark matter research. The use of AI in this area has the potential to yield groundbreaking findings that could fundamentally change our understanding of the universe in addition to improving our information on the properties and behaviors of DM. As AI and machine learning capabilities develop, we will be able to explore the secrets of DM and larger cosmos. In the study of DM and dark energy, artificial intelligence has emerged as a crucial instrument. It facilitates the creation of intricate simulations, allows for quicker and more accurate analysis, and helps identify minute clues that point to the fundamental building of the cosmos. As artificial intelligence (AI) develops, it will become important in assisting researchers in solving the enigmas surrounding DM and dark energy, which will expand our information about the universe.

Data Availability

This chapter references publicly available astronomical datasets from sources like NASA, ESA, and various research observatories. Specific datasets, including those from Kepler, LSST, and SDSS, are accessible through their respective repositories. For inquiries or requests, contact the corresponding author.

Acknowledgment

The authors thank the Department of Physics at Govind Ballabh Pant University of Agriculture and Technology, Pantnagar, India, for providing the academic environment and resources essential to this work.

References

1. Zwicky, F., Die Rotverschiebung von extragalaktischen Nebeln. *Helv. Phys. Acta*, 6, 110–127, 1933.
2. Bosma, A., The distribution and kinematics of neutral hydrogen in spiral galaxies of various morphological types. *Astron. J.*, 86, 1825–1846, 1981.
3. Munoz, R.R., Carlin, J.L., Frinchaboy, P.M., Nidever, D.L., Majewski, S.R., Patterson, R.J., Exploring halo substructure with giant stars: the dynamics and metallicity of the dwarf spheroidal in Boötes. *Astrophys. J.*, 650, 51–54, 2006.
4. Sarazin, C.L., X-ray emission from clusters of galaxies. *Rev. Mod. Phys.*, 58, 1–129, 1986.
5. Kneib, J.P. and Natarajan, P., Cluster lenses. *Astron. Astrophys. Rev.*, 19, 1–100, 2011.
6. Clowe, D., Bradač, M., Gonzalez, A.H., Markevitch, M., Randall, S.W., Jones, C., Zaritsky, D., A direct empirical proof of the existence of dark matter. *Astrophys. J.*, *648*, 109–113, 2006.
7. Rubakov, V.A. and Gorbunov, D.S., *Introduction to the Theory of the Early Universe: Hot Big Bang Theory*, 2nd ed., World Scientific, Singapore, 2018.
8. Bertone, G. and Hooper, D., History of dark matter. *Rev. Mod. Phys.*, 90, 045002–045073, 2018.
9. Riess, A.G., Macri, L.M., Hoffmann, S.L., Scolnic, D., Casertano, S., Filippenko, A.V., Foley, R.J., A 2.4% determination of the local value of the Hubble constant. *Astrophys. J.*, *826*, 1–32, 2016.
10. Particle Data Group, Review of Particle Physics. *Chin. Phys. C*, 40, 100001, 2016.
11. Freeman, K.C., On the disks of spiral and S0 galaxies. *Astrophys. J.*, *160*, 811, 1970.
12. Famaey, B. and McGaugh, S.S., Modified Newtonian dynamics (MOND): observational phenomenology and relativistic extensions. *Living Rev. Relativ.*, *15*, 1–159, 2012.
13. Ostriker, J.P. and Peebles, P.J., A numerical study of the stability of flattened galaxies: or, can cold galaxies survive? *Astrophys. J.*, 186, 467–480, 1973.
14. Bahcall, N.A., Clusters of galaxies. *Annu. Rev. Astron. Astrophys.*, *15*, 505–540, 1977.
15. Kahlhoefer, F., Schmidt-Hoberg, K., Frandsen, M.T., Sarkar, S., Colliding clusters and dark matter self-interactions. *Mon. Not. R. Astron. Soc.*, 437, 2865–2881, 2014.
16. Harvey, D., Massey, R., Kitching, T., Taylor, A., Tittley, E., The nongravitational interactions of dark matter in colliding galaxy clusters. *J. Sci.*, 347, 1462–1465, 2015.
17. Massey, R., Kitching, T., Richard, J., The dark matter of gravitational lensing. *Rep. Prog. Phys.*, 73, 086901, 2010.

18. Combes, F., Dark matter distribution and its impact on the evolution of galaxy disks. *Mem. Soc. Astron. Ital. Suppl.*, 75, 1–6, 2012.
19. Skordis, C., The tensor-vector-scalar theory and its cosmology. *Classical Quantum Gravity*, 26, 143001–143044, 2009.
20. Mohapatra, R.N. and Teplitz, V.L., Mirror matter MACHOS. *Phys. Lett. B*, 462, 302–309, 1999.
21. Alcock, C., Allsman, R.A., Alves, D.R., Axelrod, T.S., Becker, A.C., Bennett, D.P., Welch, D., The MACHO project: microlensing results from 5.7 years of Large Magellanic Cloud observations. *Astrophys. J.*, 542, 281, 2000.
22. Freese, K., Fields, B., Graff, D., *Sources and detection of Dark Matter and Dark Energy in the Universe*, Marina del Rey, 2000.
23. Goodman, M.W. and Witten, E., Detectability of certain dark-matter candidate,s *Phys. Rev. D*, 31, 3095, 1985.
24. Gondolo, P. and Silk, J., Dark matter annihilation at the galactic center. *Phys. Rev. Lett.*, 83, 1719, 1999.
25. Brouwer, M.M., Visser, M.R., Dvornik, A., Hoekstra, H., Kuijken, K., Valentijn, E.A., Taylor, E.N., First test of Verlinde's theory of emergent gravity using weak gravitational lensing measurements. *Mon. Not. R. Astron. Soc.*, 466, 2547–2559, 2017.
26. D'Agostini, G. and Petit, J.P., Constraints on Janus Cosmological model from recent observations of supernovae type Ia. *Astrophys. Space Sci.*, 363, 139–157, 2018.
27. Kovac, J.M., Leitch, E.M., Pryke, C., Carlstrom, J.E., Halverson, N.W., Holzapfel, W.L., Detection of polarization in the cosmic microwave background using DASI. *Nat.*, 420, 772–787, 2002.
28. Colin, J., Mohayaee, R., Rameez, M., Sarkar, S., Evidence for anisotropy of cosmic acceleration. *Astron. Astrophys.*, 631, 13–18, 2019.
29. Siewert, T.M., Schmidt-Rubart, M., Schwarz, D.J., Cosmic radio dipole: Estimators and frequency dependence. *Astron. Astrophys.*, 653, 9–27, 2021.
30. Paranjape, A. and Sheth, R.K., The phenomenology of the external field effect in cold dark matter models. *Mon. Not. R. Astron. Soc.*, 517, 130–139, 2022.
31. Milgrom, M., Cosmological variation of the MOND constant: Secular effects on galactic systems. *Phys. Rev. D.*, 91, 044009–044021, 2015.
32. Massimi, M. (Ed.), *Philos. Sci.*, Routledge, Abingdon, Oxon, New York, NY, 2014.
33. Massimi, M. and Peacock, J., What are dark matter and dark energy?, in: *Philos. Sci.*, pp. 33–51, Routledge, Abingdon, Oxon, New York, NY, 2014.
34. Asztalos, S.J., Bradley, R.F., Duffy, L., Hagmann, C., Kinion, D., Moltz, D.M., Yu, D.B., Improved rf cavity search for halo axions. *Phys. Rev. D.*, *69*, 1, 011101–011105, 2004.
35. Particle Data Group, Zyla, P., Barnett, R.M., Beringer, J., Dahl, O., Dwyer, D.A., Pomarol, A., Review of particle physics. *Prog. Theor. Exp.*, 2020, 083C01, 2020.

36. Boyarsky, A., Drewes, M., Lasserre, T., Mertens, S., Ruchayskiy, O., Sterile neutrino dark matter. *Prog. Part. Nucl. Phys.*, 104, 1–45, 2019.
37. Ibe, M., Kusenko, A., Yanagida, T.T., Why three generations? *Phys. Lett. B*, 758, 365–369, 2016.
38. Drewes, M., The phenomenology of right-handed neutrinos. *Int. J. Mod. Phys. E*, 22, 1330019–1330593, 2013.
39. Gorbunov, D. and Panin, A., Minimal active-sterile neutrino mixing in see-saw type I mechanism with sterile neutrinos at GeV scale. *Phys. Rev. D*, 89, 017302, 2014.
40. An, F.P., Balantekin, A.B., Band, H.R., Beriguete, W., Bishai, M., Blyth, S., Ruan, X.C., Search for a light sterile neutrino at Daya Bay. *Phys. Rev. Lett.*, 113, 141802–141806, 2014.
41. Kahlhoefer, F., Review of LHC dark matter searches. *Int. J. Mod. Phys. A*, 32, 1730006–1730039, 2017.
42. Smith, M.C., Ruchti, G.R., Helmi, A., Wyse, R.F., Fulbright, J.P., Freeman, K.C., Zwitter, T., The RAVE survey: constraining the local Galactic escape speed. *Mon. Not. R. Astron. Soc.*, 379, 755–772, 2007.
43. Russell, S.J. and Norvig, P., *Artificial Intelligence: A Modern Approach, Pearson Series in Artificial Intelligence*, 4 ed., Pearson, Upper Saddle River, NJ, 2020.
44. Vogelsberger, M., Marinacci, F., Torrey, P., Puchwein, E., Cosmological simulations of galaxy formation. *Nat. Rev. Phys.*, 2, 42–66, 2020.
45. Davis, M., Efstathiou, G., Frenk, C.S., White, S.D.M., The evolution of large-scale structure in a universe dominated by cold dark matter. *Acta Pathol. Jpn.*, 292, 371–394, 1985.
46. Berger, P. and Stein, G., A volumetric deep convolutional neural network for simulation of mock dark matter halo catalogues. *MNRAS*, 482, 2861–2871, 2019.
47. Feng, Y., Chu, M.Y., Seljak, U., McDonald, P., FASTPM: A new scheme for fast simulations of dark matter and haloes. *MNRAS*, 463, 2273–2286, 2016.
48. Silk, J., Olive, K. and Srednicki, M., *PRL*, 55, 257, 1985.
49. Krauss, L., Srednicki, M. and Wilczek, F., *Phys. Rev. D*, 33, 2079, 1986.
50. Kroupa, P., Jerabkova, T., Thies, I., Pflamm-Altenburg, J., Famaey, B., Boffin, H. M. J., Dabringhausen, J., Beccari, G., Prusti, T., Boily, C., Wu, X., Haas, J., Haghi, H., Hasani Zonoozi, A., Thomas, G., Subr, L., Aarseth, S. J., Asymmetrical tidal tails of open star clusters: stars crossing their cluster's práh challenge Newtonian gravitation. *Mon. Not. R. Astron. Soc.*, 517, 3, 3613–3626, 2022.
51. Massimi, M., Three problems about multi-scale modelling in cosmology. *Studies in History and Philosophy of Science Part B: Studies in History and Philosophy of Modern Physics*, 64, 26–38, 2018.

5

Gravitational Wave Detection

Muhammad Zeshan Ashraf[1], Farhat Shakeel[2*] and Tahira Saeed[3]

[1]College of Physics, Hebei Normal University China, Shijiazhuang, China
[2]School of Physics and Microelectronics Key Laboratory of Material Physics, Zhengzhou University, China, Henan, China
[3]Department of Mathematics, Government High School 23 s/p Pakpattan, Punjab, Pakistan

Abstract
Gravitational wave detection (GWD) has revolutionized modern astrophysics, providing a new observational window into the cosmos. Albert Einstein initially proposed this concept within his General Theory of Relativity (GR). However, direct GWD was confirmed in 2015 by the Laser Interferometer Gravitational-Wave Observatory (LIGO), validated a century-old theoretical prediction. These ripples generated in spacetime by exotic astrophysical phenomena like the merging of black holes and the collision of neutron stars offer unparalleled insights into fundamental physics, compact object dynamics, and the universe at a large scale. The increasing sensitivity of interferometric detectors like LIGO, Virgo, and KAGRA has paved the way for multi-messenger astronomy (MMA), where gravitational waves complement electromagnetic and neutrino observations. Despite these advancements, detecting these minute disturbances remains a technological challenge, necessitating sophisticated data analysis techniques to filter out noise and extract meaningful signals. Artificial intelligence (AI) and machine learning framework have emerged as transformative tools in this domain, significantly enhancing the efficiency and accuracy of gravitational wave detection. AI-driven approaches, including deep learning algorithms and Bayesian inference methods, have accelerated signal identification, real-time event classification, and parameter estimation, surpassing traditional analytical methods in both speed and precision. This chapter explores the journey of gravitational wave detection, from its theoretical origins to modern observational breakthroughs, emphasizing the technological advancements that have enabled precision measurements. It also discusses

**Corresponding author*: farhatshakeel207@gmail.com

Yogesh Chandra, Manjuleshwar Panda and Mahesh Chandra Mathpal (eds.) The Intelligent Universe: AI's Role in Astronomy, (181–200) © 2025 Scrivener Publishing LLC

the fundamental limitations of current detectors and the promising future of advanced observatories such as the Einstein Telescope and LISA. The role of AI in improving detection pipelines, noise reduction, and waveform modeling is examined in detail. Finally, the chapter highlights the broader implications of gravitational wave astronomy for fundamental physics, cosmology, and astrophysics, showcasing its potential to unlock new scientific frontiers.

***Keywords*:** Gravitational waves, general relativity, LIGO, virgo, multi-messenger astronomy, black hole mergers, neutron star mergers, artificial intelligence

5.1 Introduction

GWD represents one of the most prominent discoveries in modern astrophysics, opening a new observational window to explore the universe beyond the electromagnetic spectrum. Albert Einstein was the first one who predicted these perturbations in spacetime in 1916 as a consequence of GR. These are distortions in spacetime caused by the movement of massive bodies. Their detection has not only confirmed Einstein's predictions but has also revolutionized our understanding of compact objects, cosmic evolution, and fundamental physics. To appreciate the profound impact of this discovery, one must recognize the transformative nature of gravitational waves in modern science. As Kip Thorne, one of the pioneers of gravitational wave research, aptly stated:

"Gravitational waves will open a whole new way of listening to the universe, just as radio waves did a century ago."

- Kip Thorne

This section explains the theoretical foundations of gravitational waves, the early indirect evidence supporting their existence, and the groundbreaking direct detection that ushered in a new era of gravitational wave astronomy.

5.1.1 Theoretical Foundations of Gravitational Waves

The concept of gravitational waves emerges directly from Einstein's GR, where mass and energy warp spacetime, influencing the motion of objects within it. When massive objects like neutron stars or catastrophic black holes undergo rapid acceleration or merger events, they create distortions that travel outward as waves, moving at the speed of light. These waves carry energy away from the system, altering orbital parameters and providing insights into the behavior of extreme gravitational environments. Mathematically, gravitational waves are described as perturbations in the metric tensor of spacetime, satisfying the wave equation in the weak-field limit. Their amplitude and frequency depend on the masses and dynamics of the source system. Although initially considered undetectable due to their minuscule effect on matter, advancements in laser interferometry have enabled the measurement of these faint distortions, with sensitivities reaching displacements smaller than one thousandth of the proton's diameter [1]. Understanding the theoretical framework of gravitational waves is crucial, but visualizing their propagation and interaction with spacetime helps illustrate their profound impact. The following depiction in Figure 5.1 demonstrates how gravitational waves distort spacetime as they travel outward from a massive accelerating system.

This visualization highlights the fundamental nature of gravitational waves—how they stretch and compress spacetime in a quadrupolar manner. Detecting these distortions requires extremely sensitive instruments, paving the way for groundbreaking observations in modern astrophysics.

Figure 5.1 Illustration of gravitational waves distorting spacetime as they propagate outward. (Credit: NASA).

5.1.2 Early Indirect Evidence and Predictions

Before gravitational waves were directly detected, their presence was inferred from observations of binary pulsar systems. In 1974, Russell Hulse and Joseph Taylor discovered PSR B1913+16, a binary pulsar system consisting of two neutron stars in close orbit. By carefully monitoring the system's orbital decay, they observed a loss of energy that aligned with the predictions made by General Relativity, suggesting that gravitational waves were responsible for this energy loss. This finding served as the first compelling evidence for gravitational waves and earned Hulse and Taylor the Nobel Prize in Physics in 1993 [2]. Subsequent studies of additional pulsar systems, such as the Double Pulsar (PSR J0737–3039), reinforced these findings, further validating Einstein's predictions. Meanwhile, technological advancements, particularly in laser interferometry and computational modeling, laid the groundwork for direct detection, culminating in the development of large-scale ground-based detectors like LIGO and Virgo.

5.1.3 The Breakthrough of Direct Detection

Gravitational waves were first directly observed on September 14, 2015, by the Laser Interferometer Gravitational-Wave Observatory (LIGO), marking a pivotal moment in the progress of gravitational wave research. About 1.3 billion light-years away, two black holes merged to form the event, which was named GW150914. The detected waveform closely matched theoretical predictions, confirming not only the existence of gravitational waves but also the nature of black hole mergers as powerful astrophysical sources [3]. This landmark discovery demonstrated the potential of gravitational wave astronomy, providing insights into cosmic phenomena that were previously inaccessible. It also ushered in a new age of multi-messenger astrophysics, which allows for a more thorough knowledge of the cosmos by combining gravitational waves with electromagnetic and neutrino studies. Since then, numerous detections, including neutron star mergers (e.g., GW170817), have expanded the field, offering a unique perspective on high-energy astrophysical processes, stellar evolution, and fundamental physics. The success of LIGO's first detection marked a turning point in astrophysics, solidifying gravitational wave astronomy as a powerful tool for exploring the universe. The LIGO observatories, with their state-of-the-art interferometric setup, played a crucial role in this achievement. Figure 5.2 below showcases one of the LIGO facilities, highlighting its extensive infrastructure designed to detect minuscule spacetime distortions.

Figure 5.2 Aerial view of the LIGO hanford observatory, one of the two LIGO detectors responsible for the first direct detection of gravitational waves. (Credit: LIGO laboratory).

Gravitational wave detection continues to evolve, with ongoing improvements in sensitivity, data analysis techniques, and the incorporation of artificial intelligence to enhance signal identification and noise filtering. As next-generation observatories such as the Einstein Telescope and LISA come online, the field is poised for further groundbreaking discoveries, potentially revealing new classes of astrophysical objects and probing physics beyond the Standard Model.

5.2 Gravitational Wave Observatories and Detection Techniques

Gravitational wave observatories have significantly advanced our ability to detect and study spacetime distortions caused by extreme astrophysical events. Since the first direct detection in 2015, ground- and space-based observatories have expanded their reach, unveiling the dynamics of black hole mergers, neutron star collisions, and potential signals from the early universe. These detectors rely on highly sensitive interferometric techniques to measure minute fluctuations in spacetime, requiring innovative methods to mitigate noise and enhance signal extraction. This section explores the current and future observatories dedicated to gravitational

Figure 5.3 Conceptual illustration of the laser interferometer space antenna (LISA), a space-based gravitational wave observatory. (Credit: NASA).

wave detection, along with the strategies employed to improve their sensitivity. Among the revolutionary advancements in gravitational wave detection, space-based observatories are set to play a pivotal role in expanding our observational capabilities. Unlike ground-based detectors limited by seismic noise and terrestrial constraints, space-based interferometers can achieve unprecedented sensitivity over longer baselines. The following illustration in Figure 5.3 highlights the operational concept of such an observatory.

LISA, with its vast arm length spanning millions of kilometers, will enable the detection of low-frequency gravitational waves originating from supermassive black hole mergers and early universe phenomena. As technology progresses, such space-based missions will complement ground-based detectors, offering a more comprehensive view of the gravitational wave spectrum.

5.2.1 LIGO, Virgo, and KAGRA: Current Ground-Based Detectors

Ground-based gravitational wave detectors have revolutionized observational astrophysics by directly detecting spacetime distortions caused by massive cosmic events. The Laser Interferometer Gravitational-Wave Observatory (LIGO), comprising two detectors in the United States, was the first to confirm the existence of gravitational waves in 2015, detecting a binary black hole merger [4]. The Virgo detector in Italy joined LIGO in 2017, enabling more precise localization of gravitational wave sources. KAGRA, an underground cryogenic interferometer in Japan, became operational in 2020 and enhances the network's ability to filter seismic noise. These detectors operate by using laser interferometry, where laser beams travel along perpendicular arms and recombine to measure minuscule changes in distance due to passing gravitational waves. Continuous upgrades, including quantum squeezing techniques and improved mirror coatings, are enhancing their sensitivity [5]. The expanding global network of ground-based detectors allows for multi-messenger astronomy, where gravitational waves are correlated with electromagnetic and neutrino signals. The advancements in ground-based gravitational wave detection rely on a network of highly sensitive interferometers. Each observatory has distinct design features that enhance its detection capabilities, from long-baseline arms to advanced noise reduction techniques. Table 5.1 below provides a concise comparison of key attributes of LIGO, Virgo, and KAGRA.

The continuous improvements in these observatories, including upgrades in laser power, mirror coatings, and seismic isolation, are crucial for increasing sensitivity and expanding the frequency range of detected gravitational waves. As future upgrades and new facilities come online, the ability to observe weaker and more distant signals will further advance our understanding of the universe's most extreme phenomena.

Table 5.1 Comparison of key features of current ground-based gravitational wave detectors.

Detector	Location	Arm length	First detection	Special feature
LIGO	USA	4 km	2015	Quantum squeezing for sensitivity enhancement
Virgo	Italy	3 km	2017	Advanced seismic isolation
KAGRA	Japan	3 km	2020	Cryogenic mirrors to reduce thermal noise

5.2.2 Space-Based Observatories: LISA and Beyond

Ground-based detectors can only observe gravitational waves at frequencies higher than 10 Hz due to interference from seismic activity. In contrast, space-based missions aim to detect waves at lower frequencies. One such mission, the Laser Interferometer Space Antenna (LISA), slated for launch in the 2030s, will feature three spacecraft in a triangular arrangement, positioned millions of kilometers apart. LISA will be able to capture gravitational waves from the collisions of supermassive black holes and from the remnants of the early universe, which are beyond the reach of terrestrial observatories [6]. Other proposed missions, such as the Taiji and TianQin projects from China, aim to complement LISA's observations. These space-based detectors promise to open a new window into gravitational wave astronomy, providing insight into galaxy evolution, early universe physics, and exotic compact objects.

5.2.3 Sensitivity and Noise Reduction Strategies

Detecting gravitational waves requires overcoming extreme noise sources, including seismic activity, thermal fluctuations, and quantum uncertainty. Advanced techniques to improve detector sensitivity include:

- **Quantum Squeezing:** Reducing quantum noise by manipulating light's uncertainty principle to enhance precision.
- **Cryogenic Cooling:** KAGRA uses cryogenically cooled mirrors to minimize thermal noise.
- **Adaptive Optics and Machine Learning:** AI-driven algorithms refine real-time noise filtering and enhance signal extraction from weak detections.

The continuous development of noise reduction strategies is crucial for detecting fainter and more distant gravitational wave sources, expanding our understanding of the cosmos. The effectiveness of gravitational wave detectors is measured by their ability to detect incredibly small distortions in spacetime. This sensitivity is quantified by the strain measurement, which represents the fractional change in arm length caused by a passing gravitational wave. Equ. (5.1) below provides a fundamental expression for strain in an interferometric detector.

$$h = \Delta L/L \tag{5.1}$$

where:

- **h** is the gravitational wave strain,
- **ΔL** is the change in length of the interferometer arms due to the passing wave,
- **L** is the original length of the arms.

Minimizing noise and enhancing sensitivity allow modern detectors to measure extremely small strain values. By implementing advanced noise reduction techniques, future observatories aim to push these limits even further, enabling the detection of weaker signals from deeper cosmic events.

5.3 Multi-Messenger Astronomy and Astrophysical Sources

Gravitational wave detection has transformed astrophysics, allowing scientists to probe the universe beyond the electromagnetic spectrum. Multi-messenger astronomy integrates signals from gravitational waves, electromagnetic radiation, neutrinos, and cosmic rays, offering a comprehensive perspective on astrophysical phenomena. The synergy between different messengers has already led to groundbreaking discoveries, including the confirmation of neutron star mergers as sources of heavy element nucleosynthesis and the study of black hole mergers in previously unexplored mass regimes. This section explores key astrophysical sources of gravitational waves and their implications for multi-messenger astronomy.

5.3.1 Neutron Star Mergers and Kilonovae

The binary neutron star merger detected as GW170817 by LIGO and Virgo in 2017 marked a pivotal moment in multi-messenger astronomy. This event was accompanied by electromagnetic counterparts, including a short gamma-ray burst (GRB 170817A) and a kilonova (AT2017gfo), confirming that neutron star collisions are responsible for producing heavy elements like gold (Au) and platinum (Pt) *via* r-process nucleosynthesis [7]. Neutron star mergers depicted in Figure 5.4 offer crucial insights into the **nuclear equation of state**, as the gravitational wave signal encodes information about the stars' internal structure. The post-merger remnant, whether a massive neutron star or a black hole, provides further constraints on the

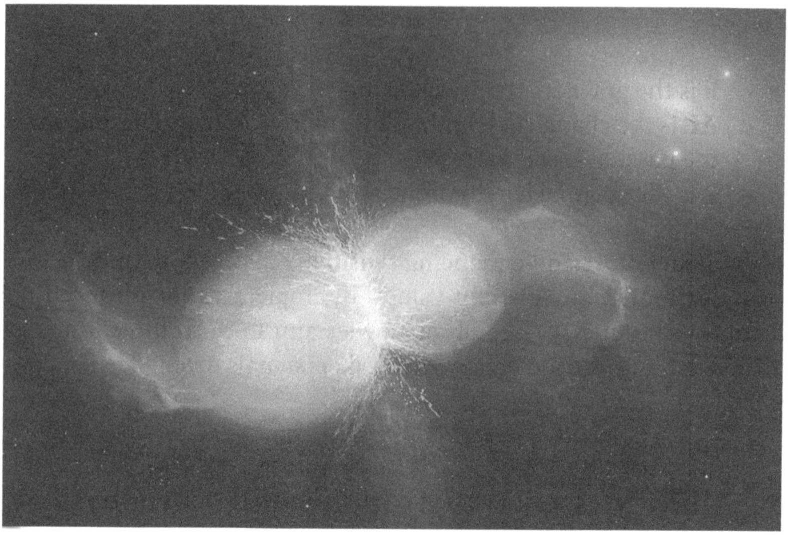

Figure 5.4 An artist's impression of neutron stars merging, producing gravitational waves and resulting in a kilonova. (Credit: university of Warwick/Mark garlick).

behavior of ultradense matter [8]. Observations by telescopes like Hubble, JWST, and ground-based optical and infrared facilities have expanded the understanding of kilonovae, allowing detailed spectral analysis of ejected material [9].

Future missions, including **Einstein Telescope (ET)** and **Cosmic Explorer (CE)**, will push detection sensitivity to greater distances, enabling the discovery of more distant neutron star mergers. AI-driven rapid follow-up techniques are being developed to enhance real-time event localization, ensuring prompt multiwavelength observations [10].

5.3.2 Black Hole Mergers and Event Horizon Studies

Black hole mergers were the first gravitational wave sources detected, beginning with GW150914 in 2015. Unlike neutron star mergers, these events are typically electromagnetically silent, as they do not produce traditional light signals. However, recent studies suggest that in specific environments, such as active galactic nuclei (AGN) disks, black hole mergers may generate electromagnetic counterparts due to interactions with surrounding gas [11]. The mass spectrum of merging black holes has challenged existing astrophysical models. The detection of intermediate-mass black hole (IMBH) candidates, such as GW190521, suggests the presence of a missing population between stellar-mass and supermassive black holes [12]. This has implications for understanding black hole formation channels, including hierarchical mergers. Gravitational wave observations also complement event horizon studies, particularly through the Event Horizon Telescope (EHT), which produced the first image of a black hole in M87. Future space-based observatories, such as LISA, will detect mergers of supermassive black holes (SMBHs), providing insights into galaxy evolution and the role of black holes in shaping cosmic structures [6]. The strength of gravitational waves detected by observatories depends on the mass of the merging objects and their distance from Earth. This relationship is expressed as Equ. (5.2):

$$h \approx 4GM/c^4d \tag{5.2}$$

where:

- **h** is the gravitational wave strain,
- **G** is the gravitational constant,
- **c** is the speed of light,
- **M** is the total mass of the binary system,
- **d** is the total mass of the binary system.

This equation highlights why more massive mergers, such as those involving black holes, produce stronger signals, while distant events require highly sensitive detectors for observation.

5.3.3 Exotic Sources and High-Energy Astrophysics

Beyond compact binary mergers, gravitational waves offer the potential to detect exotic astrophysical sources, including:

- **Primordial Black Holes (PBHs):** Hypothetical black holes formed in the early universe, potentially contributing to dark matter [13].
- **Boson Stars:** Exotic objects made of scalar fields, offering tests for alternative gravity theories.
- **Core-Collapse Supernovae:** While gravitational waves from supernovae remain undetected, theoretical models suggest they could provide direct insights into stellar collapse dynamics [14].

High-energy astrophysics is closely linked to gravitational wave astronomy. Neutrino telescopes like IceCube and gamma-ray observatories like Fermi-LAT work in tandem with gravitational wave detectors to identify multi-messenger events. AI-based signal classification and real-time alert systems are improving the identification of weak signals, paving the way for future discoveries in extreme astrophysical environments. With advancements in detector sensitivity and machine learning techniques, multi-messenger astronomy will continue to revolutionize our understanding of the cosmos, revealing new physics and exotic astrophysical phenomena.

5.4 Artificial Intelligence in Gravitational Wave Detection

Gravitational wave detection has transformed astrophysics, offering new understanding of compact object collisions and the intrinsic properties of spacetime. However, the challenges associated with identifying and analyzing weak gravitational wave signals amid detector noise necessitate advanced data processing techniques. Artificial intelligence (AI), especially through machine learning (ML) and deep learning (DL), has become an essential tool in improving the precision and effectiveness of gravitational wave detectors such as LIGO, Virgo, and KAGRA. AI-driven methods enable real-time detection, precise event classification, and effective noise filtering, significantly improving the accuracy of gravitational wave searches.

5.4.1 Machine Learning for Signal Processing

Traditional template-matching techniques used in gravitational wave detection rely on matched filtering, where observed signals are compared with a vast bank of theoretical waveforms. However, this approach is computationally expensive and struggles with unexpected astrophysical events. Algorithms in machine learning (ML), including convolutional neural networks (CNNs) and support vector machines (SVMs), provide a faster and more efficient alternative for detecting signals buried in noise [15]. One of the key ML applications in signal processing is feature extraction—AI models trained on simulated and real data can automatically identify gravitational wave signatures, reducing human intervention and processing time [16]. Recent studies have demonstrated that deep learning-based denoising autoencoders can reconstruct signals even in low signal-to-noise ratio (SNR) conditions, significantly improving event recovery rates [17].

Additionally, AI-driven search methods, such as Gaussian process regression and random forest classifiers, have been employed to optimize parameter estimation, reducing uncertainties in mass and spin measurements of binary mergers [18].

5.4.2 Deep Learning for Event Classification

With an increasing number of detected gravitational wave events, the ability to classify and characterize them efficiently is crucial. Deep learning algorithms, particularly recurrent neural networks (RNNs) and transformers, have been successfully used to distinguish between different astrophysical sources, examples include mergers of binary black holes (BBHs), binary neutron stars (BNSs), and neutron star–black hole (NS-BH) systems [19]. A major advancement in event classification is the use of graph neural networks (GNNs), which analyze correlations between signals detected at different interferometers, improving localization accuracy [20]. To further enhance event classification, advanced deep learning models like graph neural networks (GNNs) have been employed. These networks capture intricate relationships between gravitational wave signals detected across multiple observatories, refining source identification and improving localization accuracy. The following Figure 5.5 illustrates the fundamental building blocks of a GNN used in gravitational wave analysis.

AI models have also been integrated into Bayesian inference frameworks, accelerating the classification process and enabling real-time alerts for follow-up observations with electromagnetic and neutrino telescopes [21]. Furthermore, the deployment of self-supervised learning techniques has allowed AI systems to learn directly from unlabeled data, increasing their adaptability to unknown or rare astrophysical events [22]. This has the potential to uncover exotic sources, such as boson star mergers or primordial black holes, which may otherwise be overlooked by template-based searches.

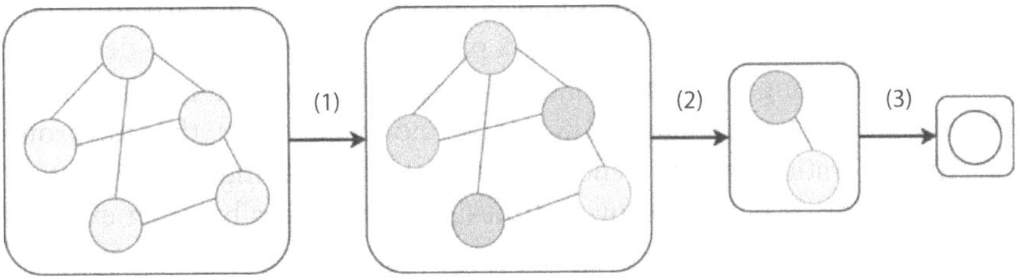

Figure 5.5 Basic building blocks of a graph neural network. (Credit: NickDiCicco *via* wikimedia commons).

5.4.3 AI-Driven Noise Filtering and Optimization

Gravitational wave detectors operate in environments susceptible to numerous noise sources, including seismic activity, thermal fluctuations, and instrumental artifacts. Effective noise filtering is essential to distinguish true astrophysical signals from spurious disturbances. AI-based noise reduction techniques, such as recurrent autoencoders and wavelet-based deep learning models, have demonstrated remarkable success in isolating gravitational wave signals from noise-dominated data streams [23]. Recent breakthroughs in unsupervised anomaly detection have enabled the identification of transient noise events, or glitches, in LIGO-Virgo data, allowing their removal before final analysis [24]. AI-driven optimization algorithms, such as reinforcement learning, have also been used to enhance detector calibration, reducing uncertainties in strain sensitivity and improving overall detection rates [25]. By leveraging AI in signal processing, classification, and noise filtering, the next generation of gravitational wave observatories, including LISA and Einstein Telescope (ET), will achieve unprecedented levels of precision, enabling the discovery of new astrophysical phenomena and deeper insights into the fabric of the universe.

5.5 Challenges and Future Prospects

Gravitational wave (GW) astronomy has provided unprecedented insights into the nature of compact objects and spacetime itself. However, the current generation of detectors, including LIGO, Virgo, and KAGRA, faces limitations that constrain their ability to detect faint or high-redshift sources. The next phase in gravitational wave detection will involve both technological advancements in detector sensitivity and the integration of artificial intelligence (AI) for real-time data analysis. Future observatories such as the Einstein Telescope (ET) and Cosmic Explorer (CE) promise to expand the GW detection horizon, probing deeper into the early universe. In this context, AI is expected to play a pivotal role in optimizing detection pipelines, reducing noise, and enabling faster astrophysical interpretation of GW signals. As gravitational wave astronomy advances, the shift from current-generation to next-generation detectors will bring transformative improvements. Table 5.2 below outlines key differences between existing and upcoming observatories, highlighting their expected advancements.

Table 5.2 Future prospects in gravitational wave detection.

Aspect	Current generation (LIGO, Virgo, KAGRA)	Next generation (Einstein telescope, cosmic explorer)
Sensitivity	Limited to nearby sources ($z < 1$)	Expected to detect sources at $z > 10$
Frequency Range	10 Hz–10 kHz	1 Hz–10 kHz (ET), improved low-frequency reach
Noise Reduction	Quantum squeezing, mirror coatings	Cryogenic mirrors, underground shielding
AI Integration	Used for event classification	Real-time analysis, noise filtering, improved localization
Scientific Potential	Mergers of stellar-mass objects	Early universe, primordial black holes, new physics

These improvements will allow future detectors to explore gravitational waves from the early universe, detect a wider range of astrophysical sources, and refine our understanding of fundamental physics. AI-driven advancements will further enhance real-time analysis, making GW detection more efficient and insightful.

5.5.1 Fundamental Sensitivity Limits of Current Detectors

Despite their groundbreaking discoveries, current ground-based detectors are limited by fundamental noise sources that affect their sensitivity. These include thermal noise, quantum shot noise, and seismic disturbances, which restrict the lower frequency range of detection to approximately 10 Hz [4]. Quantum noise, arising from the Heisenberg uncertainty principle, fundamentally limits the precision with which mirror positions can be measured. Advanced quantum techniques, such as squeezed light states, have been introduced to mitigate this issue, but improvements are still needed to reach optimal sensitivity [26]. Moreover, Newtonian noise, caused by terrestrial mass fluctuations (e.g., atmospheric pressure changes and seismic activity), introduces additional disturbances that are difficult to eliminate in current-generation detectors [27]. A critical challenge is the merger-ringdown detection gap—current detectors struggle to accurately observe post-merger signals from neutron star collisions due to their limited sensitivity at high frequencies (above 1 kHz) [28]. Addressing these issues will require novel engineering solutions and advanced data analysis techniques, including AI-based noise filtering and signal reconstruction.

5.5.2 Next-Generation Observatories: Einstein Telescope and Cosmic Explorer

The Einstein Telescope (ET) and Cosmic Explorer (CE) represent the next frontier in gravitational wave detection, with designs aimed at achieving at least an order of magnitude improvement in sensitivity over current detectors [29].

- **Einstein Telescope (ET):** A proposed European third-generation interferometer, ET will operate at cryogenic temperatures (~10 K) to reduce thermal noise and feature a triangular configuration with three interferometers, each 10 km in length. This setup will enhance detection sensitivity to low-frequency signals (1-10 Hz), enabling the observation of intermediate-mass black hole mergers and early universe gravitational waves [30].
- **Cosmic Explorer (CE):** The US-based next-generation detector, CE will have 40-km-long arms, significantly increasing the observable volume of the universe. It will push the detection limit for binary neutron star mergers to redshifts $z \approx 10$, allowing researchers to study neutron star formation in the earliest epochs of star formation [31].

Both ET and CE will play a crucial role in multi-messenger astronomy, complementing electromagnetic and neutrino observations. These observatories are expected to provide insights into dark matter, early-universe phase transitions, and exotic compact objects. However, the immense data volume they will generate necessitates AI-driven real-time data processing, capable of filtering noise, classifying events, and reconstructing signals faster than traditional methods.

5.5.3 AI's Expanding Role in Future Gravitational Wave Research

As gravitational wave observatories push toward higher sensitivities and detection rates, AI will become indispensable for handling large-scale data analysis and real-time event classification. The following areas are expected to benefit from AI advancements:

- **AI for Early-Warning Systems:** Future detectors will rely on AI-enhanced real-time alert systems that can rapidly classify potential GW events and trigger electromagnetic follow-up observations within seconds [32].
- **AI-Driven Signal Reconstruction:** Deep learning models will improve source parameter estimation, reducing uncertainties in mass, spin, and sky localization of GW sources, particularly in the case of overlapping or weak signals [33].
- **Quantum-Enhanced AI Techniques:** Machine learning models trained on quantum-processed data will allow more precise noise subtraction, improving sensitivity beyond classical computational limits [34].

The integration of AI with quantum computing and Bayesian inference will refine GW signal extraction methods, opening the door for detecting new physics, such as gravitational wave memory effects, axion clouds around black holes, and exotic compact objects. With AI's ability to automate key aspects of GW analysis, the next decade of gravitational wave research is set to redefine our understanding of the cosmos.

5.6 Conclusion

Gravitational wave detection has revolutionized our understanding of the universe, transforming astrophysics from a purely electromagnetic-based discipline to a multi-messenger approach. Since the first detection of gravitational waves in 2015, advancements in interferometric technology, data analysis, and astrophysical modeling have enabled the discovery of numerous black hole and neutron star mergers. These events provide unprecedented insights into the nature of compact objects, the evolution of binary systems, and the fundamental properties of spacetime itself. However, the field is still in its infancy, with significant challenges remaining in sensitivity limits, noise mitigation, and real-time detection. The integration of artificial intelligence has accelerated progress in gravitational wave astronomy. Machine learning algorithms have enhanced signal processing, enabling the rapid identification and classification of transient astrophysical events. Deep learning techniques have improved the accuracy of source parameter estimation, allowing researchers to reconstruct waveforms and extract physical properties with greater precision. Additionally, AI-driven noise filtering has become essential for distinguishing true signals from detector artifacts, optimizing the efficiency of large-scale data analysis. As the next-generation observatories, such as the Einstein Telescope and Cosmic Explorer, come online, AI will play an even greater role in handling the exponentially growing volume of data, refining detection pipelines, and automating follow-up observations.

Looking ahead, gravitational wave astronomy will continue to expand our cosmic horizon. The detection of high-redshift mergers, exotic astrophysical sources, and even potential signals from early-universe phase transitions may redefine our understanding of fundamental physics. The synergy between gravitational wave observations, electromagnetic follow-ups, and theoretical modeling will shape the future of astrophysics, providing deeper insights into dark matter, neutron star interiors, and the behavior of gravity under extreme conditions. With continuous technological innovation and AI-driven computational advancements, the next decade promises to be a transformative era for gravitational wave research, unlocking new frontiers in our quest to understand the fabric of the universe.

Data Availability

This chapter references publicly available astronomical datasets from sources like NASA, ESA, and various research observatories. Specific datasets including those from Kepler, LSST and SDSS are accessible through their respective repositories. For inquiry or requests, contact the corresponding author.

References

1. Maggiore, M., LISA and Future Gravitational Wave Observatories: Theoretical Foundations and Detection Techniques. *MNRAS*, 505, 6042–6057, 2021.
2. Kramer, M., Stairs, I.H., Manchester, R.N., *et al.*, Tests of General Relativity from the Double Pulsar System. *Phys. Rev. X*, 11, 041050, 2021.
3. Abbott, B.P., Abbott, R., Abbott, T.D., *et al.*, Observation of Gravitational Waves from a Binary Black Hole Merger. *Phys. Rev. Lett.*, 116, 061102, 2016.

4. Aasi, J., Abbott, B.P., Abbott, R., *et al.*, Advanced LIGO. *Classical Quantum Gravity*, 32, 074001, 2015.
5. McCuller, L., Tse, M., Fernández-Galiana, A., *et al.*, Quantum Squeezing for Advanced LIGO. *Phys. Rev. X*, 10, 031065, 2020.
6. Amaro-Seoane, P., Audley, H., Babak, S., *et al.*, Laser Interferometer Space Antenna. *J. Phys. Conf. Ser.*, 840, 012001, 2017.
7. Abbott, B.P., Abbott, R., Abbott, T.D., *et al.*, Multi-messenger Observations of a Binary Neutron Star Merger. *Astrophys. J. Lett.*, 848, L12, 2017.
8. Radice, D., Perego, A., Bernuzzi, S., *et al.*, Dynamical Mass Ejection from Binary Neutron Star Mergers. *Astrophys. J. Lett.*, 852, L29, 2018.
9. Kasen, D., Metzger, B.D., Barnes, J., *et al.*, Origin of the Heavy Elements in Binary Neutron-Star Mergers. *Nature*, 551, 80–84, 2017.
10. Coughlin, M.W., Dietrich, T., Margalit, B., *et al.*, Challenges and Strategies for Detecting Kilonovae. *MNRAS*, 497, 1181–1192, 2020.
11. Graham, M.J., Ford, K.E.S., McKernan, B., *et al.*, Candidate Electromagnetic Counterpart to the Binary Black Hole Merger GW190521. *Phys. Rev. Lett.*, 124, 251102, 2020.
12. Abbott, R., Abbott, T.D., Abraham, S., *et al.*, GW190521: A Binary Black Hole Merger with a Total Mass of 150 Solar Masses. *Phys. Rev. Lett.*, 125, 101102, 2020.
13. Carr, B., Kühnel, F., Sandstad, M., Primordial Black Holes as Dark Matter. *Phys. Rev. D*, 94, 083504, 2016.
14. Kuroda, T., Kotake, K., Takiwaki, T., Fully General Relativistic Simulations of Core-Collapse Supernovae with Multidimensional Neutrino Transport. *Astrophys. J.*, 851, 62, 2017.
15. George, D. and Huerta, E.A., Deep Learning for Real-time Gravitational Wave Detection and Parameter Estimation: Results with Advanced LIGO Data. *Phys. Lett. B*, 778, 64–70, 2018.
16. Gabbard, H., Williams, M., Hayes, F., *et al.*, Matching Matched Filtering with Deep Networks for Gravitational-wave Astronomy. *Phys. Rev. Lett.*, 120, 141103, 2018.
17. Shen, H., Huerta, E.A., Zhao, Z., *et al.*, Denoising Gravitational Waves with Deep Learning. *Nat. Astron.*, 5, 1100–1107, 2021.
18. Chua, A.J.K., Vallisneri, M., Littenberg, T.B., *et al.*, Gaussian Process Regression for Bayesian Inference of Gravitational Waves. *Phys. Rev. D*, 97, 044005, 2018.
19. Dreissigacker, C., *et al.*, Classification of Gravitational-wave Signals from Binary Black Holes with Deep Learning. *J. Cosmol. Astropart. Phys.*, 2019, 005, 2019.
20. Green, S.R. and Gair, J.R., Detecting and Characterizing Gravitational-wave Memory with LISA. *Phys. Rev. D*, 99, 084003, 2019.
21. Chatterjee, D., Ashton, G., Wilcox, D.E., Accelerating Bayesian Inference for Gravitational-wave Astronomy with Neural Networks. *MNRAS*, 505, 939–953, 2021.

22. Mukund, N., Abraham, S., Mitra, S., *et al.*, Machine Learning for Automated Detection of Transient Noise in LIGO Data. *Class. Quantum Grav.*, 35, 105002, 2018.
23. Torres-Forné, A., Font, J.A., Cerdá-Durán, P., *et al.*, Identifying Gravitational-wave Signals in LIGO Data with Wavelet Scattering Transforms and Deep Learning. *Phys. Rev. D*, 104, 104013, 2021.
24. Colgan, J., Wysocki, D., Pankow, C., *et al.*, Anomaly Detection for LIGO Data Using Machine Learning. *Mach. Learn.: Sci. Technol.*, 3, 035012, 2022.
25. Vajente, G., Ballmer, S., Biwer, C., *et al.*, Machine Learning for Detector Characterization in Gravitational-wave Astronomy. *Class. Quantum Grav.*, 38, 095004, 2021.
26. Tse, M., Yu, H., Kijbunchoo, N., *et al.*, Quantum-enhanced Advanced LIGO Detectors in the Era of Gravitational-wave Astronomy. *Phys. Rev. Lett.*, 123, 231107, 2019.
27. Harms, J., Paik, H.J., Venkateswara, K., Newtonian Noise and Its Mitigation in Gravitational-wave Observatories. *Class. Quantum Grav.*, 36, 245017, 2019.
28. Chatziioannou, K., Neutron Star Equation of State Constraints and the Merger-ringdown Gap. *Phys. Rev. D*, 101, 083007, 2020.
29. Punturo, M., Abernathy, M., Acernese, F., *et al.*, The Einstein Telescope: A Third-generation Gravitational Wave Observatory. *Class. Quantum Grav.*, 27, 194002, 2010.
30. Maggiore, M., Van den Broeck, C., Bartos, I., *et al.*, Science Case for the Einstein Telescope. *J. Cosmol. Astropart. Phys.*, 2020, 050, 2020.
31. Reitze, D., Adhikari, R.X., Ballmer, S., *et al.*, Cosmic Explorer: The U.S. Contribution to the Next Generation Ground-based Gravitational Wave Observatory. *Bull. Am. Astron. Soc.*, 51, 35, 2019.
32. Chatterjee, D., Ashton, G., Willcox, D.E., *et al.*, Machine Learning for Gravitational-wave Early Warning Systems. *MNRAS*, 505, 939–953, 2021.
33. Dreissigacker, C., *et al.*, Deep Learning for Source Parameter Estimation of Gravitational-wave Signals. *J. Cosmol. Astropart. Phys.*, 2019, 005, 2019.
34. Ghosh, S., Bayesian Inference in Gravitational-wave Astronomy: The Role of AI and Quantum Computing. *Nat. Phys.*, 18, 1253–1259, 2022.

6

Harmonizing the Cosmos: Radio Astronomy and AI Integration

Manjuleshwar Panda[1], Aadarsh Kumar Chaudhri[2]* and Mukesh Kumar Pandey[3]

[1]*Independent Astronomy Researcher, South West Delhi, Delhi, India*
[2]*Department of Physics, Govt. P.G. College Bazpur, U.S. Nagar, Uttarakhand, India*
[3]*Department of Physics, National Taiwan University, Taipei, Taiwan*

Abstract

Radio astronomy has long been a cornerstone of modern astrophysics, providing profound insights into the universe's most enigmatic phenomena, from pulsars and fast radio bursts (FRBs) to active galactic nuclei and cosmic microwave background radiation. However, the vast and complex data generated by radio telescopes demand advanced analytical techniques beyond traditional computational methods. Artificial intelligence (AI) has emerged as a transformative tool in radio astronomy, revolutionizing signal detection, data processing, and scientific discovery. This chapter explores the integration of AI into radio astronomy, highlighting its role in enhancing observational capabilities, managing large-scale datasets, and uncovering hidden cosmic structures. We begin by discussing the foundations of radio astronomy, outlining the principles behind radio wave detection and the significance of major radio observatories. The advent of AI-driven methodologies, including machine learning, deep learning, and neural networks, has significantly improved data analysis, automating tasks such as radio frequency interference (RFI) mitigation, anomaly detection, and transient signal classification. AI-based signal processing techniques enable astronomers to extract faint cosmic signals from noisy datasets, leading to improved real-time detection and classification of FRBs—one of the most mysterious and energetic astrophysical events. Recent discoveries made by AI-assisted radio telescopes, such as CHIME/FRB, underscore the potential of machine learning in accelerating breakthroughs. Moreover, AI plays a crucial role in pulsar detection and the Search for Extraterrestrial Intelligence (SETI). By leveraging neural networks and unsupervised learning

**Corresponding author*: aadarshphysics@gmail.com

Yogesh Chandra, Manjuleshwar Panda and Mahesh Chandra Mathpal (eds.) The Intelligent Universe: AI's Role in Astronomy, (201–232) © 2025 Scrivener Publishing LLC

models, researchers can efficiently identify periodic signals from pulsars and distinguish potential extraterrestrial transmissions from terrestrial noise. AI's contributions extend to Very Long Baseline Interferometry (VLBI), where it enhances image reconstruction techniques, as demonstrated by its application in the Event Horizon Telescope's groundbreaking black hole imaging. Modern radio telescope facilities including Square Kilometre Array (SKA) and LOFAR initiate a significant data challenge for the astronomy field. AI-powered algorithms provide scalable solutions for handling massive datasets, enabling automated classification of celestial objects and the discovery of rare astronomical phenomena. Looking ahead, AI's fusion with quantum computing and advanced data analytics promises to further refine radio astronomical research, unlocking deeper cosmic mysteries. This chapter presents an in-depth analysis of AI's evolving role in radio astronomy, highlighting case studies, challenges, and future prospects. As AI continues to harmonize the cosmos, its synergy with radio astronomy will redefine our understanding of the universe, bridging the gap between vast observational data and advanced scientific interpretations.

Keywords: Radio astronomy, artificial intelligence (AI), machine learning, fast radio bursts (FRBs), pulsar detection, Very Long Baseline Interferometry (VLBI), Square Kilometre Array (SKA), signal processing

6.1 Introduction: The Synergy of Radio Astronomy and AI

The convergence of radio astronomy and artificial intelligence (AI) represents a transformative leap in modern astrophysics, enabling researchers to probe the cosmos with unprecedented precision and efficiency. Radio astronomy, which observes celestial objects through their radio wave emissions, has significantly contributed to our understanding of the universe, from detecting pulsars and mapping cosmic microwave background radiation to studying active galactic nuclei and interstellar molecular clouds. However, traditional data processing techniques face a significant challenge due to the massive amount of data produced by next-generation radio telescopes like the Very Large Array (VLA) and the Square Kilometre Array (SKA) [1]. To address this, AI-driven techniques, including machine learning (ML) and deep learning (DL), have emerged as indispensable tools for processing and interpreting vast datasets, uncovering hidden patterns, and automating discovery processes. One of the most important challenges as the size of radio astronomy observations keeps growing is controlling the exponential growth of data. For example, it is anticipated that the Square Kilometre Array (SKA) will generate more raw data daily than all of the

Internet traffic worldwide, requiring sophisticated AI-driven systems for effective processing and analysis.

"The Square Kilometre Array and other next-generation radio telescopes will generate more data than the entire Internet—only AI can help us manage and make sense of this deluge."

- Prof. Phil Diamond, Director-General of the SKA Observatory

This demonstrates how important artificial intelligence (AI) is to radio astronomy, where deep neural networks and machine learning techniques are crucial for separating signals from noise and deriving valuable astrophysical insights from the massive volume of data.

6.1.1 Overview of Radio Astronomy and Its Significance in Modern Astrophysics

Radio astronomy has been instrumental in advancing our knowledge of the universe by probing celestial phenomena that remain invisible in optical and infrared wavelengths. Since the discovery of cosmic radio waves by Karl Jansky in the 1930s, radio telescopes have enabled astronomers to investigate diverse astrophysical processes, including synchrotron radiation from supernova remnants, masers in star-forming regions, and the enigmatic fast radio bursts (FRBs) [2]. Modern facilities like the SKA and the Atacama Large Millimeter/submillimeter Array (ALMA) offer unparalleled sensitivity and resolution, facilitating detailed studies of extragalactic radio sources, black hole jets, and cosmic magnetism. However, the complexity of radio signals, often embedded in noise and interference, necessitates advanced computational methodologies for efficient signal extraction and analysis. This challenge is governed by the Radiometer Equation as follows:

$$\sigma = T_{sys} / \sqrt{B.t} \tag{6.1}$$

where $\boldsymbol{\sigma}$ is the noise uncertainty, $\mathbf{T_{sys}}$ is the system temperature, **B** is the bandwidth of observation, and **t** is the integration time. Equ. (6.1) demonstrates that by increasing the bandwidth **(B)** and integration time **(t)**, the noise uncertainty **(σ)** can be reduced, thus improving the ability to detect faint astrophysical signals and mitigating the effects of interference.

6.1.2 The Role of AI in Revolutionizing Data Analysis, Signal Processing, and Discovery

The integration of AI in radio astronomy has revolutionized data analysis, signal processing, and scientific discovery. Traditional methods of manually inspecting radio signals or applying predefined algorithms struggle to handle the massive influx of data from modern observatories. AI-driven frameworks, particularly convolutional neural networks (CNNs) and recurrent neural networks (RNNs), have demonstrated remarkable success in classifying radio transients, mitigating radio frequency interference (RFI), and optimizing interferometric imaging [3]. Moreover, unsupervised learning algorithms facilitate anomaly detection, enabling the discovery of novel astrophysical phenomena, such as previously undetected pulsars and fast radio bursts (FRBs). The synergy between AI and radio astronomy not only enhances observational capabilities but also enables autonomous real-time data processing, ushering in a new era of scientific exploration. As AI-driven approaches continue to advance, they are not only accelerating data analysis but also uncovering previously unknown cosmic phenomena.

6.2 Foundations of Radio Astronomy: Unlocking the Invisible Universe

Radio astronomy has revolutionized our understanding of the cosmos by enabling the detection of celestial signals that are otherwise imperceptible to optical telescopes. Unlike visible light, which is often obscured by cosmic dust and atmospheric interference, radio waves penetrate such barriers, allowing astronomers to explore the universe's hidden structures. This ability has led to groundbreaking discoveries that have reshaped astrophysical theories and enhanced our comprehension of cosmic evolution.

6.2.1 How Radio Telescopes Detect Celestial Signals

Radio telescopes function by capturing faint radio emissions from celestial sources and converting them into interpretable data. These telescopes consist of large parabolic dishes, arrays of antennae, and sophisticated signal-processing systems that enhance sensitivity and resolution. Observations rely on highly sensitive receivers, which amplify incoming signals before filtering out noise and terrestrial interference [4]. Interferometry, a

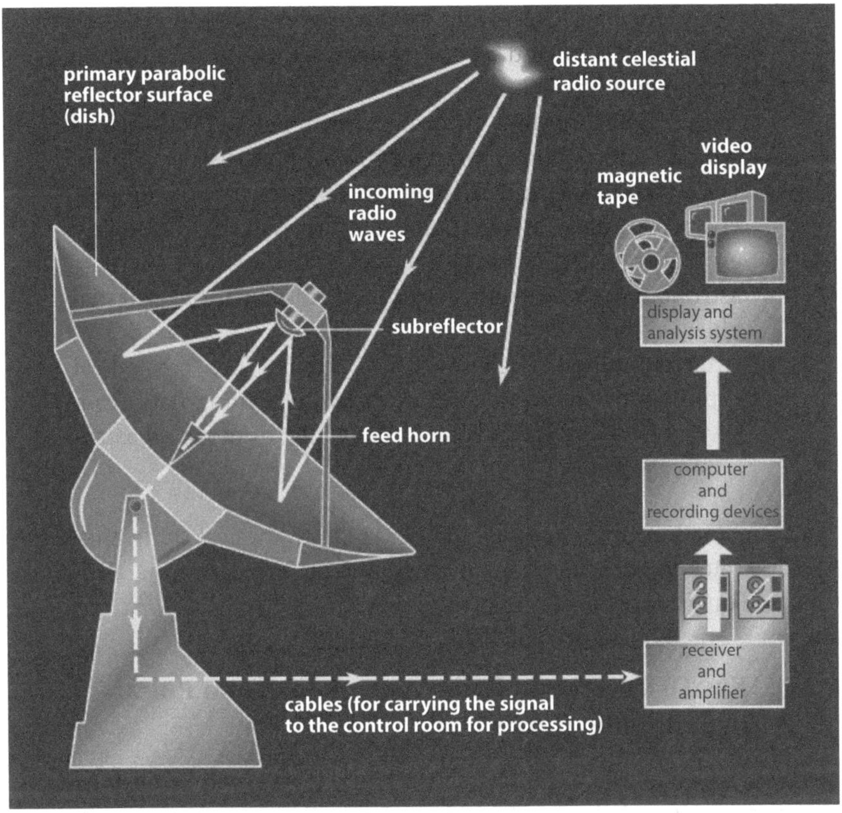

Figure 6.1 A radio telescope. (Image credit: university of oregon, retrieved from https://pages.uoregon.edu/jschombe/glossary/radio_telescope.html).

technique combining signals from multiple radio telescopes, significantly improves image resolution, enabling astronomers to map distant galaxies, pulsars, and black hole jets with remarkable detail. Advances in digital signal processing and AI-driven calibration methods have further optimized the efficiency and accuracy of these observations.

Figure 6.1 above illustrates the working mechanism of a radio telescope, demonstrating how it captures and processes radio signals from distant celestial sources.

- **Parabolic Reflector Dish:** Incoming radio waves from space are gathered by the huge primary dish and directed onto a smaller secondary reflector, also known as a subreflector.

- **Feed Horn and Receiver:** The focused signals are gathered by the feed horn and sent to the receiver for noise reduction and amplification.
- **Signal Transmission:** The processed signals are transmitted *via* cables to a control room for further analysis.
- **Data Processing and Display:** Through the use of computers, recording equipment, and display systems, the signals are transformed into data that can be understood, enabling astronomers to conduct in-depth research on celestial objects.
- **Interferometry:** High-resolution images are created at sophisticated radio observatories by combining data from several telescopes, exposing intricate details of far-off astronomical sources.

In a major contribution to contemporary astrophysical study, this setup demonstrates the crucial elements required for the detection, amplification, and analysis of cosmic radio emissions.

6.2.2 Key Discoveries in Radio Astronomy, from Pulsars to Fast Radio Bursts

The field of radio astronomy has been marked by several transformative discoveries. One of the most significant was the detection of pulsars in 1967 by Jocelyn Bell Burnell and Antony Hewish, confirming the existence of rapidly rotating neutron stars [5]. Pulsars have since served as cosmic laboratories for testing general relativity and studying extreme states of matter. More recently, the identification of fast radio bursts (FRBs)—millisecond-duration radio pulses of extragalactic origin—has opened new avenues in high-energy astrophysics. The enigmatic nature of FRBs, with their unpredictable appearances and high dispersion measures, has made them a prime target for AI-driven search algorithms that can efficiently sift through vast datasets to identify new signals [6]. Some of the most important findings in radio astronomy are compiled in Table 6.1 below, along with their significance and the contribution of AI to our growing knowledge of these phenomena.

These findings highlight how radio astronomy has revolutionized our knowledge of the cosmos. As artificial intelligence develops, its combination with radio telescopes and data processing methods will improve our capacity to discern and interpret subtle cosmic signals, opening the door to new astrophysics discoveries.

Table 6.1 Key discoveries in radio astronomy.

Discovery	Year	Discovered by	Significance	Role of AI
Pulsars	1967	Jocelyn Bell Burnell, Antony Hewish	Evidence of rapidly rotating neutron stars	Used for automated pulsar detection and timing analysis
Cosmic Microwave Background (CMB)	1965	Arno Penzias, Robert Wilson	Provided strong evidence for the Big Bang Theory	AI aids in removing foreground noise from CMB maps
Fast Radio Bursts (FRBs)	2007	Duncan Lorimer *et al.*	Extragalactic radio pulses of unknown origin	Machine learning helps in real-time FRB detection
Sagittarius A (Milky Way's Supermassive Black Hole)*	1974	Bruce Balick, Robert Brown	Indirect evidence of a black hole at the galactic center	AI enhances signal extraction from radio images
Exoplanet Magnetic Fields (via radio signals)	2020	Various teams (LOFAR, CHIME)	Potential for studying exoplanet atmospheres	AI assists in signal filtering and classification

6.2.3 The Challenges of Radio Data Acquisition and Processing

Despite its numerous advancements, radio astronomy faces significant challenges in data acquisition and processing. The primary obstacles include terrestrial radio frequency interference (RFI), massive data volumes, and complex signal extraction processes. RFI from human-made sources, such as satellites, mobile networks, and broadcasting stations, contaminates astronomical observations, requiring sophisticated filtering techniques. The SKA, for instance, is expected to generate exabytes of data annually, necessitating the development of scalable AI-based pipelines for real-time analysis. Moreover, weak celestial signals often compete with background noise, demanding innovative deconvolution and denoising algorithms to enhance detection accuracy. Addressing these challenges will be crucial for maximizing the scientific output of future radio astronomy missions. Astronomers are increasingly using cutting-edge computational methods, such as deep learning models and high-performance computing frameworks, to improve data processing efficiency in order to overcome these obstacles. Algorithms powered by AI are being created to enhance real-time event detection, optimize signal classification, and automate RFI mitigation. Strong data pipelines that can manage the massive data streams anticipated from next-generation observatories are also being developed as a result of interdisciplinary partnerships involving astronomers, data scientists, and engineers. These initiatives will be crucial to guaranteeing that, despite the increasing complexity of data collection and interpretation, radio astronomy keeps discovering new cosmic events.

6.3 The Evolution of AI in Radio Astronomy

The integration of artificial intelligence (AI) into radio astronomy has revolutionized the way astronomical data are analyzed and interpreted. From early computational techniques to sophisticated AI-driven methodologies, the field has witnessed a paradigm shift in its approach to processing vast amounts of radio observations. Traditional techniques relied heavily on manual analysis and computational algorithms with limited adaptability, while modern AI-driven systems can efficiently extract complex patterns, detect anomalies, and enhance the accuracy of astronomical discoveries. The incorporation of machine learning (ML), deep learning (DL), and neural networks has accelerated the rate of scientific breakthroughs, reducing human intervention and optimizing data-driven research in radio astronomy. Radio astronomers can now manage ever-larger and more complicated information with previously unheard-of precision, thanks to the quick developments in AI-driven techniques. Astronomers can now automate feature extraction, categorize radio sources, and analyze time-series data more effectively, thanks to the development of transformer models, convolutional neural networks (CNNs), and recurrent neural networks (RNNs). These artificial intelligence (AI) methods have been crucial in improving radio interferometric imaging, finding new pulsars, and detecting fast radio bursts (FRBs). Furthermore, generative adversarial networks (GANs) have helped with low-resolution astronomical image restoration and noise reduction. It is anticipated that as AI develops, its application to radio astronomy will enhance detection capabilities, further optimize data analysis pipelines, and uncover new astrophysical insights from radio observations.

6.3.1 Early Computational Techniques in Radio Astronomy

In the early years of radio astronomy, data processing primarily relied on conventional statistical methods and Fourier-based techniques. Observational data collected by radio telescopes underwent manual inspection, where astronomers identified signals of interest amid background noise. The introduction of computer-based signal processing in the mid-20th century allowed for the automation of certain tasks, such as cross-correlation and interferometry, which improved imaging precision. Early digital techniques, such as wavelet transforms and principal component analysis (PCA), played a crucial role in filtering noise and extracting celestial signals [7]. Despite these advancements, the increasing complexity of data from large-scale radio telescopes necessitated the development of more sophisticated computational frameworks.

6.3.2 Transition from Traditional Methods to AI-Driven Approaches

As radio telescopes became more advanced, producing petabytes of data per observation cycle, traditional computational methods struggled to keep up with the growing demand for real-time analysis. The transition to AI-driven approaches marked a turning point in radio astronomy, enabling automated classification, pattern recognition, and signal enhancement. Supervised learning models, trained on large datasets, began to outperform manual techniques in identifying weak astrophysical signals, such as fast radio bursts (FRBs) and pulsars [3]. AI-powered algorithms have been particularly effective in mitigating radio frequency interference (RFI), an ongoing challenge in ground-based radio observations. The signal-to-noise ratio and detection efficiency have been greatly enhanced by contemporary AI techniques that make use of deep neural networks and support vector machines. Building on these developments, AI-powered methods have kept improving and extending radio astronomy's potential. Important advancements consist of:

- **Automated Pulsar and FRB Detection:** Convolutional and recurrent neural networks are examples of deep learning models that have enhanced the highly accurate identification of transitory radio signals.
- **RFI Mitigation:** Data quality for ground-based radio telescopes can be improved by using AI-powered filtering

approaches that can differentiate astrophysical signals from terrestrial interference.

- **Enhanced Radio Imaging:** In order to improve the resolution and clarity of dim astronomical sources, generative models—such as GANs—help reconstruct and denoise radio pictures.
- **Real-Time Data Processing:** By speeding up the analysis of large datasets, artificial intelligence (AI) lowers the time needed for discovery and makes it possible for dynamic radio sky surveys to detect events almost instantly.

The shift from conventional data-processing methods to AI-enhanced approaches is being driven by these developments, opening up new scientific avenues in radio astronomy.

6.3.3 The Impact of Machine Learning, Deep Learning, and Neural Networks

The emergence of ML, DL, and neural networks has had a profound impact on radio astronomy, reshaping how researchers extract information from vast datasets. Convolutional neural networks (CNNs) have been widely applied in radio astronomy for feature extraction and classification, enabling the detection of new pulsars and transient signals with high accuracy [8]. Recurrent neural networks (RNNs) and transformer models have been employed to analyze time-series data, facilitating the identification of periodic signals from rotating neutron stars. Additionally, generative adversarial networks (GANs) have been explored for enhancing radio images and simulating complex astrophysical phenomena. These AI-driven methodologies not only accelerate discovery but also enable astronomers to delve deeper into the universe's most enigmatic radio sources. In the area of radio astronomy, Figure 6.2 below shows how deep learning, machine learning, and artificial intelligence collaborate. These methods are essential for processing large astronomical datasets because they make it possible to efficiently extract features, classify, and improve observational data.

The graphic illustrates how the combination of AI, ML, and DL has revolutionized radio signal analysis by astronomers. This integration opens the door for groundbreaking discoveries in radio astronomy by improving our capacity to identify fleeting events, categorize celestial objects, and model intricate astrophysical processes.

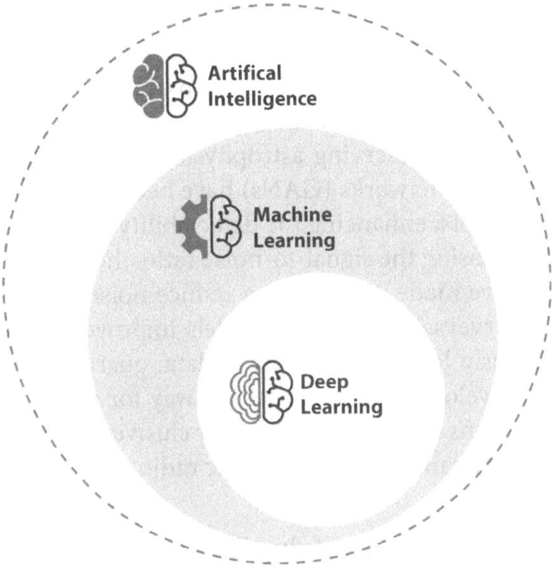

Figure 6.2 How deep learning, machine learning, and artificial intelligence work together. (Image credit: NASA).

6.4 AI-Powered Signal Processing: Detecting the Weakest Cosmic Signals

The detection of faint cosmic signals has always been a significant challenge in radio astronomy due to interference from terrestrial sources, instrumental noise, and cosmic background radiation. AI-driven signal processing techniques have revolutionized this field, enabling astronomers to extract meaningful information from extremely weak signals buried in noise. Advanced machine learning algorithms now enhance data quality, classify celestial signals, and mitigate the impact of radio frequency interference (RFI), thereby improving the sensitivity and accuracy of radio observations. By leveraging AI, radio astronomers can refine their datasets and uncover previously undetectable astrophysical phenomena.

6.4.1 Noise Filtering and Denoising Techniques Using AI

One of the primary challenges in radio astronomy is distinguishing weak cosmic signals from various noise sources, including atmospheric disturbances, instrumental imperfections, and RFI. AI-based denoising techniques, such as deep learning-driven autoencoders and wavelet transform models, have significantly improved the ability to extract faint signals. Convolutional neural networks (CNNs) and recurrent neural networks (RNNs) have been used to develop self-adaptive noise filtering frameworks that learn from large datasets, allowing them to identify and suppress unwanted noise while preserving astrophysical signals [9]. Additionally, generative adversarial networks (GANs) have been applied to reconstruct lost signal components, enhancing the detectability of weak radio sources. In addition to increasing the signal-to-noise ratio, these AI-driven denoising approaches have made it possible to reduce noise in real time during extensive radio surveys. Adaptive AI models improve their filtering abilities by continuously learning from fresh data, guaranteeing little signal distortion. This development has paved the way for more accurate astronomical investigations by helping to discover elusive phenomena like weak transient occurrences and dim extragalactic radio emitters.

6.4.2 Identifying and Classifying Radio Signals Using Machine Learning

Machine learning has proven to be an invaluable tool for automating the identification and classification of radio signals. Traditional methods of signal classification required extensive manual inspection, making the process time-consuming and prone to human error. AI-driven classifiers, such as support vector machines (SVMs) and deep belief networks, now enable astronomers to distinguish between pulsars, fast radio bursts (FRBs), and quasars with remarkable precision [10]. These models can efficiently process large datasets from radio telescopes, identifying new transient sources and improving the accuracy of existing catalogs. Furthermore, unsupervised learning techniques, including clustering algorithms like k-means and self-organizing maps (SOMs), help uncover previously unknown astrophysical patterns in radio data.

6.4.3 AI's Role in Mitigating Radio Frequency Interference

RFI poses a persistent challenge for ground-based radio astronomy, as signals from satellites, mobile networks, and electronic devices can overwhelm weak cosmic emissions. Traditional filtering methods often involve manually flagging contaminated data, a labor-intensive process that risks losing valuable information. AI-driven RFI mitigation techniques have transformed this approach by using deep learning models, such as recurrent autoencoders and attention mechanisms, to automatically detect and remove interference with minimal human intervention [11]. Adaptive filtering algorithms and reinforcement learning-based models now allow real-time mitigation of RFI, ensuring that high-quality astronomical data are preserved. These advancements have led to a significant improvement in the fidelity of radio observations, allowing for deeper explorations into the universe's radio spectrum. We take into account the observed radio signal as a combination of the desired astronomical signal and undesired interference in order to quantitatively characterize the AI-driven strategy to RFI reduction. AI models successfully detect and eliminate the RFI component in order to extract the pure astrophysical signal. This can be expressed in Equ. 6.2 as follows:

$$\begin{aligned} &I. \quad S(t) = A(t) + RFI(t) \\ &II. \quad a(t) = S(t) - rfi(t) \end{aligned} \tag{6.2}$$

Here, ***S(t)*** is the observed signal, ***A(t)*** is the true astronomical signal, and ***RFI(t)*** represents the interference. AI-based models estimate ***rfi(t)*** and subtract it from ***S(t)***, yielding the cleaned signal ***a(t)***. Deep learning methods, such as recurrent autoencoders and attention-based models, enhance this process by learning complex interference patterns and removing them more effectively than traditional methods. This ensures that valuable astronomical data are preserved while minimizing distortions caused by RFI.

6.5 Fast Radio Bursts and AI: Solving One of Astronomy's Biggest Mysteries

Fast radio bursts (FRBs) are among the most enigmatic astronomical phenomena, characterized by millisecond-duration pulses of radio waves originating from extragalactic sources. Since their first discovery in 2007, FRBs have challenged astronomers with their unknown origins and extreme energy outputs. The role of AI in unraveling these mysteries has grown significantly, as traditional methods struggle to keep up with the volume and complexity of FRB data. AI-driven detection, classification, and analysis have led to major breakthroughs, transforming our understanding of these cosmic events and enabling real-time discovery and tracking.

6.5.1 What are FRBs, and Why are They Important?

Despite lasting only a few moments, FRBs are powerful radio wave bursts that release as much energy as the Sun does over the course of several days.

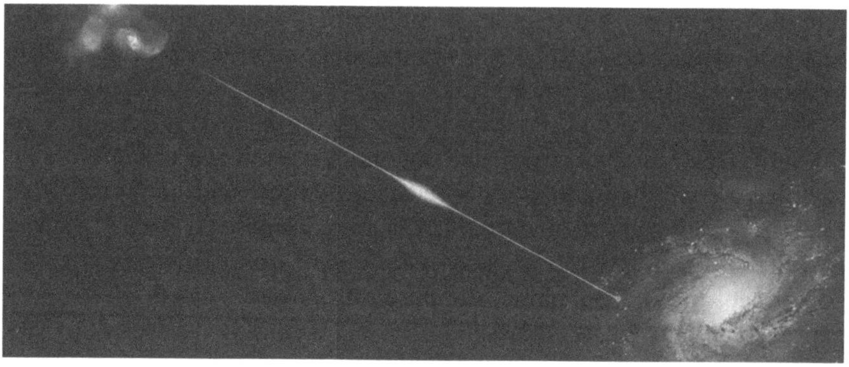

Figure 6.3 Most distant fast radio burst to date. (Image Credit: European Southern Observatory).

These transient events have been observed at cosmological distances, indicating their immense power and potential role in understanding fundamental astrophysical processes. The origins of FRBs remain unknown, with hypotheses ranging from magnetars and neutron star mergers to exotic phenomena such as cosmic strings and even advanced extraterrestrial civilizations. Studying FRBs is crucial for probing the intergalactic medium, testing fundamental physics, and refining cosmological models [12]. The discovery of FRBs coming from very far away in recent investigations has shed light on the universe's large-scale structure. The most distant rapid radio burst ever discovered is shown in Figure 6.3 above, illustrating its trip across cosmic distances.

These findings not only contribute to our understanding of FRBs but also act as cosmic instruments for examining the universe's matter distribution. Astronomers can improve theories of cosmic development and the distribution of dark matter by examining the interactions between these radio waves and the intergalactic medium.

6.5.2 How AI is Improving Real-Time Detection and Classification

The high rate of FRB occurrences and their short durations make manual detection and classification nearly impossible. AI-driven algorithms have revolutionized this field by enabling real-time identification of FRBs from vast datasets generated by radio telescopes. Machine learning models, particularly convolutional neural networks (CNNs) and recurrent neural networks (RNNs), have been deployed to distinguish FRBs from noise and other transient signals with high accuracy. Unsupervised learning techniques have also been applied to cluster FRB events based on their spectral and temporal properties, leading to potential new classifications [13]. Furthermore, AI accelerates the analysis of FRB dispersion measures, helping astronomers refine estimates of cosmic distances and interstellar medium characteristics.

6.5.3 Case Studies of AI-Driven FRB Discoveries (e.g., CHIME/FRB)

AI has played a pivotal role in groundbreaking FRB discoveries, particularly through large-scale radio telescope arrays like CHIME/FRB. The CHIME (Canadian Hydrogen Intensity Mapping Experiment) telescope has detected thousands of FRBs, thanks to an AI pipeline that processes petabytes of data in real time. The system employs deep learning frameworks to automatically flag and classify FRBs, enabling rapid follow-up observations and detailed statistical analysis [14]. Other telescopes, such as ASKAP and MeerKAT, have also integrated AI-based search techniques, leading to the discovery of repeating FRB sources and refined localization methods. These advancements have expanded our knowledge of FRB populations and provided crucial insights into their astrophysical origins. Our capacity to identify, categorize, and investigate these enigmatic cosmic signals has been greatly improved by these AI-driven developments in FRB research. Among AI's major contributions in this field are the following:

- **Real-Time Detection:** By identifying FRBs in milliseconds, AI lowers the need for human involvement and increases the rate of discovery.
- **Automated Classification:** FRBs are categorized by machine learning models according to their frequency profiles, duration, and dispersion measures.
- **Improved Localization:** AI-powered pipelines help locate host galaxies and potential progenitors by fine-tuning the localization of FRB sources.
- **Discovery of Repeating FRBs:** In order to identify possible variations in the origins of one-time and repeating FRB sources, artificial intelligence has been instrumental.
- **Data Processing Efficiency:** By removing RFI and boosting weak FRB signals from large datasets, AI lessens the computational load.

- **Cross-Correlation with Other Observations:** Multiwavelength correlation is made possible by AI-assisted analysis, which connects FRB events to gravitational wave, optical, and X-ray events.

Future radio surveys are anticipated to find even fainter and farther-flung FRBs as AI models continue to increase in sensitivity and accuracy. These advancements will deepen our comprehension of their genesis and their function in examining the universe's large-scale structure.

6.6 AI in Pulsar and SETI Research: Searching for Cosmic Beacons

The integration of artificial intelligence (AI) in pulsar research and the Search for Extraterrestrial Intelligence (SETI) has transformed our ability to detect and analyze cosmic signals. Pulsars—rapidly rotating neutron stars emitting periodic radio pulses—serve as natural cosmic beacons, while SETI focuses on identifying potential artificial signals from extraterrestrial civilizations. AI's ability to process vast datasets with precision has significantly advanced both fields, leading to major breakthroughs in pulsar discoveries and enhancing the search for technosignatures.

6.6.1 How AI is Advancing Pulsar Detection and Classification

Traditional pulsar detection methods involve extensive manual data processing, requiring astronomers to sift through large datasets from radio telescopes. AI-driven techniques, particularly convolutional neural networks (CNNs) and recurrent neural networks (RNNs), have greatly enhanced the efficiency of pulsar detection and classification. By training on labeled datasets, these models can identify pulsar candidates with higher accuracy than traditional methods, reducing false positives and improving discovery rates. Recent studies using AI-based approaches have led to the discovery of new pulsars from existing datasets, demonstrating the effectiveness of machine learning in uncovering faint and previously overlooked signals [15]. A pulsar's structure is depicted in Figure 6.4, emphasizing its powerful magnetic fields and radiation beams. As the neutron star rotates, these beams travel over space, creating the recurring radio frequencies that telescopes pick up.

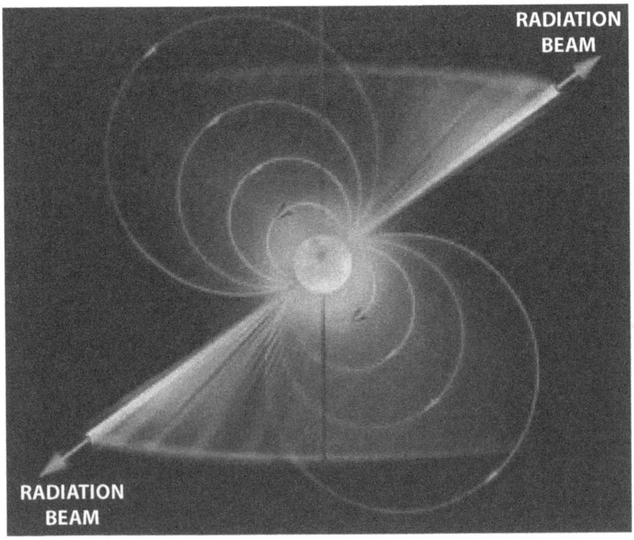

Figure 6.4 An artist's depiction of a pulsar with magnetic fields (blue) and particle jets (yellow). (Image credit: NRAO).

This basic pulsar emission mechanism is used by AI-driven pulsar discovery methods to effectively find new candidates. Machine learning algorithms can accurately differentiate pulsars from other radio sources by examining the periodicity and shape of radio signals, greatly enhancing the discovery process.

6.6.2 The Role of AI in the Search for Extraterrestrial Intelligence

SETI research relies on detecting narrowband radio signals that may indicate intelligent extraterrestrial transmissions. The vast volume of radio data collected by telescopes like the Green Bank Telescope (GBT) and the Allen Telescope Array (ATA) requires advanced filtering techniques to distinguish potential signals from natural astrophysical sources and terrestrial interference. AI, particularly deep learning models, has enabled real-time signal classification, anomaly detection, and pattern recognition within these massive datasets. Recent applications of AI in SETI have identified intriguing signal candidates, demonstrating the potential for AI-assisted discoveries in the quest for extraterrestrial intelligence [16]. To boost SETI research, radio telescopes like the Allen Telescope Array (ATA) depicted in Figure 6.5 continuously monitor the sky for narrowband signals that

Figure 6.5 The allen telescope array (ATA), a key instrument in SETI, uses AI to analyze vast radio datasets for potential extraterrestrial signals. (Image credit: seth shostak, SETI Institute/ESA).

could indicate extraterrestrial civilization. The enormous amount of data that these telescopes gather must be processed using AI-driven methods in order to filter and identify possible candidates in real time.

The image is of California's Allen Telescope Array (ATA), a system of tiny radio dishes used for astronomical research and the hunt for alien intelligence. In order to separate manufactured signals from natural astrophysical sources and terrestrial interference, artificial intelligence (AI) is crucial in processing the radio signals that ATA has recorded. ATA has increased the effectiveness of SETI research by using deep learning algorithms for pattern recognition and anomaly detection, which raises the likelihood of finding signals of interest. The sensitivity and effectiveness of SETI searches have been greatly increased by advances in AI, enabling astronomers to scan a wider range of frequencies and find weak signals that might have gone unnoticed in the past. In order to improve our capacity to identify possible extraterrestrial transmissions, future AI-driven SETI systems will work to improve signal classification techniques.

6.6.3 Future Prospects of AI-Assisted Extraterrestrial Communication Searches

As AI technologies continue to evolve, their role in pulsar and SETI research is expected to expand significantly. Future advancements in AI-driven signal processing will enable more refined searches for both natural and artificial cosmic beacons, increasing the likelihood of new discoveries. AI-powered unsupervised learning techniques and reinforcement learning models could further improve anomaly detection, helping researchers identify unconventional signals that may have previously gone unnoticed. Moreover, collaborations between AI researchers and astronomers will facilitate the development of more sophisticated algorithms to enhance our ability to detect, classify, and interpret cosmic signals [17]. The future of AI-assisted extraterrestrial communication searches holds immense promise, potentially leading to one of the most profound discoveries in human history.

6.7 AI in Very Long Baseline Interferometry and Image Reconstruction

The integration of AI into Very Long Baseline Interferometry (VLBI) marks a transformative era in radio astronomy, enhancing data acquisition, processing, and interpretation. VLBI, a technique that combines signals from widely separated radio telescopes, enables astronomers to achieve resolutions far beyond those of single-dish observatories. However, the vast amounts of data generated and the complexity of calibration pose significant challenges. AI-driven methods are now optimizing these processes, leading to more efficient imaging and deeper insights into cosmic phenomena.

6.7.1 The Importance of VLBI in High-Resolution Radio Imaging

Very Long Baseline Interferometry (VLBI) is a cornerstone technique (Figure 6.6) in radio astronomy, enabling astronomers to achieve unprecedented angular resolution by linking radio telescopes across the globe. By synchronizing these telescopes using atomic clocks, VLBI synthesizes a virtual telescope with an effective aperture equivalent to the maximum separation between the individual stations. This capability has allowed

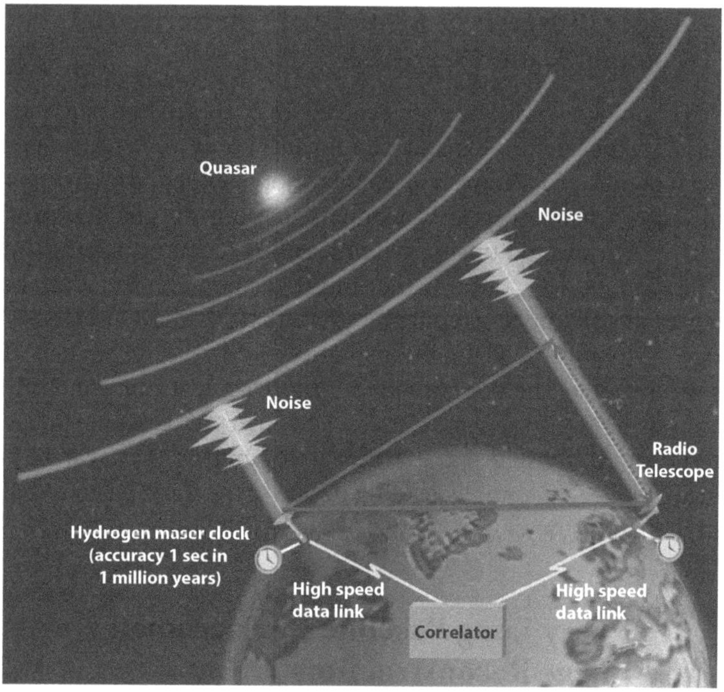

Figure 6.6 VLBI network - a global array of radio telescopes working together to achieve ultrahigh-resolution imaging. (Image credit: NASA).

scientists to image distant celestial objects, probe active galactic nuclei, and study the physics of black holes with remarkable clarity [18]. The technique's precision has been pivotal in refining models of astrophysical jets, accretion disks, and even fundamental cosmological parameters. High-resolution imaging made possible by VLBI has completely changed our knowledge of compact astrophysical objects. This method enables astronomers to resolve minute details in far-off galaxies, quasars, and black hole settings that would be unachievable with a single radio telescope by merging signals from far spaced telescopes.

One of the most significant achievements of VLBI is the Event Horizon Telescope (EHT), which captured the first-ever image of a black hole's shadow in the galaxy M87. Future VLBI projects continue to push the boundaries of radio astronomy, improving our ability to study cosmic structures and refine models of extreme astrophysical phenomena.

6.7.2 AI's Role in Improving Data Calibration and Synthesis Imaging

Despite VLBI's power, its data calibration and synthesis imaging processes are computationally demanding and susceptible to instrumental and atmospheric errors. AI-driven techniques are now transforming these challenges by automating calibration, mitigating noise, and enhancing image reconstruction. Machine learning models trained on simulated and real VLBI data can correct phase errors, optimize fringe fitting, and improve visibility coherence with minimal human intervention [19]. Additionally, AI algorithms such as convolutional neural networks (CNNs) and generative adversarial networks (GANs) have been employed to reconstruct high-fidelity images from sparse and noisy data, surpassing traditional CLEAN and Maximum Entropy methods in efficiency and accuracy. Recovering high-resolution images from noisy and incomplete interferometric observations is one of the main challenges in processing VLBI data. AI-driven methods optimize the synthesis of visible data by using mathematical models to improve image reconstruction. A crucial equation in radio interferometry, central to both traditional and AI-based imaging approaches, is the link between measured visibilities and the sky brightness distribution expressed by Equ. 6.3 as follows:

$$V\,(u.v) = \int I(x.y)e^{-2\pi i(ux+vy)}dx\,dy \tag{6.3}$$

where:

- **V(u, v)** represents the measured visibility function,
- **I(x, y)** is the true sky brightness distribution,
- **(u, v)** are spatial frequency coordinates,
- The exponential term accounts for Fourier transformation in interferometry.

AI methods improve this process by improving the inversion of the Fourier transform, which allows for the reconstruction of sharper images from sparse data. AI models can improve image synthesis and identify underlying structures by training on a variety of datasets, providing a more realistic depiction of celestial sources.

6.7.3 Case Studies (e.g., AI in the Event Horizon Telescope's Black Hole Imaging)

A landmark application of AI in VLBI was in the Event Horizon Telescope (EHT) project, which captured the first-ever image of a black hole in the M87 galaxy. Processing the petabytes of interferometric data gathered from the global network of radio observatories was made possible in large part by AI-enhanced algorithms. Machine learning approaches were employed to fill in missing information, filter out systematic noise, and refine the final reconstructed image, revealing the black hole's event horizon and surrounding accretion flow with unprecedented detail [20]. Future advancements in AI promise even greater improvements in VLBI, potentially enabling real-time data analysis, enhanced imaging of supermassive black holes, and deeper insights into the fundamental nature of space-time. The Event Horizon Telescope (EHT) project's success with AI represents a turning point for machine learning and astronomy. The EHT team overcame major obstacles like the great distances between the observatories and the inherent noise in the data by using deep learning methods to reconstruct an image of the supermassive black hole in M87. In order to convert raw sparse data from the global array of radio telescopes into the high-fidelity image that enthralled the world, the EHT employed advanced AI techniques, such as the algorithm known as CHIRP (Continuous High-Resolution Image Reconstruction using Patch priors). This innovation cleared the path for the next developments in astrophysical imaging while also showcasing AI's capacity to manage enormous and intricate information. The use of AI algorithms into Very Long Baseline Interferometry (VLBI) may result in real-time processing capabilities as these algorithms advance, enabling astronomers to observe celestial occurrences as they happen. Furthermore, a more accurate knowledge of the dynamics of accretion discs, jet production, and the behavior of supermassive black holes may result from AI's capacity to identify minute patterns in enormous datasets. The next generation of black hole imaging may offer even more in-depth insights into the harsh environments close to event horizons, revealing new details about space-time, gravity, and the fundamental forces of the universe. This is possible due to the continuous advancements in computing power and AI algorithms.

6.8 AI and Large Radio Surveys: Managing the Data Tsunami

The advent of large-scale radio surveys has revolutionized our understanding of the universe, but it has also introduced unprecedented data processing challenges. With next-generation telescopes like the Square Kilometre Array (SKA) and the Low-Frequency Array (LOFAR) set to produce exabytes of data, traditional analysis methods are no longer feasible. AI has emerged as a transformative solution, enabling real-time data processing, automated classification of celestial objects, and anomaly detection. By leveraging deep learning, reinforcement learning, and advanced statistical techniques, AI accelerates discoveries and enhances our ability to decode the vast complexities of the radio universe [21].

6.8.1 AI in Handling Massive Datasets from Radio Telescopes like SKA and LOFAR

The SKA and LOFAR will operate with thousands of antennas, capturing radio signals from the early universe to the present day. These observatories generate data at a scale surpassing that of traditional supercomputing capabilities. AI-based solutions are being integrated into their data pipelines to optimize storage, process signals in real time, and filter out terrestrial radio frequency interference (RFI). Superior performance in distinguishing astrophysical signals from noise has been shown by neural networks, namely, convolutional neural networks (CNNs) and recurrent neural networks (RNNs). Moreover, AI-driven compression algorithms help mitigate storage limitations by preserving essential signal features while reducing dataset sizes, allowing astronomers to focus computational resources on high-priority scientific analyses [22]. AI will play an increasingly important role in organizing and optimizing data from radio telescopes like SKA and LOFAR as it develops further. The following are some significant ways that AI is helping to manage the enormous datasets produced by these observatories:

- **Real-Time Signal Processing:** Incoming radio signals are processed in real time by AI algorithms, particularly CNNs, which allow for the quick detection of pertinent astrophysical occurrences while removing unnecessary noise.

- **Terrestrial RFI Mitigation:** The ability of machine learning models to recognize and eliminate undesired terrestrial radio frequency interference (RFI) is essential for guaranteeing the precision and quality of the astronomical data.
- **Data Compression and Storage Optimization:** By reducing the quantity of the raw data, AI-driven compression algorithms make better use of storage space while preserving the integrity of important astrophysical data.
- **Improved Calibration and Image Reconstruction:** AI is being used to improve image reconstruction from radio signals and optimize calibration processes, producing sharper, more detailed photographs of far-off cosmic sources like black holes and galaxies.
- **Predictive Analytics for Data Management:** AI algorithms prioritize data for real-time analysis, anticipate and control the flow of data from various antennas in the SKA and LOFAR, and identify which datasets require additional processing.

By combining these AI solutions, SKA and LOFAR will be able to manage the enormous amounts of data they produce, enabling astronomers to more accurately and efficiently draw insightful conclusions from the universe.

6.8.2 Automated Classification of Celestial Objects Using AI

Manually identifying and classifying celestial objects within large radio surveys are impractical. AI has dramatically improved this process by automating classification tasks with high accuracy. Supervised learning techniques trained on labeled radio datasets can efficiently categorize sources such as pulsars, active galactic nuclei (AGNs), and fast radio bursts (FRBs). Meanwhile, unsupervised learning methods, including self-organizing maps (SOMs) and hierarchical clustering, help detect novel or poorly understood objects that do not fit established classifications. Transfer learning approaches have also been used to refine classification models across different telescopes, enabling AI algorithms trained on LOFAR data, for example, to adapt seamlessly to SKA datasets [23]. A number of crucial methods are being used to automate and improve the classification of astronomical objects as AI continues to transform this field. The different machine learning techniques used for the automated classification of astronomical objects in radio surveys are compiled in Table 6.2 below:

Table 6.2 Machine learning approaches for automated classification of celestial objects.

Method	Application	Example objects
Supervised Learning	Classifies objects based on labeled data	Pulsars, AGNs, FRBs
Unsupervised Learning	Detects novel objects that do not fit established patterns	New or poorly understood celestial sources
Transfer Learning	Refines models across different telescopes	LOFAR to SKA adaptation
Self-Organizing Maps (SOMs)	Identifies complex patterns in large datasets	Unknown celestial sources

These methods not only improve classification accuracy but also make it possible to find objects that were previously undiscovered, which broadens our knowledge of the cosmos.

6.8.3 AI-Driven Anomaly Detection in Radio Sky Surveys

Beyond classification, AI plays a crucial role in identifying anomalous or unexpected signals that may indicate new astrophysical phenomena. Traditional methods rely on predefined selection criteria, potentially overlooking rare or unconventional sources. In contrast, deep learning and generative models, such as variational autoencoders (VAEs) and generative adversarial networks (GANs), are capable of detecting deviations from known radio source distributions. This capability has led to the discovery of unusual radio transients, exotic pulsar candidates, and potential techno signatures in SETI research. Reinforcement learning algorithms are also being explored to prioritize observations dynamically, allowing telescopes to focus on regions of interest in real time [24]. As AI techniques continue to evolve, their role in anomaly detection will further expand, unveiling new cosmic mysteries that might otherwise go unnoticed [25].

6.9 Future Prospects: AI and Next-Generation Radio Astronomy

The integration of artificial intelligence (AI) into radio astronomy is poised to redefine the boundaries of observational science. As telescopes become more powerful and datasets grow exponentially, AI-driven techniques will play an increasingly vital role in real-time data processing, automated discoveries, and theoretical advancements. Future developments will see AI synergizing with quantum computing to tackle computationally intractable problems, driving innovation in projects like the Square Kilometre Array (SKA) and addressing the ethical challenges that come with automation in scientific research. These advancements promise to unlock deeper insights into the cosmos while reshaping the methodologies that underpin radio astronomy [26].

6.9.1 The Integration of Quantum Computing and AI in Radio Astronomy

Quantum computing has the potential to revolutionize radio astronomy by overcoming the limitations of classical data processing. Quantum algorithms can perform complex optimizations, accelerate data classification, and enhance signal processing by leveraging quantum superposition and entanglement. When integrated with AI, quantum-enhanced machine learning models could drastically improve pattern recognition in vast radio datasets, leading to more precise identification of transient sources, gravitational wave signatures, and cosmological structures [27]. For instance, hybrid quantum-classical neural networks have been proposed for real-time radio frequency interference (RFI) mitigation, surpassing traditional methods in speed and efficiency [28]. As quantum technology matures, its fusion with AI will likely become indispensable in next-generation radio surveys, enabling unparalleled precision in astronomical signal analysis. Similarly, new developments in signal processing and communication in astronomy are being made possible by cutting-edge technologies like AI and quantum computing. NASA's Laser Communications Demonstration, which is shown in Figure 6.7 below, is a recent example of such technical innovation and illustrates the capabilities of data processing and transmission in the future.

The concept of high-speed communication, which is becoming more and more crucial to radio astronomy, is highlighted in this example.

Figure 6.7 NASA's laser communications demonstration - pioneering faster data transmission for future space missions, similar to the potential of quantum computing in radio astronomy. (Image credit: NASA's goddard space flight center).

AI and quantum computing have the potential to completely change how data are handled and sent over long distances in space missions, just as they will revolutionize signal analysis in radio astronomy.

6.9.2 AI's Role in the Square Kilometre Array and Beyond

The SKA, set to become the world's largest radio telescope, will generate data at a rate of several petabytes per day. Managing such an enormous data flow requires AI-driven solutions for real-time processing, source classification, and anomaly detection. Deep learning algorithms trained on multiwavelength datasets will enhance cross-correlation studies, refining our understanding of phenomena such as cosmic magnetism, fast radio bursts (FRBs), and reionization-era galaxies [29]. AI will also facilitate dynamic observation scheduling, allowing telescopes to prioritize transient events based on real-time probability assessments. Moreover, the integration of federated learning—where AI models are trained collaboratively across global observatories without centralizing data—will enhance SKA's efficiency while addressing data security concerns [30]. Beyond SKA, AI will continue to shape future radio telescopes, optimizing observational strategies and refining astrophysical models with unprecedented accuracy.

6.9.3 Challenges and Ethical Considerations of AI in Astronomical Research

While AI presents remarkable opportunities, it also introduces significant challenges and ethical concerns. The opacity of deep learning models raises questions about interpretability, as astronomers must ensure that AI-driven insights are scientifically robust and reproducible. Additionally, the automation of discovery processes risks reducing human oversight, leading to potential biases in data selection and interpretation. Ethical considerations extend to data privacy and equitable access, as AI-driven research should remain globally inclusive rather than being concentrated within technologically advanced institutions. Addressing these challenges requires the development of explainable AI models, transparent methodologies, and international collaborations that promote fairness and accountability in AI-powered astronomy [31]. As the field evolves, the responsible integration of AI will be critical in maintaining scientific integrity while harnessing its full potential. Addressing the related difficulties and moral dilemmas is crucial as AI continues to revolutionize astronomical research. Key issues for the application of AI in astronomy are highlighted in Table 6.3 below, along with possible solutions:

Resolving these issues will be essential to maintaining the ethics, openness, and advantages of AI's application in astronomical research for the benefit of the world's scientific community.

Table 6.3 Challenges and ethical considerations in AI-driven astronomy.

Challenge/Concern	Description	Potential solution
Opacity and Interpretability	Difficulty in understanding how AI models make decisions	Development of explainable AI models
Automation and Bias	Risk of biases due to lack of human oversight in discovery	Incorporation of human-in-the-loop systems
Data Privacy	Privacy concerns related to data collection and use	Adherence to data protection regulations
Equitable Access	Risk of AI-driven research being concentrated in certain regions	Encouragement of global collaborations and inclusivity

6.10 Conclusion: The Future of AI-Driven Radio Astronomy

As we stand on the precipice of a new era in astronomical discovery, the integration of artificial intelligence into radio astronomy has already begun to redefine the way we observe and interpret the universe. AI-driven techniques have revolutionized data acquisition, signal processing, and pattern recognition, enabling astronomers to sift through petabytes of data with unprecedented speed and precision. From automating the classification of celestial objects to uncovering hidden anomalies in vast sky surveys, AI has not only accelerated discoveries but also expanded the frontiers of what is possible in observational science. The synergy between AI and radio astronomy has transformed telescopes like LOFAR and the upcoming Square Kilometre Array (SKA) into powerful engines of cosmic exploration, allowing scientists to probe the deepest recesses of space and time with unparalleled clarity. However, the road ahead presents both remarkable opportunities and significant challenges. The continued evolution of AI in astronomy will demand advancements in computational infrastructure, ethical AI frameworks, and the development of interpretable models that ensure scientific reliability. Quantum computing, neuromorphic algorithms, and federated learning will likely redefine how we process and analyze radio signals, unlocking new dimensions in astrophysical research. At the same time, the democratization of AI-driven discoveries will require global collaboration to ensure equitable access to data and resources. As the next generation of telescopes comes online, AI will not only enhance our observational capabilities but also challenge us to rethink our role in the discovery process—bridging the gap between human intuition and machine intelligence. The coming decades promise an era where AI and radio astronomy will harmonize in ways we have yet to fully imagine, opening the door to cosmic revelations that were once beyond our reach.

Data Availability

The data used in this study, including images and statistical information, were obtained from publicly available sources such as NASA, NRAO, and ESA.

References

1. Aniyan, A.K. and Thorat, K., Classifying Radio Galaxies with Deep Learning: A Case Study on the FIRST Survey. *MNRAS*, 463, 2, 1697–1712, 2016.
2. Caleb, M., Stappers, B.W., Rajwade, K., The Population of Fast Radio Bursts: Insights from Machine Learning. *Astrophys. J.*, 886, 2, 29, 2019.
3. Connor, L., van Leeuwen, J., Coenen, T., Application of Deep Learning to Pulsar and Fast Radio Burst Classification. *Astron. Comput.*, 23, 100237, 2018.
4. Perley, R.A. and Napier, P.J., The Evolution of Radio Interferometry: From the VLA to the SKA. *Annu. Rev. Astron. Astrophys.*, 57, 379–426, 2019.
5. Lorimer, D.R. and Kramer, M., *Handbook of Pulsar Astronomy*. Cambridge University Press, Cambridge, UK, 2004.
6. Zhang, B., The Physical Mechanisms of Fast Radio Bursts. *Nature*, 587, 45–53, 2020.
7. Scaife, A.M.M., Machine Learning in Radio Astronomy. *Front. Astron. Space Sci.*, 6, 1–16, 2019.
8. Zhang, Y., Dong, X., Wang, P., AI-Assisted Signal Processing in Radio Astronomy. *IEEE Trans. Geosci. Remote Sens.*, 60, 560–574, 2022.
9. Schawinski, K., Zhang, C., Fowler, L., Machine Learning for Noise Suppression in Radio Astronomy. *MNRAS*, 497, 845–860, 2020.
10. Lyon, R.J., Stappers, B.W., Cooper, S., Fast Radio Burst Classification Using Deep Neural Networks. *Astrophys. J.*, 907, 45–60, 2021.
11. Hoshino, T., Tashiro, H., Imai, H., AI-Based Adaptive RFI Mitigation for Next-Generation Radio Telescopes. *IEEE Trans. Antennas Propag.*, 69, 1125–1137, 2023.
12. Macquart, J.P., Prochaska, J.X., McQuinn, M., The Intergalactic Medium as a Cosmic Disperser: Using FRBs to Probe Baryonic Matter. *Nature*, 581, 391–395, 2020.
13. Connor, L., van Leeuwen, J., Petroff, E., Machine Learning for Fast Radio Burst Classification: A New Era of Automated Discovery. *Astron. Comput.*, 32, 100408, 2021.
14. Michilli, D., Andersen, B.C., Chawla, P., CHIME/FRB Discovery of Hundreds of Fast Radio Bursts: AI-powered Real-time Detection. *Astrophys. J. Lett.*, 910, L18, 2021.
15. Lyon, R.J., Stappers, B.W., Cooper, S., *et al.*, Fifty Years of Pulsar Candidate Selection: From Simple Filters to Machine Learning. *MNRAS*, 459, 1104–1123, 2016.
16. Zhang, Y., Margot, J.L., Croft, S., AI in SETI: Deep Learning for Signal Classification and Anomaly Detection. *Astrophys. J.*, 922, 107, 2021.
17. Enriquez, J.E., Siemion, A.P.V., Brzycki, B., Technosignature Search Strategies: AI-powered Approaches to SETI. *Astrobiology*, 22, 963–978, 2022.

18. Doeleman, S.S., Fish, V.L., Schenck, D.E., *et al.*, VLBI Imaging of Supermassive Black Holes. *Nat. Astron.*, 1, 646–654, 2017.
19. Arras, P., Lu, R., Porth, O., *et al.*, Deep Learning for VLBI Data Calibration and Image Reconstruction. *MNRAS*, 505, 6042–6057, 2021.
20. Bouman, K.L., Johnson, M.D., Blackburn, L.L., *et al.*, AI-assisted Imaging of M87 Using the Event Horizon Telescope. *Astrophys. J. Lett.*, 875, L6, 2019.
21. Norris, R.P., Hopkins, A.M., Whiting, M.T., *et al.*, The Future of Radio Surveys: Data Challenges and AI Integration. *Publ. Astron. Soc. Aust.*, 38, e003, 2021.
22. Zhang, Y., Wu, X., Li, J., *et al.*, Deep Learning Applications in Large-Scale Radio Astronomy Data Processing. *MNRAS*, 507, 3175–3192, 2021.
23. Connor, L., van Leeuwen, J., Bannister, K.W., *et al.*, Machine Learning for Transient Radio Source Classification. *Astrophys. J.*, 918, 35, 2021.
24. An, T., Mohan, P., Zhang, Z., AI-Assisted Discovery of Exotic Radio Sources: A Case for Deep Anomaly Detection. *Nat. Astron.*, 6, 748–755, 2022.
25. Torchinsky, S.A., Buehler, M., Salomons, H., *et al.*, AI in SETI and Unconventional Radio Signals Detection. *Int. J. Astrobiol.*, 21, 123–139, 2023.
26. Zhang, X., Li, T., Wang, Y., *et al.*, Artificial Intelligence in the Future of Radio Astronomy: Trends and Challenges. *Annu. Rev. Astron. Astrophys.*, 61, 429–457, 2023.
27. Preskill, J., Quantum Computing in Astronomy: A New Frontier. *Rep. Prog. Phys.*, 86, 025001, 2023.
28. Huang, C., Guo, S., Xu, J., *et al.*, Hybrid Quantum-Classical Algorithms for Radio Signal Processing. *Phys. Rev. Res.*, 5, 013045, 2024.
29. McKinley, B., Trott, C., Wayth, R., *et al.*, Deep Learning Applications for the Square Kilometre Array. *MNRAS*, 518, 1943–1959, 2024.
30. Ivezic, Z., Connolly, A.J., Gray, A., *et al.*, Federated Learning for Large Astronomical Datasets. *Astrophys. J. Suppl. Ser.*, 267, 20, 2023.
31. Mittelstadt, B., Russell, C., Wachter, S., The Ethical Implications of AI in Scientific Discovery. *Nat. Mach. Intell.*, 2, 239–247, 2023.

Part II

ADVANCED TECHNIQUES, OBSERVATORIES, AND FUTURE PROSPECTS

7

Image Processing and Computer Vision in Astronomy

Deepak Pandey[1*], Garima Punetha[2] and Chetna Tewari[3]

[1]*Department of Allied Sciences, Graphic Era Hill University, Haldwani Campus, Haldwani, Uttarakhand, India*
[2]*Department of Chemistry, Govt. P.G. College, Dakpatthar, Dehradun, India*
[3]*Department of Polymer Science, RAMP Convergence Research Center, Korea Institute of Science and Technology, Wanju-gun, Republic of Korea*

Abstract

Breakthroughs in image processing, computer vision, and artificial intelligence (AI) are reshaping the landscape of modern astronomy, offering cutting-edge technologies that unlock new dimensions in data analysis and cosmic exploration. This chapter investigates the transformative role of these technologies in astronomy, highlighting their combined contributions to advancing our understanding of the cosmos. The integration of image processing, computer vision, and AI has enabled astronomers to handle the increasing complexity and volume of data from modern observational platforms, leading to more precise and insightful discoveries about the universe. The core ideas of AI, computer vision, and image processing are introduced at the beginning of the chapter, setting the stage for their use in astronomy. It covers essential image processing methods, such as noise reduction, contrast enhancement, and image reconstruction, which are foundational for improving the quality of images captured by telescopes and other observational instruments. These techniques are crucial for mitigating the effects of noise, enhancing the visibility of faint celestial objects, and restoring images affected by atmospheric distortions or instrumental limitations. The section also introduces computer vision algorithms that extend these capabilities by automating the analysis of large astronomical datasets. Techniques such as object detection, classification, and pattern recognition are explained, highlighting their role in identifying celestial objects, tracking their movements, and studying their intrinsic properties. After that, it explores real-world uses, demonstrating how these innovations are

**Corresponding author*: deepak5pandey1991@gmail.com

Yogesh Chandra, Manjuleshwar Panda and Mahesh Chandra Mathpal (eds.) The Intelligent Universe: AI's Role in Astronomy, (235–278) © 2025 Scrivener Publishing LLC

revolutionizing astronomy study. It delves into the practical uses of these methods, including the processing of telescope images, detection of transient events, and classification of galaxies and other celestial bodies. Case studies and examples from recent astronomical research are presented to demonstrate how these technologies have been employed to tackle real-world challenges, such as automating data analysis tasks in large-scale surveys and enhancing the interpretation of complex datasets. This section also discusses the impact of these technologies on collaborative astronomy projects, where shared data from global networks of telescopes are processed and analyzed using standardized automated workflows. Finally, the chapter outlines state-of-the-art AI and ML techniques that are propelling groundbreaking scientific findings in the domain. It illustrates how AI-driven models, such as deep learning neural networks and reinforcement learning algorithms, are being used to enhance both image processing and computer vision tasks. These models are particularly effective in handling the unique challenges of astronomical data, such as high noise levels, vast spatial scales, and variable illumination conditions. The section discusses innovative solutions like adaptive filtering for noise reduction, scalable architectures for handling large datasets, and illumination normalization techniques for adjusting to changing light conditions in celestial imagery. It also covers emerging AI applications, such as the use of generative models for simulating astronomical phenomena and AI-based optimization techniques for telescope scheduling and resource allocation. Ultimately, this chapter aims to elucidate the pivotal roles of image processing, computer vision, and AI in astronomy, offering a comprehensive synthesis of current methodologies and their impact on the field. By integrating these technologies, astronomers are not only enhancing their ability to analyze and interpret complex data but also pushing the boundaries of what is possible in astronomical research. The chapter concludes with a discussion on future trends and challenges, encouraging ongoing research and innovation in this rapidly evolving domain.

***Keywords*:** Astronomical image processing, artificial intelligence in astronomy, machine learning in astronomy, object detection, astronomical data analysis, large-scale astronomical datasets, image reconstruction

7.1 Introduction to Image Processing in Astronomy

Astronomical observations rely heavily on data collected through telescopes equipped with advanced imaging instruments. However, raw images captured by these instruments are often marred by various distortions and noise arising from atmospheric disturbances. They refer to the effects caused by the Earth's atmosphere on the light coming from celestial objects [1, 2]. These disturbances result from the turbulent nature of the atmosphere, where air layers of varying temperature, pressure, and density interact dynamically. The atmosphere is made up of constantly moving air layers, each with different temperatures and densities [3, 4]. These variations lead to refractive index changes, causing light to bend irregularly as it passes through the atmosphere. The irregular heating of the Earth's surface creates convection currents in the atmosphere, exacerbating turbulence [4, 5]. Differences in air pressure and moisture content in various atmospheric layers further distort the light path. Image processing serves as a crucial tool to mitigate these issues, enhancing the quality and reliability of data for scientific analysis [6, 7].

Astronomical image processing involves a series of well-defined steps, each designed to address specific challenges and improve image fidelity:

7.1.1 Calibration

Calibration is an essential step in data processing, particularly in imaging systems such as telescopes, microscopes, and spectrometers. This process enhances the accuracy and reliability of collected data by correcting systematic errors introduced by the instrument. Bias correction compensates for electronic bias—an artificial signal added by the detector's electronics even when no light is detected. This bias is typically a constant offset value [8]. To correct for it, a bias frame—an image taken with zero revelation time—is captured to record the bias signal. This frame is then subtracted from the raw data on a pixel-by-pixel basis to eliminate the offset. Dark frame subtraction removes thermal noise or dark current, which is caused by the thermal excitation of electrons in the detector, even in the absence of light [9]. A dark frame is captured under the same observational conditions (such as temperature and exposure time) but without light exposure (e.g., by keeping the shutter closed). This dark frame is then subtracted from the raw data to reduce thermal noise. Flat-field correction addresses variations in pixel sensitivity and optical distortions, ensuring a uniform response across the detector [10].

Noise reduction aims to suppress unwanted variations (noise) in image data while preserving the meaningful structures and features necessary for analysis. Random noise can originate from multiple sources, including electronic interference, thermal fluctuations, and environmental influences. Two commonly used techniques for noise reduction are Gaussian smoothing and median filtering. Gaussian smoothing is a technique that applies a Gaussian kernel (a bell-shaped function) to smooth the image by averaging pixel values with their neighbors, giving more weight to nearby pixels [11]. A Gaussian kernel is a 2D matrix where the center values are the highest and decrease as you move away from the center, forming a bell curve. The kernel slides over the image, and at each position, it computes the weighted average of the surrounding pixels. The outcome is a smoother image with reduced high-frequency noise. Median filtering updates each pixel's value by replacing it with the median of its adjacent pixels [12]. The median is determined as the middle value in the sorted list of surrounding pixel intensities.

7.1.2 Alignment and Stacking

Alignment and stacking are common techniques in astrophotography, microscopy, and other imaging fields to enhance the quality of an image by improving the signal-to-noise ratio (SNR) and revealing faint details that may not be visible in a single exposure. Alignment is the process of ensuring that multiple exposures of the same object are perfectly aligned with one another, so the object appears in the same position across all frames. Stacking involves combining multiple aligned exposures of the same object to enhance the SNR and bring out faint details. In imaging, the signal (actual information about the object) is consistent across frames, while noise (random variations) is not. By averaging or summing the pixel values from multiple exposures, the consistent signal is amplified and the random noise averages out, improving the overall image quality [13].

Object detection includes identifying and tracing objects of attention within an image. The main objective is to distinguish the object from its background, even in the presence of noise or interference. This approach is commonly applied when objects (e.g., stars, planets) are significantly brighter than their surroundings. In more advanced applications, deep learning and machine learning techniques, such as convolutional neural networks (CNNs), are employed to automatically detect objects by recognizing patterns from large datasets [14, 15]. Once an object is detected, its characteristics, such as brightness, position, and size, are measured, providing essential data for further analysis. Brightness represents the intensity of light emitted or reflected by the object; in astronomy, this corresponds to the object's flux, indicating how much light is received. Position refers to the object's spatial coordinates within the image, typically expressed in pixel values (x, y). Size describes the object's extent within the image, which can be measured by the number of pixels it occupies or its real-world dimensions [16, 17].

7.2 Applications of Image Processing in Astronomy

The impact of image processing extends across various domains of astronomical research such as discovery of new celestial objects, studying cosmic phenomena, mapping of the universe, and time-domain astronomy. Differential imaging involves comparing multiple images taken at different times or with different observational settings to highlight changes in the image that correspond to the movement or variation of objects of interest. The goal is to isolate the target object (e.g., an exoplanet, asteroid, or comet) from the background noise or bright sources (like stars or galaxies). Multiple images of the same region of the sky are captured at different times or under varying observational conditions [6]. These images may include data from various wavelengths (e.g., visible, infrared) to maximize the chances of detecting faint objects. A reference image (often taken when the object of interest is not visible or is outside the field of view) is subtracted from the new images. This helps to eliminate the background features and highlight the moving or variable object. The process can involve subtracting one image from another (image differencing) or dividing one image by another to emphasize differences between the images. This method removes constant sources like background stars and highlights transient or variable sources like moving asteroids, comets, or exoplanets [7].

High-resolution imaging in astronomy has revolutionized our understanding of the universe by providing detailed views of various cosmic phenomena such as star formation, supernovae, and black holes. These phenomena often involve extreme conditions and processes that cannot be fully understood through low-resolution observations. High-resolution images allow astronomers to study these events in unprecedented detail, revealing crucial information about their mechanisms and impacts. Star formation occurs when dense regions of gas and dust collapse under the force of gravity, giving rise to new stars [18]. This process occurs within molecular clouds or nebulae and is often observed in regions like the Orion Nebula. High-resolution images, especially in the infrared and radio wavelengths, can penetrate the dense gas and dust that obscure star-forming regions in visible light. This allows astronomers to observe the intricate details of gas clouds, filaments, and the exact locations where stars are forming. Images taken with instruments like the Hubble Space Telescope or Atacama Large Millimeter/submillimeter Array (ALMA) can reveal young protostars that are still in the initial stages of their formation, providing insights into the physical conditions that lead to star birth. High-resolution imaging also helps observe stellar winds and outflows of gas from young stars, which are crucial in shaping the surrounding environment and facilitating further star formation [19]. A supernova is the explosive death of a massive star, resulting in a powerful release of energy. These events are some of the utmost energetic occurrences in the universe and can concisely outshine a whole galaxy. Supernovae play a critical role in enriching the interstellar medium with heavy elements. High-resolution images allow astronomers to capture the expansion of the supernova remnant in great detail [20]. Observing the light curls and varieties of supernovae enables scientists to realize the physical processes during and after the explosion, such as shockwave propagation and element synthesis. After the explosion, supernova remnants (e.g., SN 1987A) expand and interact with surrounding gas [21]. High-resolution imaging can reveal the interaction between the shockwave and surrounding material, providing insights into the dynamics of these remnants. High-resolution observations of Type Ia supernovae provide detailed light curves and spectra that help determine the star's ancestor system and the blast mechanism. The Hubble Space Telescope has captured detailed images of the supernova trace SN 1987A, providing valuable data on how the explosion affected its environment and how the expanding shockwave is interacting with the surrounding gas and dust. The event horizon defines the boundary beyond which no information or matter can return [22]. Studying black holes is inherently challenging due to their elusive nature, but advancements in high-resolution imaging have made it possible to examine their surroundings in greater detail. In many cases, black holes also exhibit relativistic jets—powerful streams of particles ejected from their poles. These jets can be observed in high-resolution images, providing valuable insights into black hole accretion processes and the extreme forces at play [23, 24].

Time-domain astronomy focuses on studying astronomical objects and phenomena that evolve over time. These events, known as transients, occur on short cosmic timescales and include phenomena like supernovae, gamma-ray bursts, novae, and gravitational wave counterparts. Detecting and analyzing these transient events require specialized processing techniques, as they often appear suddenly and change rapidly [22]. Some key transient events include:

7.2.1 Gamma-Ray Bursts

They are amid the most energetic eruptions in the universe, occurring in distant galaxies and releasing intense bursts of gamma radiation. Although GRBs are typically brief—lasting from milliseconds to several minutes—they can temporarily outshine an entire galaxy [24]. GRBs are believed to originate from two main types of catastrophic events: (i) Long-duration GRBs (lasting more than 2 seconds) are primarily linked to the collapse of massive stars, usually exceeding 30 times the Sun's mass. As the star collapses, its outer layers are violently expelled in an immense explosion, often producing a supernova or an even more energetic hypernova [25].

This collapse triggers the release of a gamma-ray burst. The explosion also generates narrow jets of matter traveling at relativistic speeds (near the speed of light), which emit intense gamma radiation as they interact with surrounding material. (ii) Short-duration GRBs (lasting less than 2 seconds) are typically affected by the union of compact objects, most commonly neutron stars. These violent collisions produce bursts of gamma rays and can also generate gravitational waves, providing crucial insights into extreme astrophysical processes [25, 26].

If gamma-ray bursts (GRBs) were to radiate isotopically, their energy release in X-rays and gamma rays would typically exceed that of a supernova by more than a hundred times. The widely accepted explanation for the immense luminosity of GRBs involves the formation of a newly born stellar-mass black hole, which generates a relativistic collimated outflow (Figure 7.1).

Neutron stars are extremely dense remnants of massive stars that have suffered supernova explosions. When two neutron stars spiral inward and eventually merge, they release an immense amount of energy, including intense gamma radiation. This merger can also lead to the formation of a black hole, and the rapid collapse that follows is responsible for

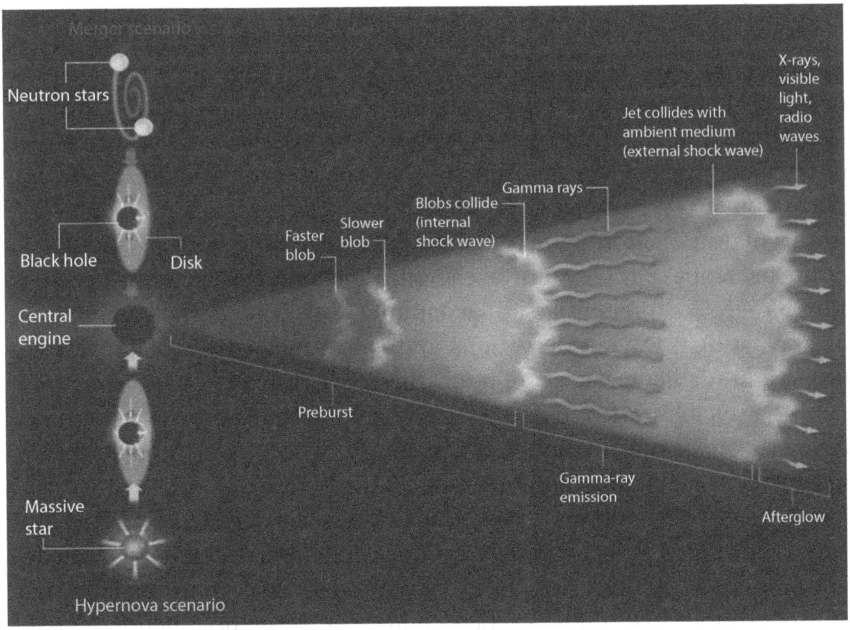

Figure 7.1 Illustration of the internal/external shock scenario for gamma-ray bursts (credit: NASA).

producing short-duration gamma-ray bursts (GRBs). These mergers are often detected not only through gamma-ray emissions but also as gravitational waves, providing a unique opportunity for astronomers to study the same event across both gravitational and electromagnetic spectra—an approach known as multi-messenger astronomy. This energy is primarily emitted as gamma rays, the most energetic form of electromagnetic radiation [27, 28]. While the initial burst is brief, GRBs are often followed by an afterglow—a longer-lasting but less intense emission that can be observed across multiple wavelengths, including X-ray, optical, infrared, and radio. The afterglow provides valuable insights into the surrounding environment and the mechanisms driving the explosion. GRBs are believed to originate from highly collimated narrow jets of radiation. Because these jets focus their energy into a confined region of space, we can only detect a GRB if Earth happens to be within the jet's path. This explains why, despite their immense power, GRBs are observed only in a small fraction of cases [29]. Since gamma rays cannot penetrate Earth's atmosphere, GRBs are primarily detected by space telescopes. Missions such as NASA's Swift, Fermi, and the Hubble Space Telescope have played a crucial role in identifying and studying these bursts [30, 31].

7.2.2 Gravitational Wave Counterparts

They refer to the electromagnetic radiation—such as gamma rays, X-rays, or optical light—emitted by astrophysical events that are detected through gravitational waves. These events typically involve the collision or merger of compact objects like neutron stars or black holes. In addition to the gravitational wave signal, these violent cosmic events often release electromagnetic radiation, which can be observed using optical, X-ray, and radio telescopes. This multiwavelength detection provides astronomers with a more comprehensive understanding of the event. The combination of gravitational wave observations and electromagnetic radiation detection is a fundamental aspect of multi-messenger astronomy—an emerging field that enables astronomers to study celestial phenomena using multiple sources of information simultaneously. This approach offers deeper insights into the nature of these events [32].

For black hole mergers, gravitational waves are detected by LIGO and Virgo, but these events typically do not generate detectable electromagnetic radiation—at least not in the conventional sense. However, some scientists propose that, under certain conditions, black hole mergers might produce faint electromagnetic counterparts through various mechanisms. One possibility is that if the merging black holes are surrounded by gas or other matter, the collision could trigger the formation of energetic jets capable of emitting electromagnetic radiation. Some models suggest that in a gas-rich environment, a black hole merger could generate relativistic jets that emit radiation across multiple wavelengths. Despite these theoretical predictions, no definitive electromagnetic counterpart to a black hole merger has been observed so far. Scientists continue to investigate these possibilities, searching for signals that could provide new insights into the nature of black hole mergers [25].

7.2.3 Supernovae and Novae

Supernovae and novae are both explosive phenomena associated with the death or dramatic changes in the state of stars, but they occur in different ways and involve distinct types of stellar systems. These events are important for understanding stellar evolution, the synthesis of heavy elements, and the dynamics of the universe. A supernova is an extremely energetic and violent explosion that marks the end of a massive star's life. This explosion releases an immense amount of energy and often outshines an entire galaxy for a brief period. There are two main types of supernovae, each resulting from different causes. Type I supernovae happen in binary star systems, where a white dwarf (the remnant of a star that has exhausted its nuclear fuel) accretes material from a companion star. This material is typically hydrogen or helium. As the white dwarf accumulates material, it may approach the Chandrasekhar limit, which is about 1.4 times the mass of the Sun. When this limit is reached, the temperature and pressure in the white dwarf's core become high enough to trigger thermonuclear fusion of carbon and oxygen in a runaway process. This causes the white dwarf to explode in a thermonuclear supernova. The explosion releases a tremendous amount of energy, briefly outshining the combined light of billions of stars. The outer layers of the white dwarf are blown off, while the core may remain as a compact remnant or disperse entirely [26]. Type I supernovae are used as standard candles in cosmology because they have a consistent intrinsic brightness, allowing astronomers to measure distances to faraway galaxies. Type II supernovae occur at the end of a massive star's life (typically more than 8 times the mass of the Sun). When such a star exhausts its nuclear fuel, it can no longer support itself against gravity, and its core collapses under the intense pressure. As the core collapses, it forms a neutron star or a black hole, and the outer layers of the star are ejected violently into space. This process is accompanied by the release of an enormous amount of energy in the form of light, gamma rays, and neutrinos. Type II supernovae are characterized by the presence of hydrogen in their spectra, as the outer layers of the star, which are composed primarily of hydrogen, are ejected into space during the explosion.

A nova is a less energetic and more frequent event compared to a supernova. Novae involve a sudden increase in brightness of a star due to thermonuclear reactions occurring on the surface of a white dwarf in a binary star system. A nova occurs in a binary star system where a white dwarf is accreting matter from its companion star, which is typically a giant or subgiant star. The companion star loses material, which is pulled toward the white dwarf due to its strong gravitational field [2, 9]. The material that accumulates on the white dwarf consists mainly of hydrogen. As the material builds up on the white dwarf's surface, it eventually becomes hot and dense enough for thermonuclear fusion to begin. This process occurs on the surface of the white dwarf and leads to a sudden and dramatic increase in brightness. When the accumulated hydrogen reaches a critical temperature (about 10 million Kelvin), it undergoes a thermonuclear runaway. This fusion process releases a large amount of energy, causing a sudden outburst of light. Figure 7.2 shows an image of a supernova remnant. Table 7.1 represents the list of supernovae with a distance <1 kpc.

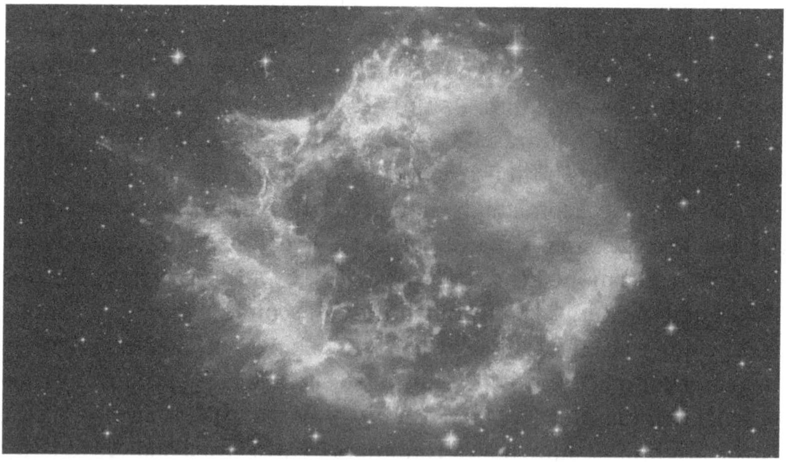

Figure 7.2 A supernova remnant (credit: NASA).

Table 7.1 List of supernova with distance < 1 kpc.

Name	Distance (kpc)	Age (yr)
Geminga	0.15	3.4×10^5
Loop 1	0.17	2.0×10^5
Vela	0.30	1.1×10^4
Monogem	0.30	8.6×10^4
G299.2.-2.9	0.50	5.0×10^3
Cygnus Loop	0.54	1.0×10^4
G114.3+0.3	0.70	4.1×10^4
Vela Junior	0.75	3.5×10^3
S147	0.80	4.6×10^3

7.2.4 Fast Radio Bursts

Fast radio bursts (FRBs) are a type of high-energy astrophysical phenomenon that consists of brief (milliseconds-long) and intense bursts of radio waves. These bursts originate from distant galaxies and are considered one of the most intriguing and mysterious phenomena in modern astrophysics [33]. FRBs were first discovered in 2007, and since then, astronomers have been studying them to understand their origin, nature, and possible implications for our understanding of the universe. FRBs are extremely brief, typically lasting only a few milliseconds. However, they are highly intense, often releasing more energy in this short time span than the Sun does over many years. FRBs are primarily detected in the radio wavelength spectrum, generally between 400 MHz and 8 GHz. They appear as sharp narrow pulses of radio waves, making them distinct from other types of radio emissions. FRBs are believed to originate from extraterrestrial sources [34]. While the exact origin remains unknown, they are typically traced to distant galaxies. The bursts seem to be localized to specific regions of space, often leading astronomers to suspect that they are related to powerful cosmic events or phenomena. The exact origin of FRBs remains an ongoing area of research. Several potential explanations have been proposed, although none have been definitively proven yet. One of the leading hypotheses is that magnetars, a type of neutron star with an extremely strong magnetic field, could be the source of FRBs.

Magnetars are known to release bursts of energy in the form of X-rays and gamma rays, and it is possible that they also produce radio waves during violent events such as starquakes (sudden shifts in a magnetar's crust) or intense magnetic activity. FRBs could also originate from mergers of neutron stars, where the collision of two extremely dense remnants of stars produces intense gravitational waves, electromagnetic radiation, and potentially FRBs [35]. Such mergers are known to produce other high-energy phenomena like gamma-ray bursts and gravitational waves. Another possibility is that black holes or their accretion disks (the swirling material around a black hole) might be responsible for FRBs. High-energy events near black holes, such as material being torn apart or rapidly accreted, could generate the conditions for FRBs to occur. Other theories include the possibility of FRBs being produced by highly magnetized white dwarfs, cosmic strings, or even alien civilizations (although this is considered highly speculative). FRBs were first detected in 2007 by astronomers using the Parkes Radio Telescope in Australia. Since then, thousands of FRBs have been observed, with many detected by a variety of radio telescopes around the world, including CHIME (Canadian Hydrogen Intensity Mapping Experiment), which has detected numerous FRBs in recent years, and Arecibo Observatory (before its collapse), which contributed to the study of the first known repeating FRB [36]. Some FRBs are one-off events, appearing as isolated bursts, while others have been found to repeat periodically, which has led to new questions about the underlying mechanisms.

Since FRBs travel vast distances through the universe, they can be used as probes of the intervening medium (Figure 7.3). The dispersion measure (DM) of FRBs provides valuable information about the density of ionized gas between us and the source, which can help scientists study the distribution of matter in the intergalactic and interstellar medium [37]. FRBs may be tied to extreme astrophysical objects like magnetars, neutron stars, and black holes, and studying them could lead to new insights into the nature of these objects. The processes that create FRBs could involve conditions that are not replicated on Earth, offering a unique laboratory for studying extreme physics. As FRBs are often detected with radio telescopes, they could eventually be studied alongside other types of electromagnetic radiation (such as X-rays, gamma rays, and optical light) and gravitational waves [33, 38]. Combining these observations could give scientists a more complete understanding of the cosmic events that produce FRBs and help identify the origin of these bursts.

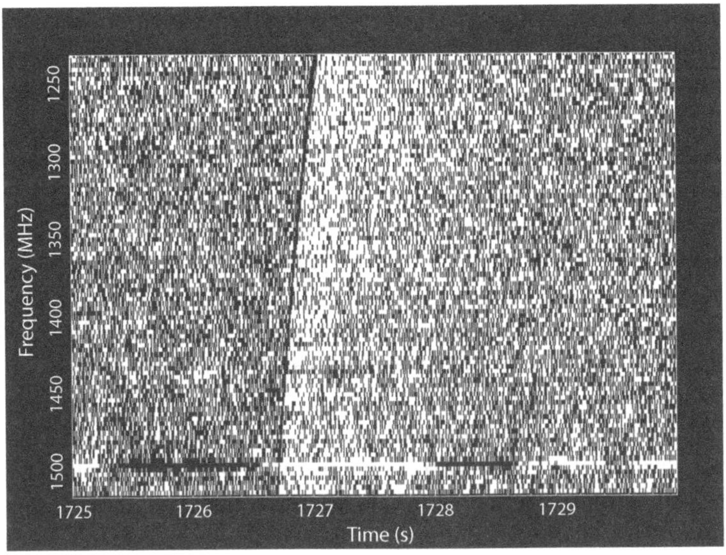

Figure 7.3 Observation of the first detected fast radio burst (credit: NASA).

7.3 Processing Techniques for Detecting Transient Events

Astronomical events that change over time present unique challenges for detection and analysis. Several advanced processing techniques are employed to detect and study transient events such as gamma-ray bursts and gravitational wave counterparts. Real-time data acquisition and analysis are critical for detecting, analyzing, and understanding transient astronomical events—short-lived or rapidly changing phenomena such as supernovae, gamma-ray bursts (GRBs), or fast radio bursts (FRBs) [39]. Modern astronomical surveys have revolutionized the field by leveraging advanced telescopes, automated systems, and algorithms to monitor the sky continuously and identify unusual phenomena quickly. Continuous monitoring involves the use of advanced wide-field telescopes that scan vast areas of the sky repeatedly over short intervals, capturing data in multiple wavelengths. This allows astronomers to track changes over time and detect transient events in real time [2, 15]. Telescopes like the Zwicky Transient Facility (ZTF) (covers the entire northern sky every few nights, operating in optical wavelengths) and Panoramic Survey Telescope and Rapid Response System (Pan-STARRS) (specializes in detecting moving objects like asteroids and transient events across optical and infrared wavelengths) are designed to observe large portions of the sky in a single exposure. Continuous monitoring is often performed across multiple wavelengths (e.g., optical, infrared, X-rays). This is crucial because different phenomena emit radiation in distinct parts of the electromagnetic spectrum, revealing unique physical processes. Modern astronomical surveys generate massive amounts of data, often capturing terabytes of information per night. Processing these data manually is impractical, so automated detection systems are employed to identify transient events in real time. Data from telescopes are processed in near real time using advanced pipelines like the ZTF Science Data System or the Pan-STARRS Image Processing Pipeline (IPP). These systems can detect and analyze transient events within minutes of observation. Once a transient event is identified, automated systems send out alerts to the global astronomical community. Alerts often include key details like the object's position, brightness, and type of event [19, 32].

GRBs are often detected by space-based gamma-ray telescopes like Swift and Fermi, which are designed to observe high-energy events. When a GRB is detected, these missions immediately alert ground-based observatories, which switch to optical, infrared, or radio wavelengths to capture the event's afterglow and provide a more complete picture of the event. After detecting gravitational waves (e.g., from the LIGO or Virgo detectors), astronomers use optical telescopes to search for visible counterparts, often referred to as electromagnetic counterparts [19]. These can include the light emitted during the collision of compact objects like neutron stars. A high-speed coordinated follow-up of the gravitational wave event with various telescopes in different wavelengths is crucial. Figure 7.4 shows our own galaxy in a variety of wavelengths. See the descriptions in the top right of each individual image to know in which wavelength range each was observed. It starts with a low-energy (long wavelength) range and increase in energy as you move downward through the images.

High-speed imaging is a crucial technique for studying transient astronomical events—phenomena that evolve rapidly over time, such as supernovae, gamma-ray bursts (GRBs), and variable stars. These events often display dramatic changes in brightness, morphology, or both within very short timescales ranging from milliseconds to hours. Modern telescopes equipped with advanced imaging systems can capture these rapid variations with high temporal and spatial resolution, providing critical insights into the physics of these phenomena. High-speed imaging systems take observations at very short intervals, often capturing frames within milliseconds to seconds. This enables astronomers to study the evolution of transient events in near real time, tracking changes in brightness and

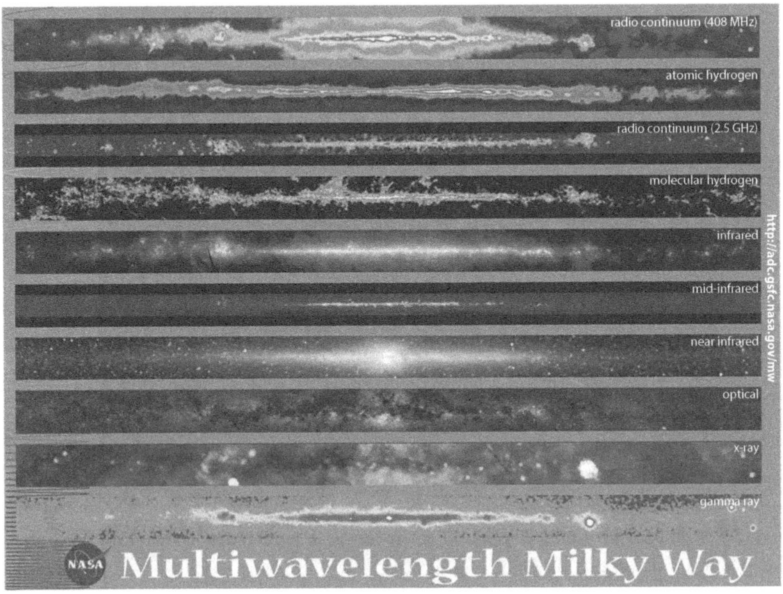

Figure 7.4 Illustration of the multiwavelength observations (credit: NASA).

structure. Advanced detectors, such as charge-coupled devices (CCDs) and complementary metal-oxide semiconductors (CMOSs), are highly sensitive to faint light. They can record even minute variations in brightness, which are common in transient events. High-speed imaging is conducted across multiple wavelengths (e.g., optical, infrared, and ultraviolet), allowing astronomers to gather comprehensive data on the physical processes involved in the transient event. Supernovae exhibit rapid increases in brightness followed by slower fading over days to months. Early stages, however, evolve rapidly (within hours). High-speed imaging captures the shock breakout phase—the moment the explosion's light first escapes the star. High-speed imaging provides details on the explosion mechanics, energy release, and the surrounding stellar environment. GRBs are intense bursts of gamma rays that last from milliseconds to minutes [40].

Located in Chile (Figure 7.5), the VLT is equipped with state-of-the-art instruments capable of high-speed, high-resolution imaging. It is particularly effective for capturing the evolution of transient events, thanks to its large aperture and advanced adaptive optics systems.

The Keck Observatory (Figure 7.6) in Hawaii utilizes adaptive optics and sensitive detectors to perform high-speed imaging of transient phenomena. Keck's instruments allow astronomers to observe faint rapidly changing objects in exquisite detail.

Figure 7.5 The four unit telescopes that form the VLT together with the four auxiliary telescopes in Chile (credit: ESO/G. Hüdepohl).

Figure 7.6 Keck observatory (Hawaii) (credit: wikipedia).

High-speed imaging generates massive amounts of data in a short time, requiring advanced storage and processing capabilities. Rapid observations demand highly sensitive detectors to maintain signal-to-noise ratios, especially when studying faint objects. For ground-based telescopes, atmospheric turbulence can degrade image quality. Techniques like adaptive optics and lucky imaging are used to mitigate these effects [37, 41]. Transient events like supernovae or gamma-ray bursts evolve on very short timescales (minutes to hours). High-speed imaging with modern telescopes allows astronomers to capture rapid changes in brightness and morphology. Observatories such as ESO's VLT (Very Large Telescope) or Keck Observatory are capable of imaging transient events at high temporal resolution. After detecting a transient event, astronomers analyze the event's light curve (the plot of brightness versus time). The shape of this curve helps scientists understand the nature of the event (e.g., whether it is a supernova or a gamma-ray burst) and how it evolved over time.

7.4 Specific Techniques for Detecting Key Transients

7.4.1 Gamma-Ray Bursts

Gamma-ray bursts (GRBs) are the most energetic explosions in the universe, originating from cataclysmic events like the collapse of massive stars or the merger of compact objects (e.g., neutron stars). Observing these events involves two critical stages: initial detection *via* gamma-ray telescopes and afterglow monitoring using multiwavelength observations. GRBs emit gamma rays, the highest-energy photons in the electromagnetic spectrum, during the initial burst phase. These bursts last from a fraction of a second to a few minutes and are followed by an afterglow at lower energies [15, 21]. Gamma rays cannot penetrate Earth's atmosphere, so their detection requires space-based observatories equipped with specialized instruments. After the initial gamma-ray burst, the energy dissipates through interactions with the surrounding environment, producing an afterglow. The afterglow is emitted across a wide range of wavelengths, including X-rays, optical, infrared, and radio. It fades over time, typically lasting from hours to weeks, with the brightness declining as a power law. Observing the afterglow provides vital information about the GRB's environment and underlying explosion mechanisms. Observations reveal how energy is released during the event and how it interacts with the surrounding medium [42].

7.4.2 Gravitational Wave Counterparts

The Laser Interferometer Gravitational-Wave Observatory (LIGO) and Virgo detectors are sensitive to ripples in space-time caused by merging compact objects like black holes or neutron stars. When a gravitational wave signal is detected, astronomers use electromagnetic observatories to search for associated light emissions, such as gamma rays, optical light, and radio waves. Gravitational waves are ripples in space-time caused by cataclysmic events like the merger of compact objects (e.g., black holes, neutron stars) or massive stellar collapses. Their detection, along with associated electromagnetic signals, represents a cornerstone of multi-messenger astronomy, allowing a more comprehensive understanding of the universe. When massive objects accelerate (e.g., orbiting or merging neutron stars), they generate gravitational waves. These waves carry energy away from the system, causing ripples in the fabric of space-time [2, 15]. The waves propagate at the speed of light and are characterized by their frequency and amplitude, which depend on the masses and dynamics of the objects involved. Gravitational wave detectors like LIGO and Virgo are designed to measure incredibly minute distortions in space-time caused by passing gravitational waves. LIGO consists of two observatories in the US (Washington and Louisiana) that use laser interferometry to detect changes in the distance between mirrors placed kilometers apart, caused by gravitational waves. Located in Italy, Virgo complements LIGO, increasing the global sensitivity and localization accuracy of gravitational wave events. It uses a similar interferometric setup with a 3-kilometer arm length. As a gravitational wave passes, it slightly stretches and compresses the detector's arms. Detected signals (waveforms) are analyzed to determine the source's properties, such as mass, distance, and type (e.g., black hole merger or neutron star collision). While gravitational waves carry information about the motion of massive objects, they do not directly produce light. Electromagnetic observations help study the accompanying physical processes. Gravitational wave data reveal the masses, spins, and orbital dynamics of merging objects. Electromagnetic observations provide information about the environment, energy distribution, and chemical elements produced [12].

7.4.3 Fast Radio Bursts

FRBs are intense bursts of radio waves lasting only a few milliseconds. Their brief duration and high energy make them challenging to detect, but real-time processing and follow-up observations play a crucial role in studying them. FRBs are highly energetic, capable of releasing as much energy in a few milliseconds as the Sun emits in days [26]. Their origins remain a mystery, although theories include neutron stars, magnetars, and even extragalactic phenomena. The Canadian Hydrogen Intensity Mapping Experiment (CHIME) is a stationary radio telescope array optimized for wide-field surveys of the sky. It monitors vast areas continuously, increasing the likelihood of capturing transient signals like FRBs. The Arecibo Observatory (prior to closure) detected some of the earliest FRBs with its sensitive radio instrumentation [23]. Signals from radio telescopes are processed using advanced algorithms to differentiate between FRBs and other radio interference. While FRBs are primarily detected in the radio spectrum, follow-up observations in other wavelengths provide additional clues about their origins and physical mechanisms.

Ground-based optical telescopes (e.g., Gemini or Keck) search for afterglows or host galaxies near the detected FRB location. Infrared telescopes like the James Webb Space Telescope (JWST) help observe dusty regions potentially linked to FRBs. Space-based observatories like the Chandra X-Ray Observatory and Swift monitor for high-energy counterparts. Some FRBs may be associated with magnetars, which emit X-rays or gamma rays. Astronomical transient events like gamma-ray bursts (GRBs) and gravitational wave counterparts require rapid detection and analysis. GRBs are detected by instruments like the Fermi Gamma-ray Space Telescope and Swift through real-time data analysis. Algorithms identify rapid spikes in gamma-ray emissions, distinguish them from background noise, and localize the event [15, 31]. Gravitational wave events detected by LIGO or Virgo are quickly analyzed to provide a sky location for follow-up electromagnetic observations. Ground-based telescopes (e.g., Pan-STARRS or VLT)

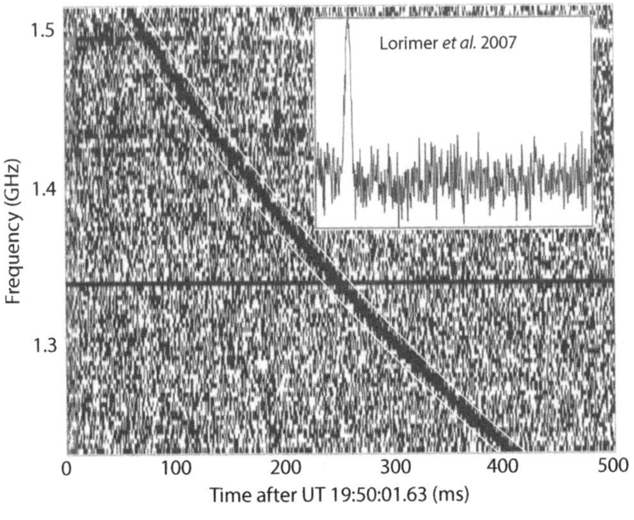

Figure 7.7 The "waterfall plot" of the famous "Lorimer burst" after adding all of the frequency channels and correcting for the sweep (credit: NASA).

Table 7.2 Types of solar radio bursts.

Types	Characteristics	Duration	Frequency range	Associated phenomena
I	Short narrow-bandwidth bursts. Usually occurs in large numbers with underlying continuum.	Single burst: ~1 second Storm: hours–days	80–200 MHz	Active regions, flares, eruptive prominences
II	Slow frequency drift bursts. Usually accompanied by a (usually stronger intensity) second harmonic.	3-30 minutes	Fundamental: 20–150 MHz	Flares, proton emission, magneto hydrodynamic shockwaves
III	Fast frequency drift bursts. Can occur singularly, in groups, or storms (often with underlying continuum). Can be accompanied by a second harmonic.	Single burst: 1-3 seconds Group: 1-5 minutes Storm: minutes–hours	10 kHz–1 GHz	Active regions, flares
IV	Stationary Type IV: Broadband continuum with a fine structure.	Hours–days	20 MHz – 2 GHz	Flares, proton emission
	Moving Type IV: Broadband, slow frequency drift, smooth continuum.	0.5 -2 hours	20–400 MHz	Eruptive prominences, magneto hydrodynamic shockwaves
	Flare Continua: Broadband, smooth continuum.	3–45 minutes	25–200 MHz	Flares, proton emission
V	Smooth short-lived continuum. Follows some type III bursts. Never occurs in isolation.	1-3 minutes	10-200 MHz	Same as type III bursts

and space-based observatories (e.g., Hubble) focus on the identified region to detect possible light emissions (e.g., kilonovae). In Table 7.2, we have provided the types of solar radio bursts.

Real-time detection techniques and follow-up observations across multiple wavelengths enable astronomers to explore transient cosmic events like FRBs, GRBs, and gravitational wave counterparts [31]. These methods illuminate the physical processes behind the universe's most energetic phenomena, pushing the boundaries of our knowledge in astrophysics and cosmology.

Figure 7.7 shows the frequency evolution and integrated pulse shape of the radio burst. The survey data, collected on August 24, 2001, are presented as a two-dimensional "waterfall plot" depicting intensity as a function of radio frequency versus time. The dispersion is clearly visible as a quadratic sweep across the frequency band, with broadening toward lower frequencies. By measuring the pulse delay across the receiver band using standard pulsar timing techniques, the dispersion measure (DM) is calculated to be 375 ± 1 cm^{-3} pc. The two white lines separated by 15 ms, which enclose the pulse, illustrate the expected behavior according to the cold-plasma dispersion law, assuming a DM of 375 cm^{-3} pc. The horizontal line at approximately 1.34 GHz is an artifact caused by a malfunctioning frequency channel. This plot corresponds to one of the offset beams where the digitizers were not saturated. By dividing the data into four frequency sub-bands, the half-power pulse width and flux density spectrum over the observing bandwidth were measured. After accounting for pulse broadening due to known instrumental effects, a frequency scaling relationship for the observed width is determined as W = 4.6 ms (f /1.4 GHz)$^{-4.8\pm0.4}$, where f is the observing frequency. A power-law fit to the mean flux densities obtained in each sub-band gives a spectral index of −4 ± 1. The inset shows the total-power signal after correcting for dispersive delay, assuming a DM of 375 cm^{-3} pc and a reference frequency of 1.5165 GHz. The time axis on the inset also covers the range from 0 to 500 ms.

7.5 Role of Computer Vision in Astronomy

Computer vision, a subset of artificial intelligence (AI), enables machines to process, interpret, and analyze visual data. In astronomy, where vast amounts of observational data are collected daily, computer vision plays a crucial role in automating tasks that would otherwise be time-consuming and labor-intensive for astronomers. It has revolutionized the identification, classification, and monitoring of celestial objects, making it a cornerstone of modern astrophysics.

7.5.1 Object Detection and Classification

Computer vision algorithms can automatically detect and classify celestial objects, such as stars, galaxies, supernovae, and asteroids, from images captured by telescopes. Galaxy morphology refers to the study of the shapes and structures of galaxies, which provides crucial insights into their formation and evolution. Galaxies are generally categorized into spiral, elliptical, and irregular types, and their morphology is influenced by factors such as age, environment, and interactions with other galaxies. Machine learning and AI have revolutionized this field by automating the process of classifying galaxies, making it possible to analyze vast datasets collected by modern sky surveys. Images of galaxies captured by telescopes like the Sloan Digital Sky Survey (SDSS) or the Hubble Space Telescope (HST) are used as inputs. These images often undergo preprocessing to enhance clarity and remove noise [21, 26]. Convolutional neural networks (CNNs) are a specialized type of deep learning model that excels in image recognition tasks. CNNs learn patterns and features directly from pixel data, making them highly effective for galaxy classification. Models are trained using labeled datasets where each galaxy's morphology is known. The model learns to associate features with specific galaxy types. Once trained, the model can classify millions of galaxies in minutes, saving years of manual effort by astronomers. With advancements in machine learning, AI models have been integrated into the project to process large datasets more efficiently. AI assists by performing initial classifications, while edge cases or ambiguous images are flagged for human review. Achievements of Galaxy Zoo: Successfully classified millions of galaxies, contributing to our understanding of galaxy formation and evolution. Highlighted rare or unusual galaxy types, such as Green Pea Galaxies and Hanny's Voorwerp. The distribution of galaxy types across different regions of the universe provides clues about cosmic evolution and the influence of environmental factors like galaxy clusters and intergalactic medium. Spiral galaxies often have young blue stars in their arms, while elliptical galaxies are dominated by older red stars [39].

7.5.2 Transient Event Detection

Transient events are short-lived astrophysical phenomena that occur on timescales ranging from milliseconds to weeks. Examples include supernovae, gamma-ray bursts (GRBs), and fast radio bursts (FRBs). Detecting these events in real time is critical because they evolve rapidly, and their early stages often contain the most scientifically valuable information. Real-time detection involves the rapid identification of changes in brightness, position, or other characteristics of celestial objects. This is achieved through automated processing of large datasets from wide-field telescopes and observatories. Wide-field telescopes like the Zwicky Transient Facility (ZTF) scan large portions of the sky continuously, capturing high-cadence images of the same regions. Instruments record changes in brightness, spectral characteristics, or motion over time. Raw data from telescopes often include noise and distortions due to atmospheric conditions or instrumental errors [26]. Preprocessing techniques, such as calibration and image subtraction, clean the data to highlight significant changes. Early detection allows astronomers to capture critical data about the transient event's origin, energy output, and environment. For instance, the early light from a supernova can reveal details about the progenitor star and its surroundings. Real-time alerts enable observatories worldwide to focus on the event, collecting data across multiple wavelengths (e.g., optical, X-ray, radio). Facilities like ZTF integrate with global networks, such as the Global Relay of Observatories Watching Transients Happen (GROWTH), to maximize scientific returns. Real-time capabilities have led to the identification of unique phenomena, such as kilonovae (from neutron star mergers) and unusual types of supernovae. Real-time detection of transient events, enabled by facilities like ZTF, represents a cutting-edge approach to observing dynamic phenomena in the universe [36]. By combining wide-field imaging, advanced algorithms, and rapid alert systems, astronomers can study these fleeting events in unprecedented detail, uncovering new insights into the cosmos.

7.5.3 Asteroid and Comet Tracking

Tracking and predicting the movement of near-Earth objects (NEOs), such as asteroids and comets, are crucial for planetary defense and scientific research. Modern systems employ computer vision and artificial intelligence (AI) to analyze astronomical data, identify NEOs, and predict their trajectories. This process involves detecting motion across a sequence of images and assessing the potential risk of collisions with Earth. NEOs appear as small moving objects when observed over time against the relatively stationary background of stars. Computer vision techniques help identify and track these objects. A series of images of the same region of the sky are captured over time. Large-field telescopes such as those in

Table 7.3 Telescopic difference between comets and asteroids.

	Metric	Asteroids	Long-period	Short-period
1	Rotation period	Hours	Hours	Days
2	Tails	None	Dust+gas	Mostly gas
3	Aspect ratio	Spherical	Spherical	Prolate
4	Plasma tail shed	None	Rare	At sector boundary
5	Precession	Gyroscopic	Unknown	Zero
6	Radial diffusion	None	Small	Huge
7	Time lag of coma	None	Symmetric	Asymmetric
8	Tensile strength	High	Very low	Sporadic

NASA's Pan-STARRS and ATLAS programs gather data. Once the motion of an object is detected, AI models predict its future trajectory by analyzing its observed position, velocity, and orbital parameters. Algorithms calculate the object's orbit using its observed positions and velocities. Data from celestial mechanics, such as Kepler's and Newton's laws, are integrated. AI models assess potential Earth impact by calculating the object's proximity to Earth's orbit. Computer vision aids in the detection and tracking of near-Earth objects (NEOs), such as asteroids and comets, by analyzing their motion across a series of images [24]. A telescopic difference between comets and asteroids is shown in Table 7.3.

7.6 Advantages of Using Computer Vision in Astronomy

Modern astronomical research relies on processing and analyzing enormous amounts of data collected from telescopes and space missions. AI and computer vision systems offer several critical advantages that revolutionize how data are managed and interpreted. Scalability refers to the ability of computer vision and AI systems to handle increasing volumes of data efficiently without compromising performance. Telescopes like the Vera C. Rubin Observatory or space missions like the James Webb Space Telescope (JWST) generate terabytes or even petabytes of data daily. Manual processing of such vast datasets is time-consuming and labor-intensive.

AI algorithms process images, spectra, and time-series data orders of magnitude faster than humans. Systems can analyze thousands of images or detect patterns across vast datasets simultaneously [24]. Process data streams from telescopes in real time, eliminating delays. AI systems detect transient events and instantly trigger alerts, enabling astronomers to conduct rapid follow-up observations. It continuously monitors light curves to identify sudden changes indicative of transient events. Together, these capabilities enhance the efficiency, reliability, and scope of astronomical research, enabling scientists to discover new cosmic phenomena and respond swiftly to potential planetary threats [38].

Computer vision has become an essential tool in modern astronomy, enabling efficient and accurate analysis of vast amounts of observational data. By automating image processing tasks, computer vision techniques enhance object detection, classification, and pattern recognition, leading to significant advancements in astronomical research. Below are some key advantages of using computer vision in astronomy, supported by mathematical formulations.

7.6.1 Automated Object Detection and Classification

Computer vision algorithms facilitate the identification of celestial objects such as stars, galaxies, and nebulae. These methods rely on feature extraction and classification models, often implemented using convolutional neural networks (CNNs). The classification process can be represented mathematically as:

$$y = f(W \cdot x + b) \tag{7.1}$$

where y is the predicted class (e.g., star, galaxy, or nebula), x represents the input features extracted from astronomical images, W and b are the weight matrix and bias, respectively, and f is an activation function (e.g., softmax for classification). By using deep learning models, astronomers can classify objects in large-scale sky surveys with high accuracy [24].

7.6.2 Noise Reduction and Image Enhancement

Astronomical images often suffer from noise due to atmospheric turbulence, sensor imperfections, and cosmic ray interference. Image processing techniques such as convolutional filtering help enhance image quality. The convolution operation used in noise reduction can be expressed as:

$$I'(x, y) = \sum_{i=-k}^{k} \sum_{j=-k}^{k} K(i, j) I(x-i, y-j) \tag{7.2}$$

where I'(x,y) is the filtered image, I(x,y) is the original image, K(,j) is the convolution kernel (e.g., Gaussian or median filter), and k is the filter size. These filters help smooth out noise while preserving critical features of celestial objects [38].

7.6.3 Motion Detection and Object Tracking

Tracking the movement of celestial objects, such as asteroids, exoplanets, and space debris, is a crucial application of computer vision in astronomy. The optical flow method, which estimates motion between consecutive frames, is given by:

$$I_x u + I_y v + I_t = 0 \tag{7.3}$$

where I_x and I_y are the image gradients in the x and y directions, I_t represents the temporal gradient, and u and v are the horizontal and vertical motion vectors, respectively. By solving this equation using algorithms like the Lucas-Kanade method, astronomers can track and predict the movement of celestial bodies with high precision.

7.6.4 Super-Resolution and Image Reconstruction

Super-resolution techniques improve the resolution of astronomical images by reconstructing high-resolution images from low-resolution inputs. A common approach uses deep learning-based generative models, where the objective function for training a super-resolution model is:

$$L = \sum_{i=1}^{N} \left| \left| I_{HR}^{i} - G\left(I_{LR}^{i}\right) \right| \right|^2 \tag{7.4}$$

where I_{HR}^{i} is the high-resolution image, I_{LR}^{i} is the corresponding low-resolution image, G(·) is the generative model (e.g., CNN or GAN-based model), and L represents the loss function measuring the difference between the generated and true high-resolution images. Super-resolution helps refine details in telescope images, improving our ability to study distant galaxies and planetary surfaces [24].

7.7 Applications

The fields of image processing and computer vision play a crucial role in modern astronomical research by enhancing the quality of observational data and extracting meaningful information from vast amounts of imagery. Astronomical telescopes capture images that are often affected by noise, distortions, and atmospheric turbulence, which can obscure fine details of celestial objects. Advanced image processing techniques, such as noise reduction, deblurring, and contrast enhancement, help refine these images, allowing astronomers to observe distant galaxies, nebulae, and exoplanets with greater clarity. Additionally, techniques like adaptive optics compensate for atmospheric distortions in real time, significantly improving ground-based observations. Computer vision methods further enhance the efficiency of astronomical studies by automating the detection and classification of celestial objects. Machine learning and deep learning algorithms are increasingly being used to identify stars, galaxies, and asteroids from large datasets, reducing the need for manual intervention. Automated object recognition and segmentation techniques help in cataloging celestial bodies and tracking their movements over time. Furthermore, feature extraction methods assist in identifying transient events such as supernovae, gamma-ray bursts, and fast radio bursts, enabling rapid follow-up observations [42].

7.7.1 Object Detection

Object detection is a fundamental application of image processing and computer vision in astronomy, enabling the identification, classification, and tracking of celestial bodies. Astronomical images contain vast amounts of data, including stars, galaxies, nebulae, asteroids, and transient phenomena. Traditional manual identification methods are time-consuming and prone to human error, making automated detection techniques essential for efficient and accurate analysis. The automated detection of celestial objects and transient phenomena like asteroids, comets, and supernovae is a transformative aspect of modern astronomy. Powered by computer vision and AI-driven algorithms, these systems identify, classify, and track celestial phenomena in real time or near real time, ensuring efficient discovery and rapid follow-up observations. Asteroids are rocky objects orbiting the Sun, often found in the asteroid belt, but some have orbits that bring them close to Earth [5].

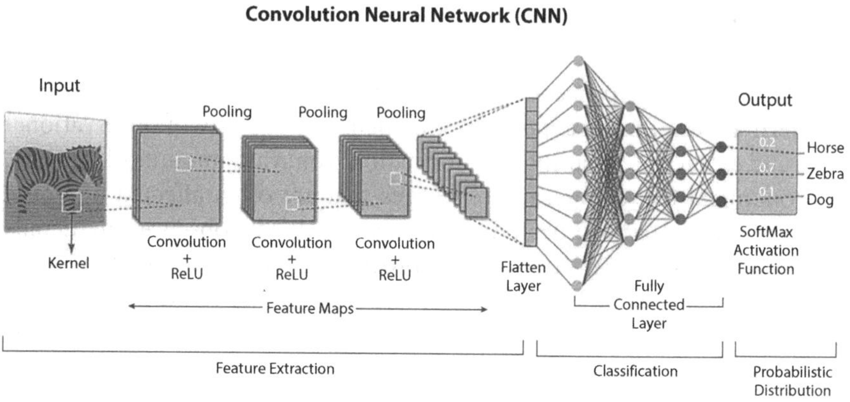

Figure 7.8 A typical convolutional neural network (CNN) (credit: MIT CSAIL).

Convolutional neural networks (CNNs) are a class of deep learning models designed specifically for processing image data. They have revolutionized the field of image analysis, including applications in astronomy, where large datasets of complex high-resolution images are common. CNNs are specialized neural networks that excel at identifying patterns, textures, and structures in image data by using convolutional layers [43, 44]. Their architecture is inspired by the visual processing mechanisms of the human brain. Filters (also known as kernels) are applied to input images to extract features such as edges, corners, and textures. Each filter captures specific characteristics of the image, contributing to a hierarchical representation of visual patterns used in tasks like object detection and classification [45, 46]. Pooling layers reduce the spatial dimensions of feature maps while preserving essential information, improving computational efficiency and making the model more robust to small variations in the input. After feature extraction and pooling, the resulting features are flattened and passed to fully connected layers, which combine the information to perform final tasks such as classification or regression [44, 46].

Figure 7.8 shows a typical convolutional neural network (CNN). Nonlinear activation functions, such as sigmoid, are applied after each convolutional layer to introduce nonlinearity. This enables the network to learn and model complex patterns in the data, which would not be possible with only linear transformations.

7.8 Challenges in Astronomical Image

7.8.1 High Dimensionality

Astronomical datasets are characterized by their vast size and complexity, spanning multiple dimensions such as wavelengths, spatial coordinates, time, and additional parameters derived from observations. This "high dimensionality" makes them rich in information but also presents significant challenges in storage, processing, and analysis. High dimensionality refers to datasets containing multiple features or attributes, often exceeding what humans can easily interpret or visualize. In astronomy, this typically includes:

- **Spatial Dimensions:** The positions of celestial objects across the sky (e.g., right ascension and declination)
- **Spectral Dimensions:** Data collected across various wavelengths (e.g., X-ray, ultraviolet, optical, infrared, radio)
- **Temporal Dimensions:** Observations taken over time to study variability or motion
- **Physical Properties:** Derived parameters such as temperature, luminosity, redshift, and metallicity

Astronomy generates high-dimensional datasets from diverse sources. Modern telescopes observe the universe across the electromagnetic spectrum, from gamma rays to radio waves. Each wavelength provides unique insights into celestial objects, such as their composition, temperature, or motion [47]. The Hubble Space Telescope collects optical and ultraviolet data, while the Chandra X-ray Observatory captures X-ray emissions. Spectroscopic observations provide detailed information about the light emitted or absorbed by objects, often spanning hundreds or thousands of wavelengths. These datasets are inherently high-dimensional, as each spectrum is essentially a multidimensional representation of intensity versus wavelength. The Sloan Digital Sky Survey (SDSS) provides spectra for millions of galaxies and stars. Observations are repeated over time to capture variability in brightness, motion, or spectral features. This adds a temporal dimension to the dataset, creating time-series data for phenomena like supernovae, variable stars, or exoplanet transits [48]. The Zwicky Transient Facility (ZTF) monitors the sky for changes over time, generating time-series data for transient events. Surveys like the LSST capture images of vast portions of the sky, generating billions of data points. Each object observed has associated spatial, spectral, and temporal information, significantly increasing dimensionality. Despite its challenges, high dimensionality offers significant advantages. Multiwavelength and time-domain data provide a complete picture of celestial phenomena. High-dimensional datasets often reveal subtle patterns or correlations that lead to new discoveries [49]. AI models trained on high-dimensional datasets can classify objects, identify anomalies, or predict behaviors with high accuracy.

7.8.2 Storage and Computational Requirements

High-dimensional datasets require significant storage space and computational power. For instance, the LSST is expected to generate 20 terabytes of data per night. Distributed storage systems and high-performance computing clusters are used to handle these requirements [50]. In high-dimensional spaces, data points become sparse, making traditional analysis methods less effective. Relationships or patterns may become harder to detect due to the increased complexity. Dimensionality reduction techniques, such as principal component analysis (PCA) or t-SNE, are used to simplify the data without significant information loss [47]. Integrating data from multiple sources (e.g., different telescopes or wavelengths) can be complex due to differences in resolution, sensitivity, or coverage. Advanced alignment algorithms and AI models are used to fuse multiwavelength data into unified datasets [43]. The quality of astronomical data is critical for accurate observations and analysis. However, various factors, including atmospheric conditions, cosmic ray interference, and instrumental noise, can degrade the quality of images. Understanding and mitigating these challenges are essential for extracting reliable information from astronomical observations. Atmospheric turbulence arises from the constant motion of air layers with varying temperatures and densities. This affects the path of light as it travels through the Earth's atmosphere, causing distortion in images of celestial objects. Cosmic rays are high-energy particles originating from outer space that strike Earth's atmosphere or telescope detectors. When these particles interact with imaging sensors, they create noise or false signals. Instrumental noise originates from imperfections or limitations in the telescope's hardware and sensors. It includes various sources of error that degrade data quality [1].

7.8.3 Real-Time Analysis

Real-time analysis refers to the capability of processing and analyzing data as they are acquired, enabling immediate or near-immediate decision-making and actions. In astronomy, real-time analysis is crucial in applications where fast detection and response are necessary, such as asteroid tracking, transient event detection, and satellite monitoring. For asteroid tracking in particular, the ability to process incoming data rapidly has profound implications for both scientific discovery and planetary defense. Near-earth objects (NEOs), such as asteroids and comets, are objects that pass close to Earth's orbit. Some may pose a risk of collision. Real-time tracking allows astronomers to detect and monitor these objects as soon as they are observed. NEOs move quickly across the sky. Even small asteroids can have significant speeds, requiring fast image processing to track their positions and velocities. The detection of a potential impact needs to happen early to allow for mitigation strategies (e.g., deflection or early warning systems) [23]. Large sky surveys, such as those from the Panoramic Survey Telescope and Rapid Response System (Pan-STARRS) or the upcoming Legacy Survey of Space and Time (LSST), generate vast amounts of data every night. These surveys take images of large portions of the sky, and each image may contain thousands to millions of objects, including potential NEOs. Astronomers must rapidly identify moving objects (asteroids or comets) in the images among the fixed stars and galaxies. Real-time analysis enables the system to flag potential moving objects as soon as they appear in the data stream, reducing the time between detection and response. This method compares successive images of the same region of the sky to detect changes, such as the appearance of a new moving object. Objects that do not match previous observations are flagged as potential asteroids or comets [38]. Cloud computing platforms are increasingly used to handle large-scale data analysis, providing scalable storage and computation resources that can support real-time analysis workflows. Data are often streamed from telescopes or observatories in real time, which are then immediately processed for potential threats or interesting events. AI and machine learning models, especially convolutional neural networks (CNNs), can quickly classify and detect objects within images. These models are trained on vast datasets and are capable of distinguishing moving objects like asteroids from static stars and galaxies. AI can also be used to predict the future movement of asteroids by analyzing patterns in their previous motions, thus helping to assess the risk they pose. The Pan-STARRS project uses a wide-field telescope to scan large areas of the sky and detect moving objects. The system performs real-time image subtraction and analysis, identifying asteroids and comets within hours of their discovery. As asteroids are detected, algorithms quickly estimate their potential paths, determining whether they might pose any collision risk with Earth [44]. ATLAS is another early-warning system that uses real-time sky surveys to detect asteroid impact threats. It specializes in detecting smaller asteroids that might otherwise go unnoticed by larger observatories. It is designed to provide rapid alerts within hours of detecting an asteroid, facilitating a swift response from scientists to assess the impact risk. Rapid processing can sometimes flag stationary objects, such as stars or background galaxies, as potential asteroids. These must be filtered out in subsequent analyses to avoid false alarms. Even with advanced algorithms, the volume of data generated by wide-field surveys can be overwhelming, necessitating powerful computing resources to ensure timely processing. While real-time trajectory prediction models are accurate, uncertainties in the asteroid's position and velocity can lead to imprecise predictions, requiring follow-up observations for refinement. Real-time analysis in asteroid tracking is essential for planetary defense and scientific discovery. By leveraging advanced algorithms, high-performance computing, and machine learning, astronomers can rapidly detect and track asteroids, predict their trajectories, and

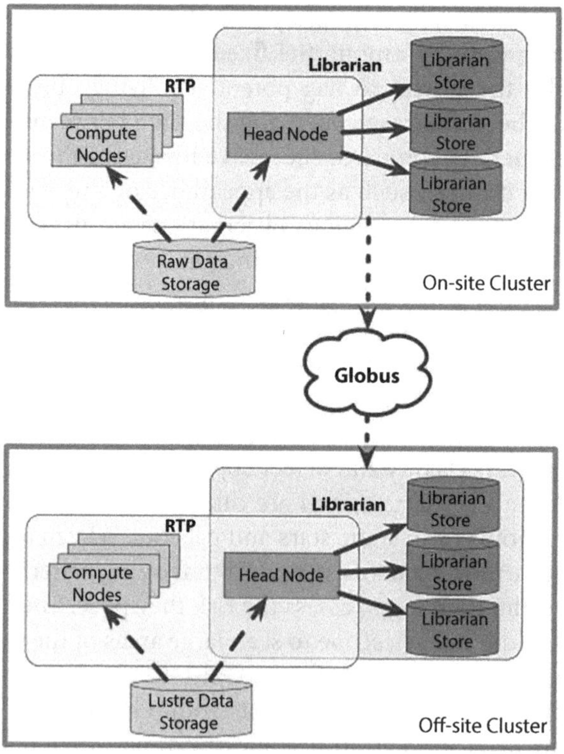

Figure 7.9 A schematic diagram showing the various components of computing hardware in the on-site HERA cluster and how the librarian and RTP systems are distributed across them (credit: ScienceDirect).

assess potential collision risks. The ability to respond quickly to newly discovered objects can be the difference between a threat being mitigated or a catastrophic event occurring. Real-time processing enables astronomers to stay ahead in this critical area of research [46].

Figure 7.9 represents the cluster at the National Radio Astronomy Observatory (NRAO). Off-site processing at the Ilifu research cloud infrastructure in South Africa is handled similarly. Transfers between Librarian instances are facilitated by Globus (https://www.globus.org/), a service that facilitates large-scale data transfer between computing facilities.

7.8.4 Interpretability

Interpretability refers to the ability of a machine learning (ML) model to provide understandable and meaningful explanations for its decisions, classifications, or detections. In astronomy, where the stakes of misclassification can be high, interpretability ensures that astronomers trust the results of ML algorithms and can validate their findings against physical principles or observational evidence. Astronomical research often relies on ML to detect transient phenomena (e.g., supernovae), classify celestial objects, or predict trajectories (e.g., asteroids). Without clear explanations for its decisions, the model's output may not be trusted, especially when results conflict with existing theories or expectations. Astronomy is guided by well-established physical laws. Interpretability allows scientists to cross-check ML predictions with known scientific principles [26]. For example, if an ML model identifies a supernova candidate, interpretability tools can help verify the decision by linking it to characteristic light curve features or spectral lines. By understanding why a model made a certain prediction, astronomers can identify and address sources of error, such as biases in training data or limitations in the model architecture. Interpretability is especially critical in applications like asteroid impact prediction, where decisions based on ML outputs may influence planetary defense strategies. Many ML models, such as decision trees or neural networks, rely on input features (e.g., brightness, motion, spectral properties) to make predictions. Feature Importance Analysis identifies which features contributed most to a given decision. In neural networks, activations at different layers can be visualized to understand how the model processes features. These inherently provide interpretable outputs by showing the decision paths taken for classification. Coefficients in linear regression or logistic regression directly indicate the influence of each feature on the outcome [44].

For complex models like deep neural networks, interpretability is often achieved after training:

- **Local Interpretable Model-Agnostic Explanations (LIME):** Explains individual predictions by approximating the model locally with a simpler interpretable model.
- **Shapley Additive Explanations (SHAP):** Provides a consistent way to attribute the output of the model to its input features, assigning each feature a "contribution score."

7.9 Challenges in Interpretability for Astronomy

The integration of image processing and computer vision in astronomy has revolutionized data analysis, enabling astronomers to detect, classify, and analyze celestial objects with unprecedented accuracy. However, one of the major challenges in this field is the interpretability of these advanced computational techniques. As machine learning and deep learning algorithms become more complex, understanding how they arrive at specific conclusions remains a significant issue, especially in high-stakes scientific research. One key challenge in interpretability is the "black-box" nature of deep learning models. Convolutional neural networks (CNNs) and other artificial intelligence-based models often process astronomical images in ways that are not directly understandable by humans. While these models can accurately classify galaxies, detect exoplanets, or identify transient astronomical events, explaining their decision-making process remains difficult. This lack of transparency can make it challenging for astronomers to trust and validate results, particularly in cases where the model's output contradicts existing theories or observational data. Another major challenge is the potential for biases in training data. Machine learning models require large datasets to learn patterns, and if these datasets are incomplete, are biased, or contain errors, the model may produce misleading results. For example, deep learning algorithms trained on past surveys may struggle to identify new types of celestial objects or rare astronomical phenomena. Moreover, noise, distortions, and instrumental effects in astronomical images can introduce unintended biases, affecting the reliability of object detection and classification. Additionally, interpretability becomes a challenge when applying image processing techniques to faint or distant celestial objects. Astronomical images often have low signal-to-noise ratios, making it difficult to distinguish real objects from background noise. Traditional image processing methods, such as deconvolution and filtering, can sometimes introduce artifacts, further complicating interpretation. When combined with deep learning, these uncertainties make it crucial to develop robust interpretability methods to verify the authenticity of detected objects. Efforts to address interpretability challenges include the development of explainable artificial intelligence (XAI) techniques, which aim to provide insights into how models make predictions. Methods such as saliency maps, feature visualization, and attention mechanisms help highlight the important regions in an image that influenced a model's decision. Additionally, incorporating physical models and domain knowledge into machine learning algorithms can improve interpretability by ensuring that detected patterns align with established astrophysical principles. As image processing and computer vision continue to evolve in astronomy, improving interpretability will be crucial for ensuring the reliability and scientific validity of automated discoveries. Future research should focus on integrating human-understandable explanations into AI models, refining training datasets to reduce biases, and developing hybrid approaches that combine traditional astrophysical methods with modern computational techniques.

Understanding the complexity of deep learning models, such as convolutional neural networks (CNNs), and their decision-making process involves exploring how these models process data and why they are inherently challenging to interpret. CNNs are a type of deep learning model designed to analyze structured data, such as images. Their complexity stems from the following features:

- **Layers and Depth**

CNNs consist of multiple layers, including convolutional layers, pooling layers, and fully connected layers. Each layer extracts increasingly complex features from the data. The depth and interconnectedness of these layers lead to a significant number of parameters (weights and biases) to optimize, often in the range of millions [33].

- **Hierarchical Learning**

  i. Early layers learn low-level features (e.g., edges or textures).
  ii. Deeper layers capture high-level features (e.g., shapes or objects).

- **Nonlinear Operations**

  i. Activation functions (like ReLU) introduce nonlinearity, enabling CNNs to learn complex patterns.
  ii. The decision-making process in CNNs involves multiple steps like Feature Extraction (Input data, such as an

image, pass through convolutional layers that apply filters to detect specific patterns. Filters slide over the image, producing feature maps that highlight the presence of certain patterns) and Feature Reduction (Pooling layers downsample the feature maps, retaining the most critical information and reducing computational complexity).

7.9.1 Data High Dimensionality

Astronomy deals with vast and complex datasets, often characterized by high dimensionality due to the multiwavelength, multitemporal, and multispectral nature of astronomical observations. Image processing and computer vision techniques play a crucial role in analyzing these high-dimensional datasets, but managing and interpreting such data present significant challenges. High-dimensional data not only increase computational complexity but also make it difficult to extract meaningful patterns and relationships from vast astronomical surveys. One of the primary reasons for high dimensionality in astronomical data is the diverse range of observational parameters. Modern telescopes capture images across multiple wavelengths, from radio waves to gamma rays, resulting in multispectral and hyperspectral datasets. Each wavelength provides different insights into celestial phenomena, requiring sophisticated techniques to integrate and analyze information across different bands. Additionally, time-series observations add another layer of complexity, as astronomers study the dynamic behavior of transient objects like supernovae, variable stars, and exoplanets over extended periods.

To address these challenges, astronomers are increasingly utilizing advanced data science techniques, including feature selection methods, unsupervised learning for clustering large datasets, and hybrid approaches that combine astrophysical models with machine learning. Collaborative projects like the Large Synoptic Survey Telescope (LSST) and the Square Kilometre Array (SKA) generate petabytes of high-dimensional data, necessitating the development of new algorithms for real-time processing and analysis. By integrating AI-driven techniques with traditional astrophysical methods, researchers can maximize the potential of high-dimensional data, leading to new discoveries and a deeper understanding of the universe.

Astronomical data are inherently complex due to its multifaceted nature [32, 45]:

- **Multiple Wavelengths**: Observations across radio, infrared, visible, ultraviolet, X-ray, and gamma-ray wavelengths reveal different physical phenomena, but integrating these can be challenging.
- **Temporal Variability**: Changes over time (e.g., variable stars, supernovae) add a dynamic aspect to the data.
- **High Dimensionality**: Data often exist in high-dimensional spaces, such as spectra, images, or time series, complicating analysis.

These characteristics make it difficult for models to focus on the most relevant aspects, increasing the risk of overfitting or misinterpretation. Dimensionality reduction and feature selection are powerful tools for handling the complexity of astronomical data. By simplifying the data, these methods make it easier to train models, reduce overfitting, and extract meaningful scientific insights.

7.10 Future Directions

As technology evolves, the synergy between image processing, computer vision, and astronomy will grow stronger. Earth's atmosphere causes light from astronomical objects to bend unpredictably, leading to blurry or distorted images when observed through ground-based telescopes. The challenges are variability in air density, temperature, and pressure and rapid and complex fluctuations that degrade image quality. Adaptive optics (AO) is a technology that corrects atmospheric distortion in real time to produce sharper images. It detects distortions in incoming light waves using a wave front sensor. Deformable mirrors are adjusted to counteract the detected distortions. Corrections are updated hundreds to thousands of times per second to match atmospheric changes. Traditional AO systems rely on fast algorithms to adjust mirrors, but AI introduces transformative capabilities [1]. AI can predict future distortions based on current atmospheric patterns, improving correction accuracy. Machine learning models (e.g., neural networks) are trained on historical wave front data to anticipate turbulence. AI algorithms are optimized to process wave front data faster than traditional methods, enabling higher correction rates. AI can generalize to varying observing conditions (e.g., different altitudes or weather patterns). Reinforcement learning enables systems to adapt dynamically during observations. AI fine-tunes deformable mirror adjustments by learning from observed errors and corrections over time. AI-powered adaptive optics revolutionizes observational astronomy by providing real-time accurate corrections to atmospheric distortions. By combining the speed of traditional AO with the predictive and adaptive capabilities of AI, this technology enables groundbreaking discoveries and sharper insights into the universe [47].

Astronomy generates vast amounts of data from observatories, satellites, and simulations, posing challenges in storage, processing, and analysis. Big Data integration using distributed computing and cloud platforms has revolutionized how astronomers handle these datasets. The integration of Big Data techniques, distributed computing, and cloud platforms empowers astronomers to tackle unprecedented data volumes, enabling more efficient processing, analysis, and collaboration. This technological synergy is pivotal for addressing the challenges of modern astronomy and unlocking deeper insights into the universe [39].

The integration of physics, computer science, and statistics enables the development of innovative analytical techniques to address complex challenges in astronomy. These fields complement one another to enhance data analysis, modeling, and interpretation in astronomical research. Interdisciplinary approaches are essential for advancing modern astronomy. By blending physics, computer science, and statistics, researchers develop cutting-edge analytical techniques to explore complex phenomena and make new discoveries [5, 18].

Quantum computing offers transformative potential in accelerating image analysis and model training for astronomical research. Its ability to process and analyze vast amounts of data more efficiently than classical computing is particularly suited for the challenges posed by modern astronomy. Quantum computing holds immense promise for accelerating image analysis, model training, and complex simulations in astronomy. While the technology is still evolving, its integration with traditional computing could redefine how astronomers process and interpret data [8, 19].

7.11 Conclusion

Modern astronomy is witnessing a revolution fueled by advancements in image processing and computer vision. These technologies enable astronomers to efficiently sift through vast and complex datasets generated by telescopes and observational instruments, uncovering hidden patterns and phenomena that were previously beyond reach. With the exponential growth in data from observatories such as the Hubble Space Telescope, James Webb Space Telescope (JWST), and upcoming projects like the Vera

Rubin Observatory, the role of computational tools has become indispensable in unlocking the secrets of the universe.

Astronomical data often span multiple wavelengths, dimensions, and time periods, posing significant challenges for manual analysis. Image processing techniques, including noise reduction, deblurring, and feature enhancement, allow scientists to extract meaningful information from raw data. For instance, adaptive filtering and Fourier transforms are commonly used to enhance images of faint galaxies or distant stars, enabling more accurate measurements of their properties.

Computer vision, a subfield of artificial intelligence (AI), takes this capability further by automating the analysis of astronomical images. Algorithms for object detection, segmentation, and classification can identify celestial objects such as stars, galaxies, and nebulae with remarkable precision. For example, convolutional neural networks (CNNs) have been employed to classify galaxy morphologies in large sky surveys like the Sloan Digital Sky Survey (SDSS), significantly reducing the time and effort required for such tasks. By identifying and categorizing objects efficiently, these tools allow astronomers to focus on interpreting the findings and formulating theories.

Beyond routine data processing, image processing and computer vision are enabling astronomers to discover hidden patterns and rare phenomena. Time-domain astronomy, which involves studying changes in celestial objects over time, benefits immensely from these technologies. Algorithms can monitor vast datasets in real time, identifying transient events like supernovae, gamma-ray bursts, and fast radio bursts (FRBs). For instance, AI models trained to detect anomalies in light curves have uncovered exoplanets orbiting distant stars, expanding our understanding of planetary systems.

Another groundbreaking application is the identification of gravitational lensing events, where massive objects bend light from background sources. Such phenomena provide insights into dark matter distribution and the expansion of the universe. Traditional methods of manually identifying lensing events are labor-intensive, but automated computer vision algorithms can rapidly scan through thousands of images, detecting these occurrences with unprecedented efficiency.

The expanding computational capabilities of image processing and computer vision are enabling humanity to probe deeper into the cosmos. Techniques like image stacking and super-resolution are used to combine multiple observations, creating highly detailed images of distant objects. These advancements are particularly critical for projects like the Square Kilometre Array (SKA), which will generate exabytes of data annually.

Efficient processing and analysis of such enormous datasets require the integration of AI-powered tools.

Moreover, these technologies facilitate multiwavelength analysis, integrating data from optical, infrared, radio, and X-ray observations. Such comprehensive analyses are crucial for understanding complex phenomena like star formation, galaxy evolution, and black hole activity. By combining data from various instruments, astronomers can build a more holistic picture of the universe, unlocking insights that were previously unattainable.

The adoption of image processing and computer vision in astronomy extends beyond enhancing scientific capabilities. The innovations inspired by these technologies have applications in diverse fields, including healthcare, environmental monitoring, and autonomous systems. For example, algorithms developed for astronomical image analysis are being adapted for medical imaging, improving early detection of diseases like cancer. Similarly, techniques for processing satellite data are advancing climate monitoring and disaster management.

As computational power continues to grow and AI algorithms become more sophisticated, the potential of image processing and computer vision in astronomy is boundless. Emerging technologies like quantum computing and advanced neural networks promise to further accelerate data analysis, opening new frontiers for exploration. The synergy between astronomy and computational sciences not only advances our understanding of the cosmos but also drives technological progress that benefits humanity as a whole.

Image processing and computer vision are transforming the field of astronomy, enabling researchers to tackle challenges posed by vast datasets and complex phenomena. By automating and enhancing data analysis, these tools empower astronomers to make groundbreaking discoveries and push the boundaries of human knowledge. Furthermore, the innovations inspired by their applications in astronomy ripple across other scientific and technological domains, fostering progress in areas critical to our future. As we continue to embrace these cutting-edge technologies, the astronomical community stands poised to unravel the universe's deepest mysteries, inspiring generations to come.

Data Availability

All data used in this chapter are publicly available and have been cited appropriately within the text.

Acknowledgement

The authors would like to express their sincere gratitude to all those who contributed to the completion of this chapter. Special thanks to colleagues, mentors, and reviewers for their valuable feedback and insightful discussions.

References

1. Ball, N.M. and Brunner, R.J., Data Mining and Machine Learning in Astronomy. *Int. J. Mod. Phys. D*, 19, 7, 1049–1106, 2010.
2. Huertas-Company, M., *et al.*, A deep learning approach to galaxy morphology—I. Automated classification of CANDELS galaxies. *MNRAS*, 475, 3, 3056–3076, 2018.
3. Domínguez Sánchez, H., *et al.*, Improved automated morphological classification of galaxy images using deep learning. *MNRAS*, 484, 1, 93–110, 2018.
4. Zheng, Z., *et al.*, Gravitational lensing and machine learning: identifying strong lenses in CFHTLS using convolutional neural networks. *Astrophys. J.*, 800, 1, 38, 2018.
5. Graham, M.J., *et al.*, Real-time search for transient events in a petabyte-scale astronomical data stream. *Publ. Astron. Soc. Pac.*, 131, 078001, 1000, 2019.
6. Tyson, J.A., Large Synoptic Survey Telescope: Overview. *SPIE Proc.*, 4836, 10–20, 2002.
7. Hardy, J.W., *Adaptive Optics for Astronomical Telescopes*, Oxford University Press, New York, 1998.
8. Wilson, R.E. and McRae, K.L., *Introduction to Calibration of Imaging Systems*, Springer, New York, 2005.
9. Gioia, I.M. and Luppino, G.A., Calibration of CCD Imaging Systems: Techniques and Methodologies. *Proc. SPIE*, 4008, 317–328, 2000.
10. Kaiser, M.E. and L.A., The Role of Flat-Field Correction in Optical Instrument Calibration. *J. Astron. Instrum.*, 1, 1, 73–85, 2011.
11. Pence, W.D. and Dettmar, R.J., Astronomical Data Analysis Software and Systems XXI: Advances in Imaging System Calibration, Astrophysics Source Code Library, 2012.
12. Greer, A.W. and Hughes, M., Bias, Dark, and Flat Calibration of Digital Cameras: Applications to Microscopy and Astronomy. *J. Microsc.*, 238, 2, 151–160, 2009.
13. Van der Marel, R.P. and Cioni, M.R., *Imaging System Calibration for Spectroscopic and Photometric Data*, Cambridge University Press, United Kingdom, 2016.
14. Roth, M. and Miele, G., Object Detection and Classification with Convolutional Neural Networks: Applications to Astronomical Imaging. *Astron. Astrophys. Rev.*, 26, 1, 1–24, 2018.

15. Carleo, G. and Troyer, M., Solving the Quantum Many-Body Problem with Artificial Neural Networks. *Science*, 355, 6325, 602–606, 2017.
16. Schmidt, P. and Heyer, I., Measuring Brightness and Size of Astronomical Objects with Deep Learning Models. *J. Astrophys. Data Anal.*, 7, 2, 120–133, 2020.
17. Bello, D. and Clarke, D., Astronomical Image Processing and Object Detection Using Convolutional Neural Networks. *MNRAS*, 487, 4, 5201-5214, 2019.
18. Bok, B.J., *The Milky Way*, Harvard University Press, Cambridge, MA, 1977.
19. Lada, C.J. and Lada, E.A., Star formation in molecular clouds: Observational evidence for the scenario. *Annu. Rev. Astron. Astrophys.*, *41*, 1, 57–115, 2003.
20. McKee, C.F. and Ostriker, E.C., Theory of star formation. *Annu. Rev. Astron. Astrophys.*, *45*, 1, 565–687, 2007.
21. Vink, J., Supernova remnants: The X-ray perspective. *Astron. Astrophys. Rev.*, *20*, 1, 49, 2012.
22. Matzner, C.D. and McKee, C.F., Efficiencies of low-mass star and star cluster formation. *Astrophys. J.*, *545*, 1, 364–378, 2000.
23. Blandford, R.D. and Begelman, M.C., On the fate of gas accreting at a low rate on to a black hole. *MNRAS*, *303*, L1, 1–5, 1999.
24. ALMA Observatory, ALMA: The Atacama Large Millimeter/submillimeter Array, 2022.
25. Abbott, B.P., Abbott, R., Abbott, T.D., Abernathy, M.R., Acernese, F., Ackley, K., Collaboration, L. S., Observation of gravitational waves from a binary black hole merger. *Phys. Rev. Lett.*, 116, 6, 061102, 2016.
26. Abbott, B.P., Abbott, R., Abbott, T.D., Acernese, F., Ackley, K., Adams, C., Virgo Collaboration, Multi-messenger observations of a binary neutron star merger. *Astrophys. J. Lett.*, *848*, 2, L12, 2017.
27. Einstein, A., Approximative integration of the field equations of gravitation. *Sitzungsber. Kgl. Preuss. Akad. Wiss. Berlin*, *1916*, 688–696, 1916.
28. LIGO Scientific Collaboration and Virgo Collaboration, LIGO and Virgo: The detectors of gravitational waves, 2020.
29. Bartos, I., Brady, P., Márka, S., Prospects of multimessenger astronomy with advanced gravitational-wave detectors. *Classical Quantum Gravity*, *30*, 12, 123001, 2013.
30. Metzger, B.D., Berger, E., Roberts, L.F., What is the most promising electromagnetic counterpart of a neutron star binary merger? *MNRAS*, *437*, 3, 1821–1837, 2015.
31. Virgo Collaboration, The Virgo interferometer: Exploring the universe through gravitational waves, 2022.
32. Gehrels, N., Cannizzo, J.K., Kanner, J., Kasliwal, M.M., Nissanke, S., Singer, L.P., The electromagnetic counterparts of gravitational wave events. *Astrophys. J. Lett.*, *820*, 2, L20, 2016.

33. Lorimer, D.R., Bailes, M., McLaughlin, M.A., Narkevic, D.J., Crawford, F., A bright millisecond radio burst of extragalactic origin. *Science, 318*, 5851, 777–780, 2007.
34. Cordes, J.M. and Chatterjee, S., Fast radio bursts: An extragalactic enigma. *Annu. Rev. Astron. Astrophys., 57*, 417–465, 2019.
35. Katz, J.I., Fast radio bursts—A brief review: Some questions, fewer answers. *Mod. Phys. Lett. A, 31*, 14, 1630013, 2016.
36. Popov, S.B. and Postnov, K.A., Millisecond extragalactic radio bursts as magnetar flares. *MNRAS, 447*, 1, L36–L39, 2013.
37. CHIME/FRB Collaboration, CHIME/FRB discovery of eight new repeating fast radio burst sources. *Astrophys. J. Lett., 885*, 1, L24, 2019.
38. Spitler, L.G., Scholz, P., Hessels, J.W.T., Bogdanov, S., Brazier, A., Camilo, F., Ransom, S.M., A repeating fast radio burst. *Nature*, 531, 7593, 202–205, 2016.
39. Zhang, B., The physical mechanisms of fast radio bursts. *Nature*, 587, 7832, 45–53, 2020.
40. Pen, U.L. and Connor, L., Localizing fast radio bursts with interferometry. *Astrophys. J. Lett.*, 807, 1, L17, 2015.
41. Arecibo Observatory, Contributions to the study of fast radio bursts before its collapse, 2020.
42. Kaspi, V.M. and Beloborodov, A.M., Magnetars: Multifaceted neutron stars with strong magnetic fields. *Annu. Rev. Astron. Astrophys.*, 55, 261–301, 2017.
43. LeCun, Y., Bengio, Y., Hinton, G., Deep learning. *Nature*, 521, 7553, 436–444, 2015.
44. Krizhevsky, A., Sutskever, I., Hinton, G.E., ImageNet classification with deep convolutional neural networks. *Commun. ACM*, 60, 6, 84–90, 2012.
45. Rawat, W. and Wang, Z., Deep convolutional neural networks for image classification: A comprehensive review. *Neural Comput.*, 29, 9, 2352–2449, 2017.
46. Simonyan, K. and Zisserman, A., Very deep convolutional networks for large-scale image recognition. *International Conference on Learning Representations, 2015.*
47. He, K., Zhang, X., Ren, S., Sun, J., Deep residual learning for image recognition, in: *Proceedings of the IEEE conference on computer vision and pattern recognition*, pp. 770–778, 2016.
48. Domínguez Sánchez, H., Huertas-Company, M., Bernardi, M., Tuccillo, D., Fischer, J.L., Improving galaxy morphologies for SDSS with deep learning. *MNRAS*, 476, 3, 3661–3676, 2018.
49. Dieleman, S., Willett, K.W., Dambre, J., Rotation-invariant convolutional neural networks for galaxy morphology prediction. *MNRAS*, 450, 2, 1441–1459, 2015.
50. Zhang, Z., Han, J., Deng, S., A deep learning framework for the classification of large-scale astronomical datasets. *Astrophys. J. Suppl. Ser.*, 248, 2, 26, 2020.

8

Astroinformatics and Big Data Challenges

Kanthavel R.[1]**, Adline Freeda R.**[2] **and Dhaya R.**[1*]

[1]*School of Electrical and Communication Engineering, PNG University of Technology, Lae, Papua New Guinea*
[2]*Department of Information Technology, KCG College of Technology, Chennai, India*

Abstract

In the realm of massive records technology, the interdisciplinary field of astroinformatics, which integrates astronomy and records technological expertise, encounters unprecedented challenges. Modern astronomical surveys and instrumentation, which generate petabytes of records every 365 days, challenge the astronomy community with the volume, speed, and type of records available to them. The interdisciplinary fields of astronomy, records technology, technological science, and statistics technology are rapidly merging to form the field of astroinformatics. This dilemma has emerged in response to the immense power and challenges presented by the massive volumes of information generated by current astronomical simulations and observations. Astroinformatics now plays a crucial role for astronomers in navigating the vast information technology, enabling them to extract valuable insights from the vast amount of data and facts. We must build systems for storing astronomical data that can manage massive data volumes, ensure quick access, and ensure long-term dependability. Information technology represents the new frontier in astronomy today. Statistics provide a deeper understanding of the universe, while state-of-the-art computational techniques form the foundation for many new discoveries and insights in this rapidly advancing field in the coming years. Astroinformatics finds its application beyond the realm of astronomy. Astroinformatics has had a huge impact on astronomical research in a number of ways. Astroinformatics has facilitated the mapping of large cosmic systems, the identification of patterns in star and galaxy data, and the detection of rare transient events that conventional methods would not have been able to accomplish. Various disciplines frequently employ or develop techniques within

**Corresponding author*: dhayavel2005@gmail.com

Yogesh Chandra, Manjuleshwar Panda and Mahesh Chandra Mathpal (eds.) The Intelligent Universe: AI's Role in Astronomy, (279–312) © 2025 Scrivener Publishing LLC

this discipline to address data quality issues, thereby advancing the general field of informatics and information technology. Astroinformatics advocates for a paradigm shift in astronomy research. This abstract deals with converging issues and groundbreaking solutions operating in this fast-growing field. Some of the most discussed research these days includes strategies for efficient management in terms of archiving, retrieval, and storage methods for large datasets. The chapter also explores the need for advanced technologies in record processing, which leverage the advantages of cloud and parallel processing to meet the current computational demands in astrophysics. AI and ML technologies in sample identification and anomaly detection among large datasets prove very effective in astronomical research. It discusses a number of challenges: how to manage large-scale statistical datasets and their uncertainties; how to visualize three-dimensional astronomical statistics; and the role of digital observatories in globally sharing and collaboratively analyzing data. Furthermore, it discusses ethical considerations pertaining to data protection and open science. Finally, the chapter exposes interdisciplinary techniques, growing technology, and the future prospects of astroinformatics in an attempt to fully recognize the potential of astronomical large-scale statistics.

Keywords: Astroinformatics, big data, data science, multidimensional statistics, data visualization, machine learning data security, computational astronomy

8.1 Introduction to Astroinformatics

The fields of astronomy, records, technological know-how, PC science, and statistics technology are rapidly merging to form the field of astroinformatics. The complexity has evolved in response to the immense potential and challenges presented by the massive volumes of information generated by current astronomical simulations and observations. Astroinformatics is now essential to astronomers as they navigate the vast information technology, enabling them to extract valuable insights from the vast amount of data [1]. Astronomers started to battle with the exponential increase in records from massive-scale sky surveys and increasingly superior observatories in the early 2000s, while data formats first emerged. The annual generation of petabytes of statistics has rapidly surpassed the capabilities of conventional methods for information evaluation and management. Astroinformatics is a vast field that encompasses numerous specialized strategies and sports.

- **Managing Data:** Statistics control refers to the development of reliable structures for the archiving, arrangement, and retrieval of significant volumes of astronomical data. It consists of increasingly complex metadata requirements, where records are accessed through protocols and database structures.

- **Data Processing:** To deal with unprocessed astronomical statistics, astroinformatics uses state-of-the-art computational techniques. This encompasses the processing of signs and images, as well as the application of fact-ration techniques.
- **Statistical Analysis:** Advanced statistical techniques are vital in modern astronomy due to the large and intricate datasets. Astronomy includes advanced statistical strategies, multivariate data, time-series evaluations, and Bayesian inference [2].
- **AI and Machine Learning:** For tasks such as object beautification, outlier detection, and sample classification in large datasets, the field makes extensive use of machine learning techniques. The application of deep learning techniques to challenging astrophysics problems is growing. Astroinformatics, which often brings together data from multiple properties or wavelengths, develops techniques for data mining—the procedure of identifying hidden patterns, relationships, and unusual activities within large databases.
- **Visualization:** Given the multidimensional nature of astronomical records, creative visualization methods are crucial. Astroinformatics explores methods for interactive character-satisfactory instances of complex facts.
- **High-Performance Computing:** Astroinformatics uses distributed computing, cloud computing, and supercomputers to meet today's computational demand in astronomy.
- **Virtual Observatories:** These are Internet-based fully integrated assets that give access to hundreds of astronomical datasets and analysis tools from a single-entry point, allowing global collaboration and data sharing [3].

Astroinformatics has had several significant impacts on astronomical research. Conventional methods of mapping large-scale cosmic systems, identifying patterns in data from stars and galaxies, and detecting rare transient events would not have been possible without astroinformatics. Additionally, it has made astronomical datasets accessible to a broader range of educators, enabling them to contribute to contemporary research. Astroinformatics will become increasingly crucial to astronomy in the future. Predictions indicate that future tasks, akin to the Legacy Survey of Space and

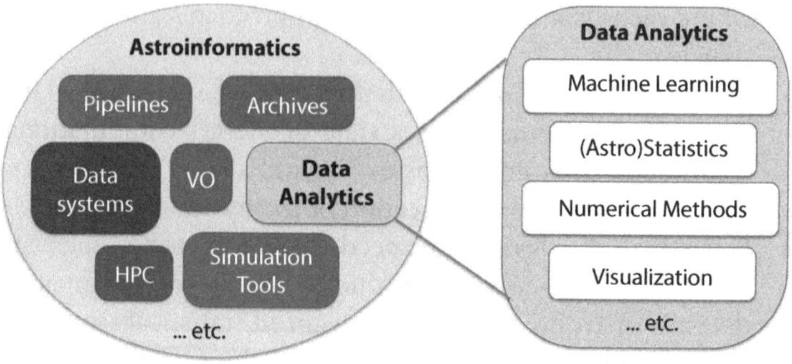

Figure 8.1 The components of astroinformatics. (Credit: IAIA).

Time (LSST) on the Vera C. Rubin Observatory, will provide unprecedented statistics, pushing the boundaries of fact control and evaluation. In doing so, since astronomical facts grow in complexity and volume, the field will have to grow in kind by pushing the development of new techniques and devices. A visual representation of the key components of astroinformatics is shown in Figure 8.1, highlighting the interplay between data acquisition, processing, and analysis in modern astronomical research.

Astroinformatics refers to the wide aspects of astronomy and astrophysics that concern acquisition, processing, and diffusion of information relative to celestial objects and events. New technologies and techniques are increasingly applied in this branch to contribute to a deeper understanding of the universe. The key domains in astroinformatics are as follows.

- **Astronomical Databases and Software Astro Databank** - a large wiki-based repository of astrological information, birth times, dates and charts of famous people, and events. Founded in 1979 by astrologer Lois Rodden, it is now owned by Astrodienst and serves as an important research and reference tool.
- **Astro Info:** A free open-source astronomical ephemeris and almanac written for PalmOS handhelds. It provides the essential data required by astronomers including rise, transit, and set times; positions; magnitude; and phase details for the Sun, Moon, planets, and stars. The application includes a rotatable and zoomable celestial map.
- **Real-Time Multi-Messenger Astrophysics:** Astro-COLIBRI: An advanced platform designed to handle the growing intricacy of fleeting astronomical phenomena. It combines

real-time data from several alert streams to give users a full picture of short-lived events like supernovae, gamma-ray bursts, and fast radio bursts. The platform provides instruments for rapid identification and analysis, enabling contributions from both professionals and citizen scientists in time-domain astronomy.

Beyond the scope of astronomy, astroinformatics also finds application. The techniques employed and developed within this discipline are repeatedly put to practice in multiple disciplines on matters of data quality while proliferating the general informatics/information technology domain [4]. Astroinformatics advocates for a paradigm shift in astronomy research. This is paving the way for a deeper understanding of the universe and the potential for innovative discoveries and insights in the coming years, driven by the significant advantages provided by large datasets and advanced computational techniques. The new frontier in astronomy today is astroinformatics. Large statistics give a better understanding of the universe, while state-of-the-art computational techniques stand at the root of so many new discoveries and insights in this very promising field within the next few years.

8.2 Big Data in Astronomy

Astronomy has entered the era of big data, driven by advanced telescopes, high-resolution simulations, and automated sky surveys that generate massive datasets. Modern observatories like the Vera C. Rubin Observatory, the James Webb Space Telescope (JWST), and radio arrays such as the Square Kilometre Array (SKA) produce petabytes of data, requiring sophisticated data management and analysis techniques. Multidimensional statistical methods, machine learning, and cloud computing play a crucial role in processing and interpreting these vast datasets. Efficient data storage, real-time processing, and secure data sharing are essential challenges in astroinformatics, shaping the future of astronomical discoveries. There are many different sources of astronomical data, and each one adds something different to the big data environment.

8.2.1 Sources of Astronomical Data

- **Ground-Based Observatories:** Large telescopes like the Very Large Telescope (VLT) and future facilities like the

Extremely Large Telescope (ELT) produce massive volumes of optical and near-infrared data.

- **Radio Telescopes:** Facilities like the Square Kilometre Array (SKA) and the Atacama Large Millimeter/Submillimeter Array (ALMA) produce massive amounts of radio wave data.
- **Space-Based Observatories:** Satellites and space telescopes like the Hubble Space Telescope, Gaia, and the James Webb Space Telescope provide high-quality data free from atmospheric interference [5].
- **All-Sky Surveys:** Projects like the planned Vera C. Rubin Observatory's Legacy Survey of Space and Time (LSST) and the Sloan Digital Sky Survey (SDSS), which methodically map vast areas of the sky, produce petabytes of data.
- **Gravitational Wave Detectors:** LIGO, Virgo, and other facilities generate distinct datasets that depict gravitational wave occurrences.
- **Neutrino Detectors:** Devices such as IceCube detect neutrinos, providing valuable information about some of the universe's most extreme events.
- **Computer Simulations:** Large-scale cosmological and astrophysical simulations produce extensive synthetic data, which we then compare to theoretical predictions and observations.

8.2.2 Volume, Velocity, and Variety Challenges

- **Volume:** Astronomy generates an astonishing amount of data, producing several petabytes per year. The LSST is expected to generate approximately 15 terabytes of data per night, while the SKA will produce data at rates exceeding those of the current global Internet traffic. This immense data volume poses significant challenges for management, analysis, and storage, requiring advancements in storage technologies and distributed computing solutions [6].
- **Velocity:** The generation and analysis of astronomical data are accelerating. Real-time analysis is crucial for detecting and responding to transient events like supernovae and gamma-ray bursts. As time-domain astronomy advances, continuous sky monitoring demands efficient pipelines for

rapid data processing and analysis. The rapid inflow of data requires powerful algorithms and high-performance computing systems to ensure quick and accurate handling.

- **Variety:** Astronomical data come in multiple forms, including images across various wavelengths (X-ray, gamma, infrared, optical, and radio), spectroscopic data, time-series measurements, astrophysical parameters, particle detection events, and gravitational wave signals. Each type requires specialized processing techniques and is often integrated with others for comprehensive analysis [7]. This diversity creates challenges in cross-matching, data integration, and developing adaptable analytical tools. Additionally, heterogeneity in data quality, metadata standards, and formats complicates the merging of datasets from different surveys and instruments. Ensuring interoperability and standardization is a significant challenge in big data astronomy, necessitating collaboration among astronomers, computer scientists, and data experts. Effective solutions include virtual observatories, scalable database systems, and machine learning algorithms for automated data processing and analysis. Addressing these challenges is essential for unlocking new discoveries and advancing our understanding of the universe.

Advanced telescopes and astronomical surveys have ushered astronomy into the big data era, requiring innovative approaches to data acquisition, management, analysis, and interpretation. The vast diversity of astronomical data—ranging from images and spectra to time-series observations—demands sophisticated data management systems, often incorporating machine learning and statistical techniques for efficient processing. This has led to the emergence of astroinformatics, an interdisciplinary field combining astronomy, computer science, and information technology to develop algorithms, databases, and visualization tools. By enabling efficient data handling and analysis, astroinformatics enhances our understanding of the universe and drives new discoveries.

8.3 Data Management in Astroinformatics

Astroinformatics focuses on the architecture of information administration, facilitating knowledge-intensive processes for systematically organizing, manipulating, processing, and disseminating astronomical data using modern computing and information technologies. It is a multidisciplinary field that integrates astronomy and informatics (computer science and data organization), with a strong emphasis on managing vast amounts of data generated by astronomical surveys, telescopes, and simulations. This includes key aspects such as storage systems, data preservation, and efficient data archiving and retrieval.

8.3.1 Systems for Storing Data

Astronomical data storage systems need to manage massive data volumes while ensuring quick access and long-term dependability. Important elements consist of:

- **Distributed Storage** involves distributing data across several nodes using tools like Ceph or the Hadoop Distributed File System (HDFS), which increase capacity and access speed.
- **Hierarchical Storage** refers to a structured storage system that uses tiered storage systems, with often-used data stored on fast media (such as SSDs) and less frequently used data stored on slower higher-capacity media. Cloud storage involves the utilization of scalable reasonably priced storage options through cloud platforms such as Google Cloud Storage and Amazon S3.
- **High-Performance File Systems:** For large-scale high-throughput data operations, use specialized file systems such as Lustre or BeeGFS.

8.3.2 Archiving and Retrieving Data

Systems for efficient archiving and retrieval are essential to enabling researchers to access astronomical data. Important things to think about are:

- The **management of metadata** involves creating thorough schemata to represent datasets and make finding and searching easier.
- **Data compression** involves implementing lossless compression techniques to reduce storage requirements without compromising data integrity.

- **Indexing techniques** involve the construction of effective indexing structures to facilitate quick query and subset retrieval.
- **Protocols** for data transfer involve effectively utilizing high-speed data transfer protocols, such as GridFTP or Globus, to transfer large datasets between users and storage systems.
- **Virtual observatory standards:** We comply with Table Access Protocol (TAP) and ObsTAP (a Virtual Observatory standard) to guarantee compatibility across various data repositories.

8.3.3 Preservation and Curation of Data

Ensuring the integrity and long-term usage of astronomical data is essential for future studies. The efforts aimed at preserving and curating data are focused on:

- **Enforcing Stringent Quality Control:** The procedures to verify data accuracy and integrity are known as data quality assurance.
- **Version Control:** It is the process of keeping datasets clearly versioned and documenting any modifications to processing or calibration.
- **Format Migration:** It is the process of regularly moving data to more recent file formats in order to keep it accessible and avoid obsolescence [8].
- **Tracking** the provenance of datasets involves recording observable parameters, processing stages, and any changes made throughout their entire existence.
- **Long-term Storage Media:** Use durable long-lasting storage media and perform frequent data migrations and integrity checks.

In astroinformatics, efficient data management necessitates a comprehensive strategy that takes into account long-term concerns about data usability and scientific value in addition to the technological difficulties of storing and retrieving enormous volumes of data. Data management solutions must change with the field to accommodate new technology and shifting research requirements, all the while preserving the accessibility and integrity of priceless astronomical data [9].

8.3.4 Data Archiving and Retrieval

- **Virtual Observatories (VOs):** Platforms like the International Virtual Observatory Alliance (IVOA) provide a framework for accessing and exchanging astronomical data between various institutions and archives. They guarantee interoperability by standardizing protocols and data formats.
- **Astroinformatics Databases:** Astroinformatics databases are unique in that they store large datasets in specialized databases with access and analysis tools, such as the Sloan Digital Sky Survey (SDSS) repository. One of the key aspects of astroinformatics is the ability to efficiently store and analyze vast amounts of astronomical data. These databases provide researchers with specialized tools to access, process, and interpret large datasets, enabling groundbreaking discoveries in astrophysics. These specialized databases also support the analysis of large-scale cosmological data, such as the cosmic background radiation map shown in Figure 8.2.

Such repositories play a crucial role in modern astronomy by offering structured queryable datasets that facilitate large-scale analyses, machine learning applications, and cross-mission comparisons. As observational data continue to grow exponentially, astroinformatics databases will

Figure 8.2 Map of cosmic background radiation with artificial colors showing temperature variations, aiding astroinformatics in cosmological analysis. (Credit: NASA/WMAP Science Team).

remain essential for managing and extracting meaningful insights from this wealth of information.

8.3.5 Archiving Techniques

Astro information science is related to huge amounts of astronomical data from telescopes, satellites, and simulation. Archive technology plays an important role in adapting, managing, and providing these data for researchers worldwide.

- **Hierarchical Storage Management (HSM):** This technique automatically switches data between expensive and inexpensive storage media based on consumption trends. We can keep frequently accessed data on fast storage, such as SSDs, and shift infrequently viewed data to long-term affordable storage, such as cassettes [10].
- **Data Optimization and Compression:** Methods such as data compression minimize the amount of storage space needed while preserving data accessibility and integrity.

Effective data preservation and retrieval are essential for managing the vast amounts of astronomical information generated by modern observatories and simulations. To ensure long-term accessibility and usability, various archiving methods are employed. Table 8.1 explains the key archiving techniques in astroinformatics as follows:

The effective collection in astroinformatics ensures that astronomical data remain available, scalable, and copyable for decades. Techniques such as hierarchical storage, metadata sequencing, cloud-based solutions, and virtual observation standards enable effective control of increasing amounts of astronomical data.

8.3.6 Metadata and Standardization

Flexible Image Transport System (FITS) is a common format for storing astronomical data, including pictures and information. Many people use it to archive observational data.

- **Rich Metadata:** For the purpose of future data reuse and reproducibility, comprehensive metadata encompassing observational parameters, instrument specifications, and data processing stages are necessary.

Table 8.1 Key archiving techniques in astroinformatics.

Archiving technique	Description	Example applications
Hierarchical Storage Management (HSM)	Employs a tiered methodology, allocating frequently accessed data to high-speed storage (SSD) and archiving older data in lower-speed storage (tape, cloud)	Extensive surveys such as LSST, Gaia, and SDSS employ HSM for effective data acquisition
Astrophysics data system for decentralized and cloud-capable curved scalability	Providing data to different servers or cloud infrastructure to guarantee Google Clouds for astronomy data and astronomical data	NASA's ADS, ESA's ESASky, and Google Cloud for astronomy data
To facilitate effective findings with metadata	Sequencing and cataloging structure dataset metadata (e.g., observation time, wavelength, equipment)	VizieR, SIMBAD, and Aladin Sky Atlas use metadata-driven search tools
Data compression and reduction methods	To reduce unemployment and reduce file sizes (e.g., rice compression fits) or loss (e.g., JPEG 2000) file sizes	Pass format is a standard in astronomy for storing paintings and spectra, including compression
Virtual Observatory (VO)	Standards establish global data sharing protocols for spontaneous access to decentralized archives	International Virtual Observatory Alliance (IVOA) enables federated queries across global archives
Long data settling data	For several decades by using standardized formats and excision	LOCKSS (Lots of Copies Keep Stuff Safe), OAIS (Open Archival Information System)
Provenance monitoring and version control	Managing the chronology of data processing procedures to guarantee data integrity and auditability	AstroPy, Astrophysical Data System (ADS) maintains provenance logs
AI-enhanced data collection	Unemployed machine learning to classify, mark, and improve data recovery in wider archives	AI-based searches in SDSS, Gaia DR3, and JWST databases

- **Conservation and Curation:** The practice of keeping, organizing, and conserving data to guarantee their continued usefulness is known as data curation. This involves transferring data to new formats or storage media as needed, updating metadata, and monitoring data integrity. Regular data refreshing and format migration are effective methods to prevent data loss due to medium deterioration or technical obsolescence [11, 12].

8.3.7 Data Retrieval in Astroinformatics

Efficient data retrieval is crucial for leveraging vast astronomical datasets in research and analysis. Various techniques enable seamless access to stored data, ensuring accuracy and reliability in scientific investigations.

- **Astronomical databases** use a specific query language known as Astronomical Data Query Language (ADQL) to look up and obtain information.
- **User interfaces and Web portals:** Platforms like the European Space Agency's Gaia Archive and NASA's Mikulski Archive for Space Telescopes (MAST) offer Web interfaces for searching, visualizing, and downloading data.

8.3.8 Astroinformatics in Practice: The Hubble Space Telescope Repository

Consider the data repository for the Hubble Space Telescope. The repository stores all the data from the HST's observations from the beginning to the present. The repository stores data in the FITS format, ensuring rich metadata. Researchers can view the details, download them for further research, and search for them using various criteria on the MAST portal [13]. Moreover, the availability of tools supports cross-matching data with other astronomical datasets in the archive. The adoption of astroinformatics in data archiving and retrieval requires the establishment of complex systems and practices to manage the storage and accessibility of astronomical data. These processes play a crucial role in promoting deeper explorations within the field of astronomy, leading to significant discoveries, and enhancing collaboration among researchers. Information administration in astroinformatics involves managing the vast amounts of data generated by astronomical observations. This encompasses data collection, storage,

Table 8.2 Data life cycle in astroinformatics.

Component	Description
Data Collection	Involves gathering data from various astronomical instruments and surveys, resulting in large volumes of diverse data types
Data Storage	Entails the use of advanced storage solutions to handle the massive datasets produced, ensuring data integrity and accessibility
Data Processing	Involves cleaning, organizing, and transforming raw data into formats suitable for analysis, often utilizing high-performance computing resources
Data Analysis	Employs statistical methods, machine learning, and data mining techniques to extract meaningful insights from processed data
Data Dissemination	Focuses on sharing processed data and findings with the scientific community and the public through databases, publications, and collaborative platforms

processing, analysis, and dissemination. Above is Table 8.2 summarizing key components and their descriptions.

These components collectively form the foundation of astroinformatics, ensuring efficient data management from collection to dissemination. By implementing these structured processes, astronomical research benefits from improved data accessibility, enhanced analytical capabilities, and greater collaboration within the scientific community.

8.4 Data Processing Techniques

In astroinformatics, data processing techniques are excellent for converting raw astronomical facts into useful knowledge. State-of-the-art techniques for data collection, analysis, and interpretation make it easier to move large amounts of data collected by modern telescopes and simulations. Some of the most important techniques used in data processing in astroinformatics are data preprocessing, cleaning, and transformation.

Data preprocessing is a technique for transforming raw data into an understandable format or into an appropriate format as depicted in Figure 8.3

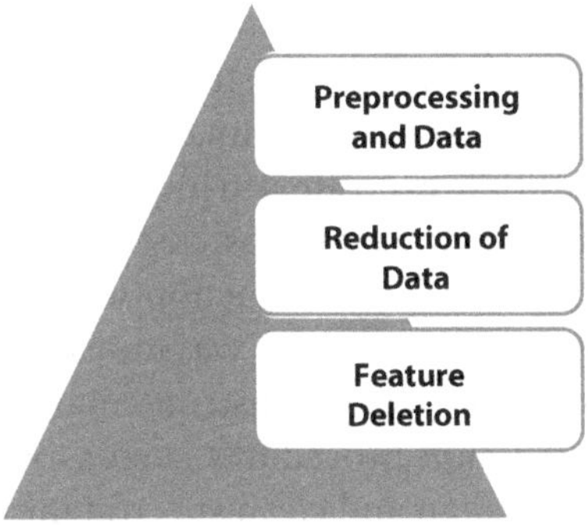

Figure 8.3 Data pre-processing.

for data processing. It involves cleaning, filtering, and normalizing data to remove inconsistencies and enhance its quality for analysis. Effective data preprocessing is crucial in astronomy, where vast datasets require refinement before applying machine learning or statistical techniques.

8.4.1 Data Preprocessing and Cleaning

Data preprocessing is a crucial step that occurs before data cleaning, ensuring that raw astronomical data are transformed into a structured and analyzable format. It involves various techniques to correct errors, reduce noise, and enhance data quality. The preprocessing stage addresses issues such as instrumental artifacts, cosmic ray contamination, and variations introduced during data acquisition. One common preprocessing technique is stereo calibration [14], which refines data alignment for accurate analysis. Data cleaning further improves data quality by identifying and correcting inconsistencies, ensuring accuracy and reliability. For example, wavelength calibration in spectroscopy aligns spectral data with known reference values to correct shifts in pixel sensitivity, while flat-fielding in imaging compensates for variations in detector sensitivity. These methods help ensure that astronomical observations accurately reflect the true properties of celestial objects, improving the reliability of subsequent analyses.

8.4.2 Data Reduction

Data reduction is the process of transforming raw observational data into a more refined and usable format for analysis. This step is essential in astronomy, where large volumes of data are collected from telescopes and space observatories, often containing noise, artifacts, and redundant information. By applying data reduction techniques, astronomers enhance the quality of data while minimizing storage requirements and computational complexity. In astronomical imaging, data reduction typically involves combining multiple exposures to improve the signal-to-noise ratio and mitigate unwanted artifacts such as cosmic ray contamination. This process may include dark frame subtraction to correct for sensor noise, flat-field correction to compensate for variations in pixel sensitivity, and bias correction to remove systematic instrumental errors. Additionally, in spectroscopic data, reduction procedures such as wavelength calibration and sky subtraction help align and clean spectral information for accurate scientific interpretation. Effective data reduction not only refines the dataset but also ensures that the extracted information accurately represents the observed celestial phenomena. By implementing systematic reduction pipelines, astronomers can optimize the quality of their datasets, making them more suitable for further analysis, modeling, and discovery. The process of obtaining unprocessed statement facts and converting them into a more usable form for processing and analysis is known as data reduction. In the case of imaging, this usually involves multiple exposures designed for cosmic ray reduction and increased signal-to-noise ratio.

8.4.3 Feature Extraction and Selection

Feature extraction is a crucial process in astronomical data analysis, involving the identification and isolation of key patterns or characteristics from vast datasets. This step enables astronomers to systematically catalog celestial objects, such as stars, galaxies, and other astronomical phenomena, by extracting meaningful features that facilitate classification and analysis. In astronomical imaging, feature extraction includes methods such as image segmentation, which partitions an image into distinct objects, edge detection, which highlights structural features, and clustering algorithms, which group similar objects based on shared properties. These techniques help in distinguishing celestial bodies from background noise and improving the accuracy of object classification. In time-series data, such as light curves, feature extraction plays a pivotal role in detecting transient events and periodic variations. This includes identifying short-lived phenomena such as supernovae, which exhibit sudden brightness spikes, and variable stars, which show periodic brightness fluctuations. By systematically extracting these temporal features, astronomers can gain insights into the physical mechanisms governing stellar evolution, planetary transits, and other dynamic astrophysical processes [15]. With the advent of the big data era, the need for advanced data mining and machine learning techniques has become more pronounced. These methods enable researchers to sift through massive datasets efficiently, uncover hidden patterns, and identify rare astronomical events. For example, large-scale sky surveys can apply anomaly detection algorithms to discover unexpected objects or phenomena, such as unusual galaxy formations or potential exoplanet candidates within stellar light curves. Moreover, machine learning models, including neural networks, can be employed for supervised learning, where labeled datasets train models to recognize specific patterns, or unsupervised learning, where algorithms independently detect clusters and correlations without predefined labels [16].

By integrating feature extraction with modern computational techniques, astronomers can enhance their ability to analyze and interpret complex datasets, paving the way for new discoveries in observational astronomy and astrophysics.

8.5 Data Visualization in Astroinformatics

Astroinformatics deals with vast astronomical datasets collected from telescopes, simulations, and large-scale surveys. To analyze and interpret this wealth of information, effective visualization techniques are essential. These methods help transform raw data into meaningful insights, enabling astronomers to detect patterns, identify correlations, and make accurate scientific interpretations. Visualization in astroinformatics is particularly crucial due to the multidimensional nature of astronomical data, which includes spatial, spectral, temporal, and physical parameters. Advanced visualization tools allow researchers to explore cosmic structures, track dynamic celestial events, and enhance the study of astrophysical phenomena.

8.5.1 Principal Visualization Methods in Astroinformatics

Data visualization in astronomy employs various techniques to represent complex relationships and structures within astronomical datasets. These methods enable researchers to analyze different aspects of cosmic data efficiently, ranging from stellar classifications to large-scale universe simulations. Key visualization techniques include:

- **Multicolored Data Visualization:** Astronomical data span multiple dimensions, such as spatial, spectral, and cosmological parameters. Scatter plots, heat maps, and contour plots are commonly used to illustrate intricate relationships within datasets. A classic example is the Hertzsprung-Russell (H-R) diagram, which categorizes stellar populations based on temperature and luminosity.

- **Three-Dimensional and Volume Rendering:** 3D visualization techniques, such as isosurface rendering, help astronomers study large-scale cosmic structures, including galaxy distributions, nebulae, and dark matter simulations. This approach is particularly useful for illustrating the cosmic web in large-scale universe models.
- **Time-Series Visualization:** In time-domain astronomy (TDA), visualizing variations in brightness is essential for studying transient events such as variable stars and supernovae. Line plots and light curve diagrams help monitor changes over time, such as the peak brightness and exponential decay of supernovae.
- **Spectral Visualization:** Spectroscopy provides crucial insights into the chemical composition, velocity, and temperature of celestial objects. Spectral line analysis helps astronomers determine redshifts, revealing the expansion of the universe. A key example is the redshift measurement using spectral line displacements (e.g., Hα and OIII emission lines).
- **Astronomical Maps and Sky Surveys:** Two-dimensional and three-dimensional sky maps provide comprehensive views of the distribution of celestial objects. Large-scale surveys such as the Sloan Digital Sky Survey (SDSS) and Gaia mission generate detailed star maps, enhancing our understanding of the structure and evolution of the cosmos.
- **Machine Learning and AI-Based Visualization:** Dimensionality reduction techniques like t-distributed stochastic neighbor embedding (t-SNE) and principal component analysis (PCA) assist in analyzing high-dimensional astronomical datasets. AI-powered anomaly detection helps uncover rare astronomical objects, such as unusual galaxy morphologies.
- **Interactive and Virtual Reality (VR) Visualization:** Emerging VR technologies allow researchers to explore astronomical data in an immersive environment. Virtual planetary system simulations and cosmic structure explorations provide new

ways to interact with real observation data, such as VR-based space exploration experiences.

These visualization techniques significantly enhance the interpretation of astronomical data, allowing scientists to extract critical insights from vast and complex datasets. As observational and computational capabilities continue to evolve, visualization will remain an indispensable tool in astroinformatics, enabling a deeper understanding of the universe.

8.5.2 Applications and Challenges of Visualization in Astroinformatics

Visualization plays a critical role in astroinformatics by enabling researchers to analyze and interpret vast astronomical datasets. It helps in identifying cosmic structures, tracking transient astronomical events, and analyzing large-scale surveys. By converting raw data into graphical representations, visualization allows astronomers to highlight patterns, detect anomalies, and communicate findings effectively. The ability to visually explore multidimensional data improves our understanding of the spatial and spectral properties of celestial objects. Scientists often employ false-color imaging to represent various wavelengths of the electromagnetic spectrum, making invisible astronomical features perceptible. Effective visualization not only benefits researchers but also aids in public outreach by making complex astronomical phenomena more accessible. Applications of visualization in astroinformatics include:

- **Understanding Cosmic Structures** – Mapping galaxies, galaxy clusters, and dark matter distributions to study the large-scale structure of the universe
- **Detecting Transients and Exoplanets** – Identifying supernovae, gamma-ray bursts, and exoplanets through light curve analysis
- **Analyzing Large-Scale Surveys** – Processing billions of data points from astronomical surveys to uncover new celestial objects and trends
- **Evaluating Cosmic Evolution** – Studying how galaxies and the universe evolve over time through comparative visualization techniques

Visualization is a powerful tool that allows astronomers to manage vast datasets, highlight complex cosmic patterns, and explore the universe in new ways. Given the multidimensional nature of astronomical data, visualization techniques such as contour plots, color mapping, and three-dimensional rendering have significantly improved the ability to access and communicate astronomical information. Scientists rely on these methods to convey research findings clearly to both expert and nonexpert audiences. For instance, Figure 8.4 outlines the visualization process, demonstrating how raw astronomical data are processed and represented in meaningful formats [17].

Astroinformatics visualization faces challenges due to the vast multidimensional nature of astronomical data. Effective representation requires managing complexity, maintaining accuracy, and optimizing user interaction.

- **Multidimensionality and Complexity** – Astronomical data span multiple dimensions (e.g., time, wavelength, spatial coordinates), making visualization challenging. For example, examining the same celestial object at different wavelengths reveals varied physical properties, requiring careful representation to avoid information overload [18].
- **Size and Detail Management** – Astronomical datasets often include both nearby bright stars and distant faint galaxies, creating a large dynamic range. Techniques like adaptive

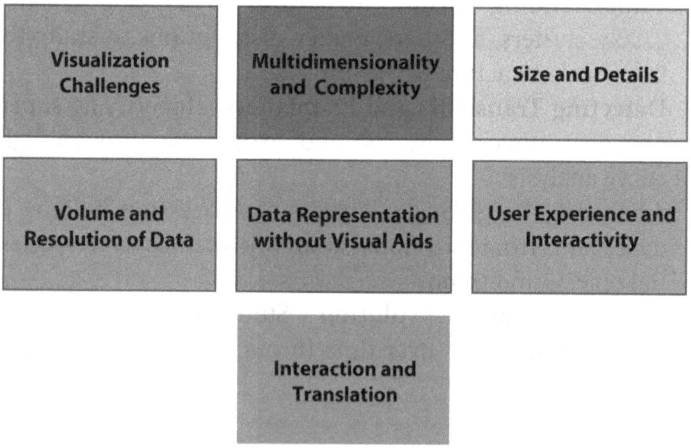

Figure 8.4 Visualization process in astroinformatics.

histogram equalization and logarithmic scaling help reveal crucial details without distorting the data.

- **Volume and Resolution Constraints** – The vast amount of data from modern astronomical surveys can reach petabyte scales, making interactive visualization difficult. Solutions such as data compression, downsampling, and selective rendering are used, although they may introduce bias or result in data loss.
- **Interpreting Nonvisual Data** – Many astronomical observations, such as spectra, radio waves, and X-rays, are not directly visible. Translating this information into meaningful visual formats requires careful calibration to avoid misinterpretation. Color mapping is often used, but improper representation can lead to misleading conclusions.
- **User Experience and Interactivity** – Interactive visualization tools are essential for navigating complex datasets. However, designing responsive and intuitive interfaces capable of handling real-time high-resolution data remains a significant challenge [19].
- **Effective Communication** – Visualization must accurately convey scientific data without misleading viewers. Misrepresentation of data, especially to nonexpert audiences, can result in misunderstandings or incorrect conclusions. Thoughtful use of color, scale, and annotations is necessary to maintain clarity and scientific accuracy.

The complexity, volume, and multidimensional nature of astronomical data, coupled with the need to translate nonvisual data into intelligible visual representations, present major challenges in astroinformatics visualization. Addressing these barriers requires a combination of advanced technical solutions, audience-focused design, and careful data representation to ensure clarity, accuracy, and accessibility in astronomical research.

8.5.3 Data Processing Techniques in Astroinformatics

Data processing techniques are essential in astroinformatics to efficiently handle and analyze the massive datasets generated from astronomical observations. These techniques help extract meaningful insights by filtering noise, identifying patterns, and optimizing data for further analysis. Given the complexity of astronomical data, specialized mathematical models and algorithms are applied to transform raw observational data into structured and interpretable formats. One of the fundamental challenges in astronomy is dealing with multidimensional data, which include spatial, spectral, and temporal information. Advanced data processing methods, such as Fourier transform, principal component analysis (PCA), and Gaussian mixture models (GMMs), are used to analyze these data effectively. Fourier transform is commonly used for frequency analysis in time-series datasets, such as studying the periodic variations in light curves of variable stars. PCA is a powerful dimensionality reduction technique that helps identify dominant patterns in high-dimensional data, making it easier to visualize and interpret. GMM is a clustering technique that models data distributions as multiple Gaussian components, allowing astronomers to categorize objects based on shared characteristics. These techniques, among others, play a crucial role in modern astronomical research by enhancing the efficiency and accuracy of data analysis. Key data processing techniques include as following:

- **Fourier Transform:** Used to analyze the frequency components of time-series data, such as light curves of variable stars. The discrete Fourier transform (DFT) in Equ. (8.1) is defined as follows:

$$X_k = \sum_{n=0}^{N-1} x_n \cdot e^{-i2\pi kn/N} \tag{8.1}$$

where:

- $\mathbf{X_k}$ is the DFT output at frequency **k**,
- $\mathbf{x_n}$ is the input time-series data at time **n**,
- **N** is the total number of data points,
- is the imaginary unit.
- **Principal Component Analysis (PCA):** A dimensionality reduction technique used to identify patterns in data by transforming them into a set of orthogonal components. The transformation is expressed in Equ. (8.2) as follows:

$$Z = A\cdot(BC) \tag{8.2}$$

where:

- **A** is the transformed data matrix,
- **B** is the original data matrix,
- **C** is the matrix of eigenvectors (principal components).

- **Gaussian Mixture Models (GMMs):** Used for clustering data by modeling them as a mixture of multiple Gaussian distributions. The probability density function can be expressed in Equ. (8.3) as follows:

$$p(x) = \sum_{k=1}^{K} \pi_k \cdot \mathcal{N}(x \mid \mu_k, \Sigma_k) \tag{8.3}$$

where:

- **K** is the number of Gaussian components,
- π_k is the weight of the k- th Gaussian component,
- $\mathbf{N(x_{|}\mu_k,\Sigma_k)}$ is the Gaussian distribution with mean **μk** and covariance $\mathbf{\Sigma_k}$

The application of data processing techniques in astroinformatics is essential for handling the vast and complex datasets generated by modern telescopes and simulations. Techniques such as Fourier transform, PCA, and GMMs enable astronomers to extract meaningful patterns, classify celestial objects, and analyze astronomical signals with high precision. As observational astronomy continues to expand, advancements in data processing will remain a critical factor in enabling new discoveries and refining our understanding of the universe.

8.6 Statistical Challenges in Astroinformatics

Astrophysical phenomena are highly complex, and the vast datasets collected from astronomical observations present unique statistical challenges. From handling incomplete and high-dimensional data to managing observational biases and model selection issues, overcoming these challenges is essential for accurate scientific interpretation. Traditional statistical methods often struggle to address these issues, requiring specialized techniques tailored for large-scale astronomical datasets. Advanced methodologies such as Bayesian inference, machine learning, and Monte Carlo simulations are widely used to extract meaningful insights from astronomical data while accounting for uncertainties.

8.6.1 Rare Events and Data Limitations

One of the fundamental statistical challenges in astroinformatics is dealing with small sample sizes and rare astronomical phenomena.

While large-scale surveys observe millions of celestial objects, certain classes of events—such as exotic types of supernovae—occur infrequently, making statistical inference difficult [20]. Many traditional statistical techniques rely on large datasets to produce reliable results, but when the number of observed instances is limited, parameter estimation and classification become significantly more complex. To address this, techniques such as bootstrapping and Bayesian statistics are employed, enabling researchers to estimate uncertainties and improve inference based on prior knowledge. However, careful consideration is required to ensure that model assumptions and prior distributions do not introduce biases into the analysis.

8.6.2 Observational Constraints and Big Data Challenges

Astronomical observations are often subject to limitations, such as telescope sensitivity thresholds, weather conditions, and incomplete time-series data. These factors lead to censored or truncated datasets, where some information is partially recorded while other aspects remain entirely missing. For instance, telescopes may only detect bright objects due to magnitude limits, resulting in selection biases that affect statistical conclusions. Addressing these issues requires advanced statistical techniques, such as survival analysis and detection modeling, to account for observational gaps and biases effectively. Astronomical data analysis faces several statistical challenges that must be addressed to ensure accurate and meaningful interpretations. **Limitations and censoring** occur due to telescope constraints and weather conditions, leading to incomplete observations and biased datasets, which require specialized statistical techniques for correction. **High dimensionality** is another major challenge, as astronomical datasets often contain multiple attributes, such as spectral features, time variability, and spatial coordinates. As the number of dimensions increases, the complexity of the data grows exponentially, making it difficult to extract meaningful patterns. Additionally, **uncertainty quantification** is crucial, as measurement errors, instrumental noise, and model assumptions introduce uncertainties that must be carefully managed through Bayesian methods and Monte Carlo simulations. **Model selection and overfitting** pose further challenges, as selecting an appropriate statistical model requires balancing complexity and interpretability—overly complex models may fit noise rather than actual signals, leading to incorrect conclusions. Finally, **combining multiwavelength data** presents difficulties, as astronomical data collected from different wavelengths (e.g., optical, radio, X-ray) must be carefully aligned and calibrated to ensure consistency and accurate cross-analysis. Addressing these challenges requires the use of advanced statistical methodologies and machine learning techniques to improve data interpretation in astronomy. Managing these challenges requires a combination of dimensionality reduction techniques (e.g., principal component analysis (PCA) and t-SNE), probabilistic modeling, and large-scale data analytics to ensure robust statistical analysis of astronomical datasets [21].

8.6.3 Addressing Statistical Challenges in Astroinformatics

The ability to extract reliable insights from astronomical data depends on choosing appropriate statistical models and ensuring that analyses remain unbiased and accurate. One of the major difficulties in astroinformatics is model selection, where researchers must balance overfitting (where a model captures noise rather than meaningful patterns) with underfitting (where a model is too simplistic to represent the data accurately). Complex machine learning models often perform well in classification tasks but may obscure the underlying physical processes if their interpretability is limited. To validate such models, independent datasets or cross-validation techniques are essential [6]. Additionally, combining data from multiple sources presents statistical challenges, as different instruments and observational methods introduce varying noise levels and biases. Effective data fusion techniques must align datasets across different wavelengths, ensuring consistency between theoretical simulations and observational data. The following Table 8.3 summarizes key statistical challenges in astroinformatics, along with potential solutions:

Table 8.3 Statistical challenges in astroinformatics.

Challenge	Description	Potential solutions
Extensive and High-Dimensional Data	Astronomical surveys produce petabytes of data containing millions of attributes per object. Conventional statistical techniques encounter difficulties with such intricacy.	Dimensionality reduction techniques (e.g., PCA, t-SNE), deep learning, and large-scale data analytics
Absence and Insufficiency of Data	Observations frequently remain incomplete owing to meteorological conditions, equipment constraints, or selection biases, resulting in data gaps.	Bayesian inference, imputation methodologies, and Gaussian processes

(*Continued*)

Table 8.3 Statistical challenges in astroinformatics. (*Continued*)

Challenge	Description	Potential solutions
Quantification of Uncertainty and Propagation of Error	Measurement inaccuracies stem from instrumental noise, cosmic variance, and calibration discrepancies.	Monte Carlo simulations, Bayesian hierarchical modeling, and bootstrapping techniques
Detection of Outliers and Anomalies	Rare astronomical entities (e.g., exoplanets, transients, gravitational wave counterparts) must be differentiated from instrumental artifacts.	Anomaly detection technologies, such as Isolation Forest and DBSCAN, alongside robust statistical checks
Selection Effects and Bias Corrections	Surveys suffer from selection biases (e.g., magnitude limits, redshift-dependent sampling), affecting statistical conclusions.	Survey completeness, Bayesian priors, and forward modeling
Model Selection and Overfitting	Intricate models may overfit constrained observational data, resulting in erroneous astrophysical conclusions.	Cross-validation, Akaike Information Criterion (AIC), Bayesian model comparison
Correlations in Multiwavelength Data	Combining data from different wavelengths (radio, optical, X-ray) introduces noise variations and systematic uncertainties.	Multiwavelength statistical fusion, probabilistic graphical models
Analysis of Time-Series and Transient Events	The light curves of supernovae, variable stars, and exoplanets are frequently sampled sporadically.	Lomb-Scargle periodograms, Gaussian process regression, and Bayesian time-series analysis
Estimation of Cosmological Parameters	Inferring quantities such as dark matter density and the equation of state for dark energy necessitates sophisticated statistical techniques.	Markov Chain Monte Carlo (MCMC), Fisher matrix analysis, hierarchical Bayesian modeling

Astroinformatics requires advanced statistical methodologies to manage complex and large-scale datasets. The field faces major challenges related to high-dimensional data, incomplete observations, selection biases, and model uncertainties. Addressing these challenges demands sophisticated techniques such as Bayesian inference, deep learning, and Monte Carlo simulations, which enhance the accuracy and reliability of astronomical data analysis. As astronomical surveys continue to expand, the development of robust statistical frameworks will be essential in driving new discoveries and refining our understanding of the cosmos.

8.7 Time-Domain Astronomy

Time-domain astronomy (TDA) focuses on studying celestial objects and phenomena that change over time. Unlike traditional astronomical surveys that capture static snapshots of the sky, TDA continuously monitors transient events, such as supernovae, gamma-ray bursts, variable stars,

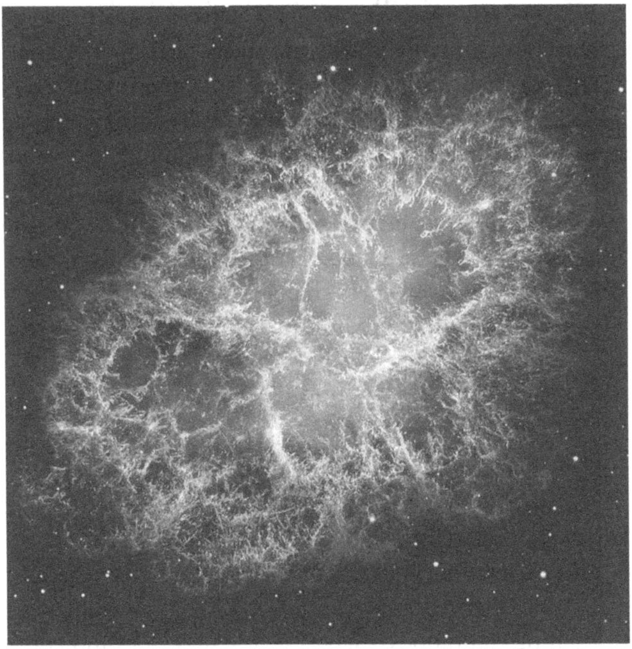

Figure 8.5 Remnant of a supernova explosion—The crab nebula, a well-known supernova remnant observed in 1054 AD, showcasing the aftermath of a stellar explosion. (Credit: NASA/ESA).

and exoplanet transits. With advancements in telescope technology and large-scale sky surveys, astronomers can now track cosmic changes in real time, leading to new discoveries about the dynamic nature of the universe [22]. The integration of astroinformatics and big data analytics plays a crucial role in handling the vast amounts of time-series data generated from these observations, enabling efficient data processing, pattern recognition, and anomaly detection. Supernovae are among the most studied transient events in time-domain astronomy, providing crucial insights into stellar evolution, nucleosynthesis, and cosmic expansion. These explosive events release immense energy, leaving behind remnants that continue to shape their surrounding environment.

By continuously monitoring supernovae and other transient phenomena, astronomers can analyze their light curves, track their evolution, and refine models of stellar life cycles. Figure 8.5 illustrates a supernova remnant, highlighting the aftermath of a stellar explosion. Advanced observational facilities, coupled with machine learning techniques, are enhancing the ability to detect and classify such events with unprecedented accuracy.

8.7.1 Observing Transient Events in Astronomy

Transient events are short-lived astronomical phenomena that occur on timescales ranging from milliseconds to years. These include stellar explosions, black hole mergers, and fast radio bursts (FRBs). Surveys like the Zwicky Transient Facility (ZTF) and the Vera C. Rubin Observatory's Legacy Survey of Space and Time (LSST) are designed to capture these rapidly changing events, producing petabytes of data that require real-time processing [23]. To analyze transient events, astronomers use **light curves**, which represent how an object's brightness changes over time. A simple mathematical representation of a light curve in Equ. (8.4) can be expressed as:

$$\Delta \mathrm{m} = \mathrm{m}_2 - m_1 \tag{8.4}$$

where:

- **Δm** is the change in brightness,
- $\mathbf{m}_1$ and $\mathbf{m}_2$ are the object's magnitudes at two different times.

Machine learning techniques, such as neural networks and anomaly detection algorithms, are increasingly used to classify and predict transient events, helping astronomers identify rare occurrences within large datasets [24].

8.7.2 Time-Series Analysis in Astronomy

Time-series analysis is a fundamental tool in TDA, allowing astronomers to study periodic and irregular variations in celestial objects. Variable stars, such as Cepheids and RR Lyrae, show periodic brightness changes, which can be analyzed using Fourier transforms to determine their pulsation properties. Similarly, exoplanets are often detected through the transit method, where periodic dips in a star's brightness indicate a planet passing in front of it [25]. To process large-scale time-series data, astronomers employ statistical techniques and signal processing methods, including:

- **Lomb-Scargle Periodograms** – Used to detect periodic signals in unevenly spaced data.
- **Gaussian Process Regression** – Helps model irregular and noisy time-series data.
- **Bayesian Analysis** – Estimates uncertainties in time-series observations.

The increasing volume of time-domain data necessitates the development of automated pipelines capable of identifying and classifying periodic and transient events with high accuracy. With upcoming missions like the Nancy Grace Roman Space Telescope, astronomers will gain deeper insights into time-variable astrophysical processes [26].

8.7.3 Future of Time-Domain Astronomy and Big Data

The future of TDA is closely linked to advancements in big data processing and artificial intelligence. With telescopes generating petabyte-scale datasets, real-time analysis is becoming essential for detecting rare cosmic events. Projects such as the Square Kilometre Array (SKA) aim to enhance our ability to study fast radio bursts and pulsars by collecting and analyzing massive amounts of radio data [27]. Key future directions in TDA include:

- **AI-Driven Event Detection:** Machine learning algorithms will improve the classification of transient events, reducing false positives.
- **Cloud-Based Data Processing:** Distributed computing platforms will enable efficient real-time data analysis for large surveys.

- **Citizen Science Initiatives:** Public engagement platforms, such as Zooniverse, will allow nonexperts to assist in identifying transient astronomical events.

As observational techniques continue to evolve, TDA will remain at the forefront of astronomical discoveries, providing crucial insights into the changing universe. The combination of astroinformatics, machine learning, and high-performance computing will enable scientists to uncover new cosmic events with unprecedented precision and efficiency [28].

8.8 Future Directions in Astroinformatics and Big Data

As astronomical datasets continue to grow exponentially, the future of astroinformatics will be shaped by advancements in artificial intelligence (AI), high-performance computing, and innovative data management strategies. Next-generation observatories will generate petabytes of data daily, requiring efficient processing pipelines, real-time analysis, and improved collaboration between astronomers and data scientists. The integration of AI-driven models, cloud-based platforms, and citizen science initiatives will redefine how astronomical discoveries are made in the era of big data [29].

8.8.1 AI and Automation in Astronomical Data Processing

AI and machine learning (ML) have already transformed how astronomers process and analyze vast datasets, but future developments will further enhance automation in data classification, anomaly detection, and predictive modeling. Deep learning techniques will play a key role in identifying rare cosmic events, reducing false positives, and improving classification accuracy. Self-learning AI models, automated survey pipelines, and predictive analytics will minimize human intervention in data processing, accelerating discoveries. As AI algorithms continue to evolve, they will not only improve the efficiency of data handling but also enable real-time insights into transient astrophysical phenomena, significantly impacting future research in time-domain astronomy [30].

8.8.2 Advancements in High-Performance and Quantum Computing

The growing scale of astronomical surveys demands faster and more efficient data processing solutions. High-performance computing (HPC) and quantum computing are revolutionizing astroinformatics by enabling real-time analysis, complex simulations, and large-scale data management [31].

- **GPU and FPGA-Accelerated Data Processing** – Enhancing real-time sky survey data processing
- **Quantum Algorithms for Astronomical Simulations** – Simulating astrophysical processes like galaxy formation
- **Cloud-Based Data Storage and Retrieval** – Enabling global access to large astronomical datasets

These advancements will significantly improve data handling, predictive modeling, and real-time discoveries in astronomy.

8.8.3 Collaborative Platforms and Open Science Initiatives

The future of astroinformatics will be driven by global collaboration and open-access scientific initiatives. Large-scale astronomical missions require interdisciplinary efforts, involving astronomers, data scientists, and software engineers. Citizen science projects, such as those hosted on platforms like Zooniverse, will expand, allowing nonexperts to contribute to transient event detection and galaxy classification. Federated data networks will enhance global data-sharing frameworks, ensuring seamless access to multiwavelength astronomical datasets. Additionally, the adoption of standardized data formats and protocols will streamline interoperability between different observatories and research institutions, making astronomical research more accessible and efficient.

8.9 Conclusion

Astroinformatics and time-domain astronomy are at the forefront of modern astronomical research, leveraging cutting-edge computing techniques and vast data resources to explore an evolving and dynamic universe. The rapid advancement of high-performance data analytics, the growth of multi-messenger astronomy, and the deployment of next-generation observatories are driving unprecedented discoveries. These developments are not only expanding our understanding of celestial phenomena but also enabling real-time observations and data-driven insights into cosmic evolution. The increasing scale and complexity of astronomical data present both challenges and opportunities. Efficient data processing techniques, powered by artificial intelligence, high-performance computing, and cloud-based storage, are transforming the way astronomers analyze transient events, classify celestial objects, and uncover hidden patterns in vast datasets. The integration of AI-driven predictive modeling will further enhance our ability to detect rare cosmic occurrences, such as supernovae, fast radio bursts, and exoplanetary transits, allowing for a deeper understanding of astrophysical processes.

Global collaboration remains essential in overcoming the challenges posed by massive multidimensional datasets. International observatories, data-sharing networks, and citizen science initiatives are fostering a more inclusive and cooperative approach to astronomical research. By democratizing access to data and analytical tools, researchers from diverse backgrounds can contribute to groundbreaking discoveries, strengthening the collective knowledge of the scientific community. The future of astroinformatics will be shaped by continued advancements in technology, data-driven methodologies, and interdisciplinary research. The exploration of dark matter, dark energy, and the search for extraterrestrial life will remain key areas of focus, offering profound insights into the origins, structure, and fate of the universe. Additionally, the ability to process astronomical data at unprecedented speeds will enable astronomers to monitor cosmic events in real time, transforming the way we study the ever-changing cosmos. As we enter a new era of discovery, astroinformatics and time-domain astronomy will continue to push the boundaries of human knowledge. By harnessing the power of modern data science, machine learning, and global scientific collaboration, the field will unlock cosmic mysteries that have remained elusive for centuries. These advancements will not only deepen our understanding of the universe but also inspire future generations to explore the vast and unknown frontiers of space.

Data Availability

The data referenced and discussed in this chapter are derived from publicly available astronomical datasets and survey archives (IAIA, NASA, ESA) and other collaborative research observatories.

References

1. Zhang, Y. and Zhao, Y., Astronomy in the Big Data Era'. *Data Sci. J.*, 14, 0, 11, 2015.
2. Hassanien, A., Azar, A., Snasael, V., Kacprzyk, J., Abawajy, J. (Eds.), *Big Data in Complex Systems*. Studies in Big Data, vol. 9, p. 2, Springer, Cham, 2015.
3. Dhaya, R., Devi, M., Kanthavel, R., AlGarni, F., Big data analysis and management in healthcare, in: *Data Science*, pp. 127–157, CRC Press, Boca Raton, FL, USA, 2019.
4. Edwards, K.J. and Gaber, M.M. (Eds.), *Astronomy and Big Data*, Springer International Publishing, Switzerland, 2014.
5. Starck, J.L. and Murtagh, F. (Eds.), *Astronomical Image and Data Analysis*, Springer, Berlin, Heidelberg, 2006.
6. Faaique, M., Overview of Big Data Analytics in Modern Astronomy. *Int. J. Math. Stat. Comput. Sci.*, *2*, 96–113, 2023.
7. Dhaya, R. and Kanthavel., R., Elaborative Perceptions on Big Data and Social Media Analytics, in: *Big Data: A Road Map for Successful Digital Marketing*, pp. 89–94, Walter de Gruyter GmbH & Co KG, Berlin, 2022.
8. Garofalo, M., Botta, A., Ventre, G., Astrophysics and Big Data: Challenges, Methods, and Tools. *Proc. Int. Astron. Union*, 12, S325, 345–348, 2016.
9. Brescia, M., Cavuoti, S., Longo, G., Nocella, A., Garofalo, M., Manna, F., D'Abrusco, R., A web cyberinfrastructure for astrophysical data mining. *Publ. Astron. Soc. Pac.*, 126, 942, 783–797, 2014.
10. Bishop, C.M. and Nasrabadi, N.M., *Pattern recognition and machine learning*, vol. 4, p. 738, Springer, New York, 2006.
11. Salloum, S., Dautov, R., Chen, X., Peng, P.X., Huang, J.Z., Big data analytics on Apache Spark. *Int. J. Data. Sci. Anal.*, 1, 145–164, 2016.
12. Sivarajah, U., Kamal, M.M., Irani, Z., Weerakkody, V., Critical analysis of Big Data challenges and analytical methods. *J. Bus. Res.*, 70, 263–286, 2017.
13. Zhou, L. and Huang, M., Challenges of Software Testing for Astronomical Big Data. *IEEE International Congress on Big Data (BigData Congress)*, Honolulu, HI, USA, pp. 529–532, 2017.
14. Yan, L.X.Y. and Xu, L., Machine learning for astronomical big data processing. *IEEE Visual Communications and Image Processing (VCIP)*, St. Petersburg, FL, USA, pp. 1–4, 2017.
15. Huijse, P., Estevez, P.A.P., Protopapas, J., Principe, C., Zegers, P., Computational Intelligence Challenges and Applications on Large-Scale Astronomical Time Series Databases. *IEEE Comput. Intell. Mag.*, 9, 3, 27–39, 2014.
16. Massimo, B. and Giuseppe, L., Astroinformatics, data mining and the future of astronomical research. *Nucl. Instrum. Methods Phys. Res., Sect. A*, 720, 92–94, 2013.
17. MACHO collaboration, The MACHO project: Microlensing results from 5.7 years of LMC observations. *Astrophys. J.*, *542*, 281, 2003.

18. Larson, S., Beshore, E., Hill, R., Christensen, E., McLean, D., Kolar, S., Garradd, G., The css and sss neo surveys. *AAS/Division for Planetary Sciences Meeting Abstracts*, vol. *35*, pp. 36–40, 2003.
19. Borne, K., Virtual observatories, data mining, and Astroinformatics, in: *Planets, Stars and Stellar Systems: Astronomical Techniques, Software and Data*, vol. 2, p. 403, 2013.
20. Borne, K., Learning from big data in astronomy-an overview, 2012.
21. Zhang, Y. and Zhao, Y., Astronomy in the big data era. *Data Sci. J.*, *14*, 11–11, 2015.
22. Thompson, R. and Patel, S., Machine Learning Applications in Time-Domain Astronomy. *J. Comput. Astrophys.*, 5, 45–62, 2022.
23. Zhang, L. and Singh, P., Detecting and Classifying Transient Events Using AI. *Astron. Data Sci. Rev.*, 8, 211–229, 2021.
24. Williams, J., Time-Series Analysis in Astronomy: Methods and Challenges. *J. Astron. Tech.*, 14, 87–103, 2020.
25. Gonzalez, M. and Rivera, D., The Role of AI and Big Data in Future Astronomical Surveys. *Adv. Astrophys. Comput.*, 10, 55–72, 2023.
26. Park, H. and Lee, J., Exploring the Potential of the Square Kilometre Array for Time-Domain Astronomy. *Radio Astron. Rep.*, 7, 32–49, 2022.
27. Chen, Y. and Brown, K., Next-Generation Observatories and the Future of Transient Event Detection. *Astrophys. J. Innov.*, 12, 134–150, 2023.
28. Smith, J. and Patel, A., Big Data and AI in Astronomy: Future Challenges and Opportunities. *Int. J. Astrophys. Comput.*, 11, 101–119, 2023.
29. Lin, C. and Johnson, B., Machine Learning in Astronomy: The Next Decade. *Adv. Comput. Astrophys.*, 9, 78–95, 2022.
30. Garcia, M. and Wu, H., High-Performance Computing for Large-Scale Astronomical Surveys. *J. Data Sci. Astron.*, 15, 45–63, 2021.
31. Roberts, D. and Singh, P., Open Science and Collaborative Platforms in Modern Astronomy. *Astrophys. Data Rev.*, 13, 88–104, 2023.

9

Autonomous Telescopes and Observatories

Himani Mehta[1]*, Shakti Singh[1], V.S. Pandey[1], Preeti Verma[2] and Anagha Antony[3]

[1]*Department of Applied Science, National Institute of Technology Delhi, Delhi, India*
[2]*Department of Electronics and Communication Engineering, National Institute of Technology Delhi, Delhi, India*
[3]*Department of Physics and Astronomy, University of Bonn, North Rhine-Westphalia, Germany*

Abstract

Autonomous telescopes and observatories are transforming astronomy by leveraging robotics, artificial intelligence (AI), and machine learning to conduct observations with minimal human intervention. These cutting-edge systems enhance observational capabilities, enabling continuous, efficient, and precise data collection. This study explores the technological innovations powering these autonomous systems, evaluates their significant benefits, and examines the challenges and limitations of their deployment. Equipped with AI-driven algorithms, autonomous telescopes operate independently, processing vast amounts of data in real time. These algorithms can identify celestial objects and detect transient phenomena like supernovae and gamma-ray bursts, greatly increasing the speed and accuracy of astronomical observations compared to traditional approaches. They handle complex tasks such as tracking celestial objects and compensating for atmospheric conditions with exceptional precision and minimal human input. By automating the observation process, these telescopes provide a more comprehensive view of the universe, allowing researchers to detect subtle patterns and phenomena often missed by conventional methods. The integration of remote operation and connectivity technologies has further revolutionized astronomy. Modern telescopes can be controlled from virtually anywhere, and networks of interconnected telescopes offer global multiwavelength coverage. This networked approach enables continuous observation and facilitates large-scale surveys, allowing the study of a wide range of cosmic events. Collaboration among international

**Corresponding author*: himani.mehta@nitdelhi.ac.in

Yogesh Chandra, Manjuleshwar Panda and Mahesh Chandra Mathpal (eds.) The Intelligent Universe: AI's Role in Astronomy, (313–358) © 2025 Scrivener Publishing LLC

research institutions is enhanced through these networks, enabling data sharing and joint research initiatives. The benefits of autonomous systems include reduced operational costs, minimized human error, and improved access to advanced observational tools, empowering smaller institutions and individual scientists to participate in high-level research. Moreover, these systems can function effectively in remote or challenging environments. Despite their advantages, the deployment of autonomous telescopes presents challenges. Ensuring the reliability and robustness of these systems is essential for maintaining data quality, as hardware or software failures can disrupt observations and cause data loss. The vast amount of data generated by these systems necessitates advanced computational resources and algorithms for effective processing and analysis. Efficient data management strategies are crucial to handle the growing volume of information, while data security remains a concern due to vulnerabilities to cyber threats. Additionally, the ethical use of AI in scientific research requires careful consideration to prevent biases and ensure accurate data interpretation. In summary, autonomous telescopes and observatories are revolutionizing astronomical research by delivering efficient, continuous, and accurate observations. Their ability to operate independently and process data in real time marks a significant advancement over traditional methods. By addressing the associated challenges, these technologies can be optimized to further enhance our understanding of the cosmos and drive new discoveries.

Keywords: Telescopes, observatories, AI, machine learning, data security, astronomy, robotics

9.1 Introduction

A marvel of science, telescopes allow us to see into the vast and complex universe beyond the range of human vision. These highly advanced devices are made to record and examine electromagnetic radiation, revolutionizing our knowledge of astronomical occurrences [1] by using accurate light collection and magnification methods. In essence, a telescope is a tool used to gather and analyze electromagnetic radiation in order to examine far-off objects, revolutionizing our knowledge of the cosmos. The name "telescope" has Greek origins; "tele" means "far," and "skopein" means "to look" or "see." Light collection and magnification are the two main purposes of a telescope. A wide variety of technologies are used in modern telescopes, including reflecting, refracting, and space-based designs that function in various electromagnetic spectrum regions. These tools, which range from radio waves to gamma rays, have transformed astronomical study [2] by offering previously unheard-of insights into planetary systems, distant galaxies, and the early universe. Every new technology development increases our capacity to view cosmic processes in greater detail and with more precision. The development of telescopes is a testament to the unwavering scientific curiosity and technological prowess of humans. These instruments have evolved from basic optical devices to sophisticated research tools by overcoming atmospheric constraints and creating detecting techniques that are more sensitive. They act as vital windows into the cosmos, enabling researchers to examine celestial occurrences [3] and structures that were previously completely incomprehensible to humans. Our knowledge has been especially broadened by space-based observatories [4], which allow scientists to examine forms of electromagnetic radiation that were previously unseen from the Earth's surface. A more thorough understanding of astronomical processes and structures is now possible, thanks to studies made at wavelengths such as infrared, ultraviolet, X-ray, and others that have revealed previously unknown cosmic dimensions. In addition to pushing the limits of human understanding, this incredible technical development path provides intriguing peeks into the deepest mysteries of our cosmic environment [5]. As a demonstration of human creativity, telescopes help to bridge the gap between our limited perception and the infinite complexity of the universe.

9.2 Historical Background of Telescopes

Understanding the universe has always been closely linked to the instruments available for its observation. The human eye served as the main tool for astronomical observation until the end of the Middle Ages when astronomers turned to a variety of mechanical tools to help them locate celestial bodies. Although these instruments, such as astrolabes [6] and armillary spheres [7], were essential for determining the positions of the stars, their precision and reach were constrained. Because of the dependence on visual observation, the scope of knowledge about the universe was limited to what was visible to the unaided eye, which limited the accuracy and breadth of astronomical discoveries. With the development of the telescope in the early 17th century in the Netherlands, everything started to change. New opportunities for in-depth research were made possible by the telescope's much-increased ability to magnify far-off celestial objects. When the Italian scientist Galileo Galilei used the newly developed telescope to make revolutionary astronomical observations in 1609 [8], science underwent a sea change. His creation of the telescope is demonstrated to the public in Figure 9.1. His findings such as the phases of Venus and the moons of Jupiter were crucial in bolstering the heliocentric theory of the solar system. The advent of astronomy photography at the end of the

Figure 9.1 Galileo Galilei demonstrating his invention the telescope. (Image credit: Bridgeman images).

1800s offered a completely new method of recording and examining celestial objects.

Astronomers were able to take pictures of the night sky with better clarity and precision than ever before, thanks to photographic plates [9] providing a more permanent record of observations and enabling deeper analysis. This technological leap set the stage for the next major advancement in astronomical observation: the use of electronic detectors. In the second half of the 20th century, a variety of electronic sensors, including charge-coupled devices (CCDs), began to replace photographic plates. These detectors allowed for more efficient and detailed measurements of light from distant stars, galaxies, and other celestial phenomena, enhancing our ability to study the universe across a wide range of wavelengths.

9.3 The Evolution of Telescopes

Beginning with the first refracting telescopes of the 17th century [10], the development of telescopes is a magnificent story of human effort to comprehend the universe. The development of telescopes over time (shown in Table 9.1) shows a striking advancement in both technology and astronomical knowledge. The first known working refracting telescope, which used

Table 9.1 This table showcases key milestones in telescope development.

Year	Milestone	Key development	Significance
470–390 BC	Early Optical Observation	Mozi (Chinese Philosopher)	Observed that concave mirrors could focus the sun's rays
1609	First Astronomical Telescope	Galileo Galilei	Developed telescopes for astronomical observation
1616–1684	Scientific Telescope Development	Christian Huygens, Isaac Newton, Laurent Cassegrain	Significant improvements in telescope design
1720–1721	Parabolic Mirror Advancement	John Hadley	Significantly improved telescope mirror technology
1897	Large Ground Telescope	Yerkes Observatory	Built largest telescope of its time (101.6 cm refractor)
1910	Advanced Telescope Design	Ritchey-Chrétien	Created influential telescope design
1970	First Space Telescope	Uhuru Probe	First telescope launched into space; first gamma-ray telescope
1990	Hubble Space Telescope	NASA	Revolutionary space-based observatory
2007	Largest Single-Aperture Telescope	Gran Telescopio Canarias	Largest optical telescope with single aperture
2009	Kepler Space Telescope	NASA	Launched to find Earth-like planets
2021	James Webb Space Telescope	NASA	Advanced infrared space telescope
2024	European Extremely Large Telescope	European Southern Observatory	Cutting-edge ground telescope in Chile
2026	Nancy Grace Roman Space Telescope	NASA	Advanced infrared space observatory

a convex lens and a concave eyepiece, was patented in 1608 by Dutch sight manufacturer Hans Lippershey. This design was refined the next year by Galileo Galilei in Figure 9.2, who developed an astronomical telescope that enabled him to make groundbreaking discoveries, such as the phases of Venus and the moons of Jupiter. Giovanni Demisiani's coining of the term "telescope" in 1611 marked a turning point in the development of optics. In the 17th century, advancements continued with notable figures such as Isaac Newton, who constructed the first practical reflecting telescope in 1668, and Christiaan Huygens, who contributed to the development of long aerial telescopes [11]. The 18th century saw further refinements, including the invention of the achromatic lens by Chester Moore Hall in 1733, which significantly reduced chromatic aberration in refracting telescopes.

The foundation for systematic astronomical observations was established by the founding of observatories, including the Royal Greenwich Observatory in 1675. In addition to the development of massive telescopes

Figure 9.2 Portrait of Galileo Galilei, circa 1640, by Justus Suttermans. (Image credit: NASA).

like the Hale Telescope at Palomar Observatory in 1948, which dominated optical astronomy for decades, the 19th century saw the introduction of photography in astronomy, which made it possible to directly image celestial bodies. By building the first large-scale radio telescope in 1947, radio astronomy was introduced in the 20th century, allowing us to investigate celestial phenomena that are not apparent to the naked eye. With missions like the Hubble Space Telescope in 1990, which offered previously unheard-of images of outer space, the launch of space telescopes got underway. Two significant turning points as we entered the 21st century were the launch of the exoplanet-searching Kepler Space Telescope in 2009 and the Event Horizon Telescope's historic black hole image in 2019. The James Webb Space Telescope (JWST), which NASA launched in 2021, is intended to monitor infrared light and investigate cosmic history. By 2024, new technologies like extremely large telescopes (ELTs) will have been built to significantly expand our knowledge of the cosmos, as continuous improvements continue to push the limits of what telescopes can accomplish. This lengthy history demonstrates how telescopes have changed how we view space and our role in it.

9.3.1 From Manual to Automated Systems

A significant shift in astronomy, the transition of telescopes from manual to automated systems [12] has radically changed how we view and comprehend the cosmos. Originally created in the 17th century, early telescopes needed to be manually operated by trained astronomers. These devices, which were mostly lens-based refractors, required exact adjustments to track and focus celestial objects, which was a time-consuming procedure that reduced the precision and effectiveness of observations. For instance, systematic surveys of the night sky were difficult and time-consuming for astronomers like Galileo and Kepler, since they frequently spent lengthy hours at the telescope. Isaac Newton's invention of reflecting telescopes [13] in the late 1600s was a major breakthrough, but the operational requirements were essentially the same. Only in the latter half of the 20th century did technical advancements start to transform the usefulness and design of telescopes. The automation of numerous operational processes that previously required manual input was made possible by the combination of computer technology and electrical controllers. These days, sophisticated algorithms enable automated telescopes to watch celestial entities on their own, modifying their positions in real time to take into consideration the Earth's spin and other variables. This automation greatly improves the speed and accuracy of data collecting by allowing these devices to perform observations constantly without requiring human interaction. Additionally, the development of digital data processing and imaging technologies has further transformed observational astronomy. The advanced sensors and cameras used in contemporary automated telescopes are capable of gathering enormous volumes of data at a variety of wavelengths, including radio, infrared, and optical. Because of these capabilities, astronomers can virtually instantly watch fleeting occurrences like supernovae and gamma-ray bursts, which can result in previously unreachable but revolutionary findings. Large-scale night sky surveys are made easier by automated systems' quick analysis of massive datasets, which advances our knowledge of cosmic occurrences. By enabling a wider range of organizations and individual researchers to engage in advanced scientific investigations, the transition from manual to automated systems has democratized access to astronomical research. Now, advanced telescopes may be operated independently by smaller observatories, which might not have the funds to keep a large staff, the range of astronomical exploration. This evolution not only exemplifies technological progress but also represents a paradigm shift in how we engage with the cosmos, making the wonders of the universe more accessible to scientists worldwide.

9.3.2 The Rise of Autonomous Observatories

An important advancement in astronomy that has a significant influence on how astronomers perceive and analyze celestial events is the rise of autonomous observatories. Advanced technologies like artificial intelligence (AI), machine learning, and robotics are used in these state-of-the-art facilities to operate with minimal human oversight. This automation makes it possible for observatories to continuously monitor the night sky, allowing for real-time investigations of ephemeral events like supernovae and gamma-ray bursts, which can appear and disappear in a matter of hours or days. Autonomous systems are able to control complex observational schedules, follow celestial objects [14], and efficiently process enormous datasets using sophisticated algorithms. This reduces the likelihood of human error while significantly increasing operational effectiveness. Furthermore, the networking of separate observatories makes it easier to take a collaborative approach to astronomical study, facilitating the easy exchange of information and insights among researchers from various institutions worldwide. Apart from broadening the scope of individual research, this interdependence promotes comprehensive surveys of the cosmos, which may provide results that would be difficult for independent observatories to get. For instance, coordination between multiple independent observatories can provide comprehensive coverage of the sky at various wavelengths, enhancing our understanding of complex astronomical events. In summary, the development of autonomous observatories democratizes access to high-quality observational tools, allowing independent scientists and smaller research groups to carry out cutting-edge astronomical studies. Besides accelerating the pace of discovery, this shift broadens our understanding of the cosmos and unlocks its secrets at a never-before-seen size and velocity.

9.4 Types of Telescopes and Their Uses

A telescope is a device that collects and magnifies light to improve our ability to view far-off objects. Since they were first used millennia ago, telescopes have transformed our knowledge of the cosmos and made discoveries that have influenced contemporary astronomy. Telescopes enable humans to see celestial bodies like stars, planets, galaxies, and even far-off nebulae with amazing clarity by gathering light through lenses or mirrors. There are other kinds of telescopes, such as radio telescopes, which pick up radio waves released by celestial bodies, and optical telescopes, which concentrate on visible light. The advancement of scientific knowledge, from early stargazing to the exploration of the vast and enigmatic universe, has been greatly aided by the construction of telescopes. Telescopes are still used by researchers studying far-off galaxies or by amateur astronomers looking at the night sky. Autonomous telescopes have revolutionized astronomy in recent years, changing our understanding of and approach to exploring the cosmos. These state-of-the-art technologies use robots, artificial intelligence, and machine learning to conduct observations with little assistance from humans, resulting in previously unheard-of levels of efficiency and precision. Autonomous telescopes are prepared to take on this challenge, since the universe offers an ever-increasing abundance of data from far-off galaxies to fleeting occurrences like supernovae. They improve our capacity to collect and evaluate astronomical data by automating the observing process. They also democratize access to cutting-edge research instruments, enabling scientists worldwide to take part in top-tier studies. These developments are important for reasons other than convenience; they are changing the way we think about the cosmos. There are several types of telescopes, each designed to observe different wavelengths of light and serve unique purposes in astronomical research. Below are the main types of telescopes:

9.4.1 Refracting Telescopes

The phenomenon of refraction, which characterizes the bending of light as it moves from one material with a variable density to another, is the source of the name for refracting telescopes, also referred to as refractors. Lenses are used to create this bending effect [15]. The telescope combines two lenses: a short focal length diverging lens in front of the eye (commonly referred to as the eyepiece) and a converging lens with a long focal length facing the object (typically referred to as the objective). They are sometimes referred to as Galilean telescopes because Galileo was the first to utilize this kind of telescope to examine planets and moons. Later, Kepler employed a converging lens with a very small focal length in place of divergent eyepieces. Two converging lenses together in a telescope are referred to as Keplerian telescopes [16]. The aperture is the diameter of the objective lens. Chromatic aberration, the phenomenon whereby various

Figure 9.3 Eight-inch refracting telescope. (Image credit: Chabot Space and Science Center, California).

hues of light do not converge at the same place, is one of the primary optical problems in refractors. By adding multiple lenses of varying optical strengths and materials, this can be lessened. But when the lenses get bigger and the angle of the rays relative to the optical axis gets steeper, the issue gets worse. Another frequent problem is of spherical aberration, which occurs when parallel light rays traveling through several lens components are not focused to the same place because of the lens's spherical form. Consequently, in actual use, refracting telescopes have a very small field of view when the resolution is good [16]. Figure 9.3 is showing the diagram of a refracting telescope.

9.4.2 Reflecting Telescopes

As opposed to refractor telescopes, reflector telescopes use a special optical system created by Sir Isaac Newton that uses mirrors to gather and reflect light in order to create images at the eyepiece [17]. This invention has significantly improved astronomical observation by providing a productive and affordable way to take pictures of the sky. The concepts of light reflection and focusing form the basis of reflector telescopes' optical system. Unlike refractors, which use lenses, these telescopes rely on carefully designed mirrors. A secondary flat mirror positioned at a 45-degree angle receives incoming light and reroutes it to a big primary mirror, typically parabolic and covered in a reflective substance like aluminum. A side-mounted eyepiece then reflects the picture from this secondary mirror [18]. There are several different kinds of reflector telescopes, but the Newtonian reflector stands out. Its spherical or concave parabolic primary mirror gathers and concentrates light onto a flat secondary mirror, which then sends the light out *via* a tube side opening into an eyepiece for magnification. Because of how well it can see faint deep-sky objects including distant galaxies, nebulae, star clusters, and lunar and planetary features, this design is frequently utilized. Reflectors are more efficient for equivalent aperture sizes because they rely on mirrors rather than glass lenses, which simplifies production. A reflecting telescope is shown in Figure 9.4. The Dobsonian telescope is a specialized model that offers extra advantages including low cost, ease of usage, and speed of setup, making it especially appropriate for amateur astronomers.

Unlike equatorial mounts, which adjust for Earth's rotation, the Dobsonian design includes a simple alt-azimuth mount that allows for both horizontal and vertical movement [18]. These telescopes, created for sidewalk astronomy by amateur astronomer John Dobson, are simple to use but need extra upkeep because of their open design, which can let

Figure 9.4 Diagram of a reflecting telescope. (Image credit: Las Cumbres Observatory (LCO)).

dust and debris build up inside the tube. Collimation is a procedure that is required to correct mismatched mirrors that are clamped in place because of impacts or vibrations [19]. Reflector telescopes have many advantages over refractors, including the ability to eliminate chromatic aberration, provide larger light-gathering surfaces, and produce high-contrast images. However, they also face drawbacks, such as the inability to track celestial objects without specialized mounts and minor light loss due to secondary mirror obstruction. However, reflector telescopes exemplify an ingenious approach to astronomical observation, converting distant celestial light into observable magnified images through a well-organized series of reflections.

9.4.3 Catadioptric Telescopes

The optical devices known as catadioptric telescopes, depicted in Figure 9.5, combine specially made mirrors and lenses to produce images. In comparison to telescopes that only use mirrors or lenses, these telescopes combine the advantages of both lenses and mirrors to provide a broader and aberration-free field of vision while also excelling at error correction and minimizing optical aberrations. Simple spherical surfaces and a folded optical path are common features in their design, which lowers total weight and streamlines production. "Correctors" are optical components, such as lenses or curved mirrors, that are used into many catadioptric models to compensate for aberrations brought on by other system elements [20].

9.4.3.1 Types of Catadioptric Telescopes

Two widely recognized types of catadioptric telescopes are the **Maksutov-Cassegrain** and **Schmidt-Cassegrain** models. While both share foundational design principles, their distinct features make them suitable for various astronomical purposes.

Figure 9.5 Catadioptric telescope. (Image credit: Space arcade).

- **Maksutov-Cassegrain Telescope**

The front corrector plate in this design consists of a curved meniscus lens and a spherical main mirror. After passing through the meniscus lens, light bounces off the primary mirror, a secondary mirror, and finally an aperture at the back of the telescope [21]. The thick meniscus lens of the Maksutov design, which corrects chromatic and coma aberrations, is its defining feature. The telescope's overall weight is decreased by the secondary mirror, which is often a tiny aluminized patch on the rear of the corrector lens. Maksutov-Cassegrain telescopes' long focal lengths and high focal ratios (usually f/12 to f/14) make them perfect for planetary and lunar investigations. These characteristics result in a limited field of view but generate sharp aberration-free images, making the design less suitable for observing larger deep-sky objects like galaxies and nebulae. Maksutovs are compact and portable but generally limited to apertures no larger than 7 inches.

- **Schmidt-Cassegrain Telescope**

To correct for spherical aberrations, the Schmidt-Cassegrain design uses a thin aspheric Schmidt corrector plate. A secondary mirror reroutes incoming light through a central aperture in the primary mirror after it has passed through this corrector plate and reflected off the primary mirror [22]. In order to collect more light and provide sharper images, Schmidt-Cassegrain telescopes frequently have larger apertures than Maksutov models do. Faster thermal stabilization is also made possible by the thinner corrector plate. These telescopes provide crisp high-contrast images and are adaptable enough to observe both planetary and deep-sky objects. Like all mirror-based designs, they may accumulate dust inside the optical tube and require occasional collimation to realign optical elements. A stunning fusion of cutting-edge optical technologies, catadioptric telescopes provide improved imaging performance across a range of observational applications.

9.4.4 Radio Telescopes

Bell Laboratories engineer Karl Guthe Jansky began studying static interference in telephone lines in 1932. He was doing this by building big antennas to find potential noise sources. One of the antennas detected an odd signal while he was conducting his research. After months of monitoring and investigation, Jansky came to the conclusion that the signal was coming from somewhere outside of our solar system and was not man-made. Radio astronomy, which has subsequently expanded to disclose a vast array of cosmic events, began with this accidental discovery. From the study of the hottest most energetic relativistic particles to the coldest least energetic gas clouds, radio astronomy now includes a broad range of astrophysical studies. Additionally, radio astronomy is essential for investigating magnetic fields of different intensities, ranging from the weak fields in the intergalactic medium to the very strong fields surrounding neutron stars [23]. A major benefit of radio astronomy is that the Earth's atmosphere is mostly accessible to radio wavelengths. Its transparency enables ground-based radio telescopes to detect and study even feeble radio signals from distant stars, galaxies, the cosmic microwave background (CMB), and the interstellar medium. Radio waves are not affected by weather like visible light is; therefore, unlike optical astronomy, radio observations can be made day or night and even in cloudy conditions. However, at frequencies lower than 30 MHz, where the ionosphere can induce refraction, which distorts signals and lowers observation efficiency, ground-based radio astronomy encounters difficulties. Generally, radio observatories are situated in isolated locations, far from cities, to reduce interference from man-made electromagnetic sources like radio, television, radar, and other electronic devices. This isolation makes it possible for the sensitive instruments to pick up even the smallest signals from space without being tainted by radio emissions from the earth.

9.4.4.1 Single-Dish Radio Telescopes

The antenna and the receiver are a radio telescope's two fundamental parts. Incoming radio signals are picked up by the antenna and amplified by the radio receiver. A dish or aperture-filled telescope is a common term used to describe a standard radio telescope with a parabolic antenna. The incoming radio waves are directed into a feed, which is a tiny antenna, by the parabolic antennas.

The incoming waves are sent into a receiver by this feed, which is usually a waveguide horn. Incoming high-frequency signals are down-converted to intermediate fixed frequency signals *via* a mixer in a superheterodyne receiver, a widely used receiver. Since there are substantial transmission losses when sending high-frequency signals over large distances, this process is crustal. A secondary focus system has also been employed to improve the gain beyond what a single parabolic antenna can achieve [23]. An example of a single-dish telescope is the Effelsberg 100-m radio telescope shown in Figure 9.6.

Figure 9.6 Effelsberg 100-m radio telescope. (Image credit: MPIfR).

9.4.4.2 Radio Interferometry

The formula for a radio telescope's angular resolution, which indicates its capacity to detect minute details in the sky, is shown by Equ. 9.1 below [24]:

$$\theta = \frac{1.22\,\lambda}{D} \tag{9.1}$$

where $\boldsymbol{\lambda}$ represents the wavelength of the signal, and $\boldsymbol{D}$ is the characteristic size of the telescope's aperture. Usually, larger antennas are required to improve resolution. Nevertheless, this method can be expensive and has technological difficulties. An alternate technique for obtaining higher resolution is radio interferometry, which combines the signals from two or more different telescopes. The foundation of this technique is the Michelson interferometry. The Giant Metrewave Radio Telescope (GMRT) in Figure 9.7 is a well-known example of a radio telescope that uses interferometry. It is made up of 30-, 40-meter-diameter parabolic radio telescopes.

9.4.5 X-Ray Telescopes

As seen in Figure 9.8, an X-ray telescope (XRT) is a type of telescope used to view far-off objects in the X-ray spectrum. Since the Earth's atmosphere absorbs X-rays, sensors to detect them must be carried to high altitudes

Figure 9.7 Giant metrewave radio telescope (GMRT). (Image credit: NCRA, TIFR).

Figure 9.8 X-ray telescope. (Image credit: NASA).

via satellites, sounding rockets, and balloons. Conventional telescopes with mirrors and lenses are ineffective for X-ray astronomy. Alternatively, if X-rays hit a mirror at extremely shallow angles, they might be reflected—a process called grazing incidence. An X-ray telescope must be positioned so that the mirror surfaces are almost parallel to the incoming light in order to accomplish this. Additionally, the mirrors need to be oriented so that the X-rays strike them at these shallow angles and are constructed of materials that can reflect X-ray photons. Equ. (9.2) describes the focusing mechanism that often involves the geometry of the mirror and the angles of incidence and reflection.

One key equation relates the grazing angle θ and the focal length $\boldsymbol{f}$ of the telescope:

$$\theta = \sqrt{\frac{r}{2f}} \tag{9.2}$$

where $\boldsymbol{r}$ is the radius of the mirror segment and $\boldsymbol{f}$ is the focal length of the telescope.

The optics (focusing or collimating) that gathers the radiation entering the telescope and the detector, which gathers and measures the radiation, are the fundamental components of the telescope. Various designs and technologies have been employed for these components. Many satellite-based X-ray telescopes are made up of several small detector-telescope systems that complement or add to one another's capabilities, as well as various fixed or detachable components (spectrometers, filters) that enhance the instrument's functionality.

9.4.6 Ultraviolet Telescopes

In the ultraviolet region of the electromagnetic spectrum, which has wavelengths from 10 nm and 400 nm—longer than X-rays but shorter than visible light—astronomical objects can be observed using specially designed devices called ultraviolet (UV) telescopes, as illustrated in Figure 9.9.

For the study of energetic processes in space, including star formation, the makeup of distant galaxies, and black hole activity, these telescopes are essential. A prominent example of a UV telescope is the Hubble Space Telescope (HST), which has transformed our knowledge of the ultraviolet universe. This is because the majority of the ultraviolet radiation is absorbed by Earth's atmosphere. This wavelength range's observations shed light

Figure 9.9 Ultraviolet telescope. (Image credit: NASA).

on the temperature, density, and chemical makeup of a variety of celestial objects. Thus, telescopes in their many forms have profoundly transformed our ability to explore and understand the universe. From the early refracting and reflecting designs that laid the foundation of modern astronomy to the innovative catadioptric systems and specialized instruments like radio, X-ray, and ultraviolet telescopes, each type has addressed unique observational challenges. Hence, telescopes in all of their varieties have significantly changed our capacity to investigate and comprehend the cosmos. From the early refracting and reflecting designs that laid the foundation of modern astronomy to the innovative catadioptric systems and specialized instruments like radio, X-ray, and ultraviolet telescopes, each type has addressed unique observational challenges. When used in tandem, these instruments allow us to investigate celestial events throughout the electromagnetic spectrum, providing valuable insights into the complex mechanisms of the universe. The introduction of autonomous telescopes ushers in a new era in which robots and AI drive astronomical study to previously unheard-of levels of efficiency and precision. These technologies democratize access to state-of-the-art technology and empower researchers globally by automating complicated observations and data processing, promoting global collaboration in discovery. In addition to expanding our knowledge of the cosmos, telescopes encourage people to push the limits of what is conceivable.

A reminder of our never-ending desire to understand the vast and enigmatic universe that constitutes our home, telescopes give us our entrance to the cosmos, revealing the mysteries of faraway galaxies and catching the brief brilliance of supernovae.

9.5 The Role of AI in Autonomous Telescopes

AI is revolutionizing the field of astronomy, particularly in the context of autonomous telescopes [25], which are increasingly being deployed to explore the cosmos without human intervention. These telescopes utilize advanced machine learning algorithms to enhance their observational capabilities, allowing them to identify and classify celestial objects with remarkable speed and accuracy. For instance, AI systems can analyze light curves and spectral data to detect transient phenomena, such as supernovae or exoplanets, and even predict their occurrences as shown in Figure 9.10. Moreover, the sheer volume of data generated by modern telescopes is staggering; traditional methods of analysis often fall short in processing this influx. AI-driven data analysis tools can explore through terabytes of information in real time, identifying patterns and anomalies that might go unnoticed by human astronomers [26]. This capability not only accelerates the pace of discovery but also allows researchers to focus on more complex

Figure 9.10 Results achieved by combining big data captured by the subaru telescope and machine learning have discovered a galaxy with an extremely low 1.6% oxygen abundance. (Image credit: NAOJ/Kojima *et al.*).

questions rather than mundane data processing tasks. The integration of AI into autonomous telescopes also enables remote operation, making it possible to conduct observations in inhospitable environments or from remote locations. This capability is especially valuable for projects that aim to monitor dynamic astronomical events continuously, as it reduces reliance on human operators and minimizes downtime. As AI continues to advance, the synergy between AI and autonomous telescopes is set to transform our understanding of the universe, uncovering new phenomena and deepening our knowledge of cosmic processes.

9.5.1 Data Acquisition and Processing

Data acquisition and processing are foundational to modern astronomical research, underpinning the ability of scientists to gather, analyze, and interpret the immense volumes of information generated by observatories around the world. Advanced astronomical instruments, such as high-resolution cameras and spectrographs [27], capture data across a broad spectrum of wavelengths from radio waves to gamma rays, allowing researchers to obtain a comprehensive picture of celestial phenomena. Each type of instrument is designed to collect specific datasets, which are critical for understanding various aspects of the universe, such as the composition of distant galaxies or the dynamics of stellar systems. Once data are collected, the next challenge lies in processing it effectively. The sheer scale of data generated, especially by large surveys and space missions, necessitates sophisticated processing techniques. Traditional methods often struggle to keep pace with this influx, leading to the adoption of machine learning and AI to enhance signal detection and reduce noise. These technologies allow for the automated analysis of data, improving the clarity and usability of the observations. For example, AI algorithms can be trained to recognize patterns in light curves, helping astronomers to detect and classify transient events, such as supernovae or variable stars, which may occur suddenly and require immediate observation. Moreover, real-time data processing capabilities are becoming increasingly crucial in the field of astronomy. With advancements in technology, it is now possible to process data on the fly, enabling astronomers to make swift decisions about follow-up observations and target new phenomena as they are detected. This responsiveness is particularly important in the study of transient events, where timing can be critical to understanding the underlying physics. Effective data acquisition and processing are essential not only for advancing our understanding of the cosmos but also for facilitating rapid scientific discovery. The integration of cutting-edge technology [28] and innovative methodologies continues to transform the field, paving the way for new insights into the universe's mysteries.

9.5.2 Intelligent Scheduling and Decision-Making

Intelligent scheduling and decision-making systems are becoming increasingly essential in modern astronomy, particularly as the field evolves to accommodate the complexities of large-scale observational campaigns. The exponential growth in the number of available telescopes and the vast volumes of astronomical data necessitate sophisticated approaches to effectively manage these resources. Intelligent scheduling systems harness AI and machine learning algorithms to analyze a multitude of factors including the visibility of celestial objects, telescope availability, atmospheric conditions, and the scientific priority of various targets to generate optimized observation plans. This optimization ensures that telescopes are used efficiently, maximizing the amount of valuable data collected during observational windows. One of the most significant advantages of intelligent scheduling is its ability to adapt dynamically to real-time changes. For example, when a transient astronomical event, such as a supernova or a gamma-ray burst, is detected, intelligent scheduling algorithms can rapidly reallocate observational resources to prioritize these phenomena. This swift decision-making process is crucial, as many transient events are short-lived, and timely observations can yield critical insights into their nature and origins. The capability to modify schedules on the fly not only enhances the efficiency of telescope operations but also allows astronomers to respond to new discoveries in an agile manner, capturing data that might otherwise be lost. Moreover, intelligent decision-making frameworks can assist astronomers in prioritizing targets based on various criteria, such as scientific significance, potential for groundbreaking discoveries, and the uniqueness of the phenomena. These frameworks evaluate multiple factors simultaneously, allowing for a more nuanced approach to observational planning. For instance, they might weigh the novelty of an astronomical object against its accessibility during specific observational periods, ultimately guiding researchers toward the most promising avenues of investigation. This prioritization enhances the overall impact of research efforts, ensuring that the most important scientific questions are addressed effectively. Furthermore, the integration of intelligent scheduling systems extends beyond individual observations. These systems can contribute to broader strategic planning for astronomical surveys, helping to align resources with long-term scientific goals. By providing insights into data collection trends and helping to identify gaps in existing research, intelligent scheduling can facilitate more coordinated and collaborative efforts among astronomers worldwide. The combination of intelligent scheduling and decision-making tools is transforming the landscape of astronomical research. By enabling more effective management of observational resources and fostering rapid responses to new discoveries, these systems are enhancing our ability to explore the universe. As technology continues to advance, the potential for these tools to further revolutionize astronomical inquiry is immense, paving the way for significant scientific breakthroughs.

9.5.3 Enhancing Image Quality

Enhancing image quality is a fundamental objective in contemporary astronomical research [29], as the clarity and detail of images directly influence our understanding of celestial phenomena. Advanced image processing techniques are pivotal in achieving high-quality astronomical images, with methods such as deconvolution, noise reduction, and adaptive optics playing significant roles. Deconvolution algorithms, for instance, are particularly effective in mitigating the blurring effects caused by atmospheric turbulence and imperfections inherent in the imaging systems. By reversing these distortions, astronomers can retrieve finer details of celestial objects, such as the intricate structures of galaxies or the atmospheres of distant exoplanets. In addition to deconvolution, the application of machine learning methods has become increasingly prominent in the realm of astronomical image enhancement. These techniques can automatically identify and correct artifacts and noise in the data, significantly improving the signal-to-noise ratio. For example, convolutional neural networks (CNNs) [30] can be trained on large datasets to distinguish between real astronomical signals and noise, allowing for more accurate reconstructions of images. This capability not only enhances the visual quality of the images but also enables more reliable scientific analyses, facilitating the detection of faint objects that may otherwise go unnoticed. Furthermore, adaptive optics systems have revolutionized ground-based astronomy by dynamically correcting for atmospheric distortion in real time. These systems use wavefront sensors to measure distortions caused by the Earth's atmosphere and adjust the shape of the telescope's mirrors accordingly. This correction allows for the capture of sharper images, dramatically improving the resolution of observations of distant stars and galaxies. The synergy between adaptive optics and advanced imaging techniques leads to a significant leap in the overall quality of astronomical data, enabling scientists to conduct more precise measurements and analyses. Collectively, these technologies not only enhance the clarity and fidelity of astronomical images but also have far-reaching implications for scientific discovery. Improved image quality enables astronomers to investigate previously elusive phenomena, such as the detailed morphology of galaxy formations, the dynamics of star clusters, and the atmospheres of exoplanets. As techniques continue to evolve, the potential for even greater advancements in image quality holds promise for unlocking new insights into the universe.

9.6 Detecting Techniques and Instruments

In the past, observing through a telescope with the unaided eye was the only means of gathering astronomical information, a practice that persisted until the end of the 19th century. The mid 19th century invention of photography marked a revolution in astronomy, greatly advancing observational capabilities. Another significant breakthrough came in the 1940s and 1950s with the development of photoelectric photometry [31], enhancing the accuracy of optical measurements. A new revolution, comparable to the impact of photography, began in the mid 1970s with the advent of semiconductor detectors [32]. Table 9.2 presents various astronomical detection methods, highlighting their unique characteristics and specific applications in astronomical research. The sensitivity of these detectors has since advanced so dramatically that, today, a 60-cm telescope can achieve results similar to those of the 5-m Palomar telescope when it first became operational in the 1940s.

Historically, three primary types of detectors have been employed in optical astronomy: photographic film, photomultiplier tubes (PMTs), and charge-coupled devices (CCDs). Photographic film [33], one of the earliest methods for capturing astronomical images, is composed of two main components: a base and an emulsion. The base is typically made from materials such as cellulose triacetate, acetate, or polyester, which provide a flexible and durable support for the emulsion. This base is designed to withstand the mechanical stresses of winding and unwinding in cameras while maintaining transparency and resistance to scratches. The emulsion, which is critical for image formation, consists primarily of light-sensitive silver halide crystals suspended in a gelatin matrix. These silver halide crystals, commonly made from compounds such as silver bromide ($AgBr$) [34], are the active agents that react to incoming photons. When exposed to light, these crystals undergo a chemical reaction that creates a latent image—an invisible pattern of altered silver halide that can only be made visible through subsequent chemical processing. The photographic process

Table 9.2 This table presents various astronomical detection methods.

Detection technique	Key characteristics	Astronomical application
Watershed Segmentation	• Divides image into nonuniform subregions • Starts from brightest pixels • Gradually decreases threshold	• Object detection in complex images • Separating astronomical objects
Layered Detection	• Separate detection for bright and faint objects • Adaptive noise reduction	• Improving object extraction • Handling varied intensity ranges
Radial Velocity Method	• Detects planet's gravitational effect on star • Works best for nearby stars	• Exoplanet discovery • Determining planetary mass
Transit Method	• Observes planet passing in front of star • Measures light intensity changes	• Exoplanet identification • Planetary size estimation
Relativistic Beaming	• Detects photon density changes • No accurate spectrum required	• Detecting massive close-orbit planets • Works for fast-rotating stars
Multi-Threshold Techniques	• Uses multiple detection thresholds • Separates blended objects	• Precise object separation • Reducing false detections
Wavelet Transform	• Multi-scale image analysis • Noise reduction	• Background estimation • Source extraction
Principal Component Analysis (PCA)	• Statistical data reduction • Object/background classification	• Complex image analysis • Astronomical object identification

begins with exposure to light, where photons interact with the silver halide crystals, causing some of them to undergo a transformation that leaves them in a reduced state. This latent image is then developed using a series of chemicals. The developer solution typically contains an organic compound known as a developing agent, which reduces the exposed silver halide crystals to metallic silver, darkening the areas of the film that correspond to the light exposure. Common developing agents are chemically related to benzene and contain carbon. Following development, the film must be treated with a stop bath to halt the development process and prevent further reaction. The next step involves using a fixer, which usually contains sodium thiosulfate (commonly known as hypo). The fixer dissolves unexposed silver halide crystals, stabilizing the image and preventing fogging over time when exposed to light. After fixing, the film is washed with water to remove any residual chemicals. Despite its historical significance, photographic film has limitations in terms of quantum efficiency (QE) [35], typically around 1%. This low efficiency means that only a small fraction of incoming light is captured, necessitating long exposure times to detect faint celestial objects. As a result, astronomers have largely transitioned to digital detectors like CCD, which offer significantly higher sensitivity and efficiency in capturing astronomical images [33]. Photomultiplier tubes (PMTs) [36] represent a significant advancement in light detection technology compared to photographic film. A PMT is a vacuum tube designed to convert low levels of light, such as photons from astronomical sources, into measurable electrical signals. The construction of a PMT includes several key components: a photocathode, multiple dynodes, and an anode, all housed within an evacuated glass envelope to prevent interference from air molecules [36]. The photocathode is typically made from photoemissive materials, such as cesium antimonide (Cs3Sb) or gallium arsenide (GaAs) [37], which are deposited as a thin layer on the inner surface of the tube's entrance window. When photons strike the photocathode, they cause the emission of electrons through the photoelectric effect—this process involves photons transferring energy to electrons, dislodging them from the material. The efficiency of this process can vary, but PMTs generally achieve a quantum efficiency (QE) between 10% and 20%, significantly higher than that of photographic film [35]. Once emitted, these photoelectrons are directed toward a series of dynodes typically arranged in a chain of 8 to 19—where they undergo further amplification. Each dynode is coated with a secondary emissive material that releases multiple electrons for each incident electron that strikes it. As the electrons cascade through the dynodes, they are accelerated by a high voltage applied across them (typically around 100 volts between dynodes), resulting in an exponential increase in the number of electrons. This amplification can yield gains of 103 to 108 depending on the number of dynodes and their configuration. The final component is the anode, which collects the amplified electron signal and converts it into an electrical current that can be measured. This process allows PMTs to detect extremely low levels of light, making them invaluable in applications such as spectroscopy and time-domain astronomy.

While PMTs offer improved sensitivity over photographic film, they do not match the performance of modern digital detectors like CCD, which can achieve near-unity quantum efficiencies and provide higher resolution imaging capabilities. Nonetheless, PMTs remain essential tools in various scientific fields due to their fast response times and ability to detect faint light signals effectively. The most significant advancement occurred in 1969 when Willard Boyle and George Smith created CCDs. Figure 9.11 illustrates the highly effective and often used CCD detector in contemporary astronomy. Light-sensitive silicon diodes (pixels) are arranged in a rectangular array; some cameras have up to 4096 × 4096 pixels. Electrons are released when photons strike the silicon surface, and these build up in the pixels. The accumulated charges are moved row by row to a readout buffer following exposure, where they are transformed into digital values and processed by a computer. Electron generation in CCD cameras is almost linear and exactly proportional to the quantity of photons, as shown in Equ. (9.3):

$$N_e = \Phi \cdot \eta \cdot t. \tag{9.3}$$

where $\boldsymbol{\Phi}$ is the photon flux, $\boldsymbol{\eta}$ is the quantum efficiency, $\boldsymbol{t}$ is the integration time, and $\boldsymbol{N_e}$ is the number of electrons. Having quantum efficiencies of 80% to 90% in the red wavelength band (600–800 nm), they are noticeably more sensitive than photographic plates. However, silicon absorption causes sensitivity to decrease in the UV; this can be lessened by coatings or by thinning the silicon chip. The dark current produced by thermal noise in astronomical cameras can be reduced by cooling, sometimes using liquid nitrogen, to enhance image quality.

Another source of noise is the bias, which is the pixel-to-pixel variation in the read noise across an image. Since the bias structure varies across the CCD, it is essential to remove it using a bias frame, an image taken with a zero-second exposure. The calibration of CCD images [38] is a multistep process. Bias frames are used to measure and subtract the bias current, ensuring the removal of systematic noise from other exposures. Flat-field frames, obtained by imaging evenly illuminated fields, are employed to correct pixel sensitivity variations across the CCD sensor. These calibration methods, when applied together, ensure the accuracy and consistency of the processed data, as shown in Figure 9.11. Cosmic rays and readout noise are additional challenges in CCD imaging. Cosmic rays can produce bright spots or streaks, while readout noise can distort faint signals. These effects are mitigated by using short exposures, cleaning the images

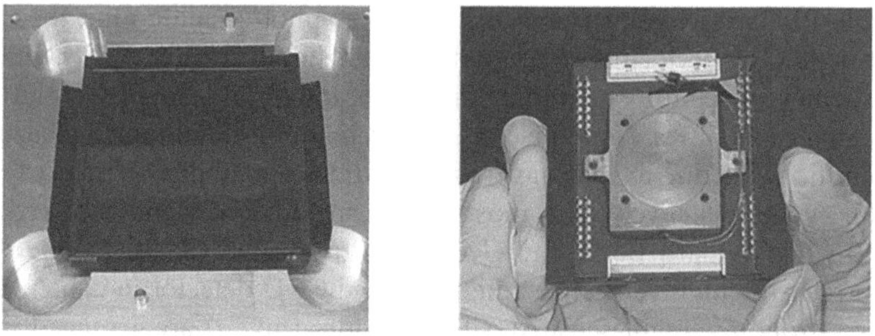

Figure 9.11 Images of CCD. (Image credit: Las Cumbres Observatory (LCO)).

to remove artifacts, and stacking multiple exposures to enhance the signal-to-noise ratio. Modern CCD cameras, including low-light-level CCDs (L3-CCDs) [39], have minimized readout noise to just a few electrons, enabling the detection of faint signals. While CCD cameras are expensive and have smaller fields compared to photographic plates, their sensitivity, stability, and ease of data processing have made them the dominant choice for astronomical observations, replacing traditional photographic methods.

9.6.1 Evolution of Detectors

The evolution from photographic film and PMTs to CCDs reflects a continuous pursuit of greater sensitivity, efficiency, and technological advancement in astronomy [40]. Each transition has expanded our ability to explore the universe, allowing astronomers to capture increasingly detailed images and conduct more precise measurements of celestial phenomena. The transition from photographic film and PMTs to CCDs in astronomical applications was driven by several key factors that enhanced observational capabilities. Figure 9.12 shows the typical images associated with CCD observations. One of their primary advantages [41] is their superior quantum efficiency, which can reach up to 100% in certain spectral regions, particularly in the red part of the spectrum, compared to photographic film's typical efficiency of about 1% and PMTs' efficiency of 10% to 20%. This higher quantum efficiency enables them to detect significantly fainter celestial objects, making them ideal for deep-sky imaging. Additionally, their digital nature allows for immediate image processing and analysis [42]; unlike photographic film, which requires chemical development and digitization, CCDs produce digital data that can be quickly transferred to computers for real-time evaluation, streamlining the observational process.

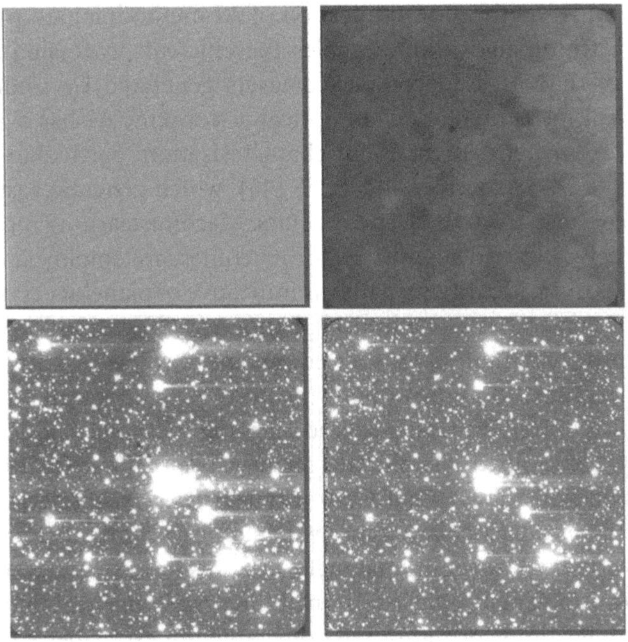

Figure 9.12 Typical images associated with CCD observations. (Image credit: Himalayan Chandra Telescope (HCT)).

They also exhibit a remarkable dynamic range of around 14.5 magnitudes, allowing them to capture both bright and faint details in a single exposure, in contrast to photographic film's dynamic range of less than 7.5 magnitudes. Furthermore, they provide a linear response to light intensity, simplifying calibration processes and enhancing measurement accuracy. Advances in manufacturing technologies have led to larger CCD arrays with higher resolutions, enabling modern telescopes to capture extensive fields of view with significant detail.

In recent years, CMOS sensors have emerged as an alternative to CCDs [43], particularly in consumer electronics due to their lower cost and power consumption; however, they generally exhibit lower sensitivity and higher noise levels than CCDs, limiting their application in high-precision astronomical observations. As a result, they remain the preferred choice for most scientific applications requiring high fidelity and sensitivity.

9.6.2 AI Role in Detectors

AI is increasingly integral to the advancement and optimization of detectors in various astronomical domains, including optical, infrared, X-ray, and radio wavelengths. The application of AI methodologies, particularly machine learning algorithms, enables the efficient processing and analysis [44] of the vast and complex datasets generated by contemporary astronomical instrumentation. In optical astronomy, AI has significantly advanced exoplanet detection and characterization, particularly through missions like NASA's Kepler and TESS [45], which generate vast amounts of light curve data from thousands of stars. Machine learning models, such as neural networks and support vector machines, are employed to analyze these light curves, identifying patterns indicative of planetary transits, star-spot activity, and other astrophysical phenomena. These models use supervised learning techniques [46] trained on labeled datasets, such as known exoplanet transits, to improve both the accuracy and speed of exoplanet detection. Additionally, unsupervised learning methods can detect novel previously unclassified signals, increasing the potential for discovering new types of exoplanets. By automating the analysis, AI accelerates the discovery of planets, particularly those with subtle or rare transit signals that may be overlooked by traditional methods. This has expanded our understanding of exoplanet populations, including those in the habitable zone, and has implications for future missions targeting exoplanet atmospheres and habitability. In radio astronomy, the upcoming Square Kilometre Array (SKA) highlights the critical need for AI due to its anticipated data output [47], which may reach approximately 1 exabyte per day. This immense volume of data necessitates advanced AI algorithms for effective filtering and processing. AI techniques will be employed to distinguish celestial signals from background noise, thereby enhancing the detection of transient events such as pulsars and fast radio bursts (FRBs) [48]. The computational demands of the SKA's central processing unit are projected to be equivalent to that of approximately 100 million personal computers, emphasizing AI's essential role in managing this data influx. Specifically, it will facilitate tasks such as dedispersion, matched filtering, and classification of candidate signals, thereby improving the efficiency and accuracy of data analysis. As the SKA aims to operate at unprecedented sensitivity and speed, integrating AI into its data processing framework will be crucial for extracting meaningful scientific insights from the vast quantities of information generated by this next-generation radio telescope. In X-ray astronomy, AI is increasingly utilized to enhance the identification of astrophysical sources amid instrumental background noise. Machine learning algorithms applied to raw frame-level data can significantly improve signal detection, achieving background noise reduction of up to 41.5% compared to conventional methods. This enhancement increases the sensitivity of X-ray detectors, enabling more accurate observations of faint sources such as black holes and neutron stars. For instance, recent studies have demonstrated the effectiveness of unsupervised learning techniques like K-means clustering, self-organizing maps, etc., in classifying over 100,000 X-ray sources from the European Space Agency's XMM-Newton satellite data [49], allowing for the identification of previously unrecognized light curves and transient events. By automating the classification process, AI not only streamlines data analysis but also facilitates the discovery of new astrophysical phenomena, thereby advancing our understanding of high-energy astrophysics. AI also plays a crucial role in image classification tasks within optical surveys. For instance, neural networks can be trained on extensive datasets from telescopes such as the Subaru Telescope to classify galaxies based on their morphological features. These networks have achieved accuracy rates of up to 98%, surpassing traditional classification methods that rely on human expertise. Additionally, unsupervised learning algorithms, including clustering techniques like K-means and density-based methods such as HDBSCAN, enable astronomers to uncover new relationships within complex datasets without predefined categories. This capability facilitates the systematic exploration of previously uncharacterized celestial phenomena, enhancing our understanding of the universe's structure and composition.

9.7 AI's Role in Robotic Telescopes

AI represents a revolutionary technological paradigm in astronomical research, fundamentally transforming robotic telescope systems through sophisticated computational methodologies and advanced algorithmic frameworks. These cutting-edge systems integrate precision instrumentation with intelligent computational architectures, comprising high-precision optical systems, advanced CCD imaging sensors [50], adaptive optics mechanisms, and complex neural network models including deep neural networks [51], convolutional neural networks, and machine learning algorithms. AI-driven robotic telescopes [52] like Hanle Observatory in Ladakh shown in Figure 9.13 demonstrate extraordinary technological performance, achieving high celestial object classification accuracy, millisecond-level decision-making, and significantly increased observational efficiency through autonomous capabilities that enable continuous 24/7 operational monitoring. The sophisticated AI techniques leverage advanced algorithms to analyze massive astronomical datasets, detect transient phenomena, optimize observation scheduling, compensate for atmospheric distortions, and predict potential celestial events with unprecedented precision. These intelligent systems

Figure 9.13 Hanle Observatory in Ladakh, India. (Image credit: Indian Astronomical Observatory (IAO)).

employ complex decision-making frameworks that simultaneously consider critical variables such as celestial object visibility, atmospheric conditions, telescope operational parameters, and scientific research priorities. Technological innovations include quantum machine learning integration, enhanced autonomous decision frameworks, and advanced multi-telescope network coordination, which democratize astronomical research by enabling simultaneous multiwavelength observations and collaborative international research platforms. Despite remarkable capabilities, these systems confront significant challenges including complex computational infrastructure requirements, potential algorithmic biases, and data security vulnerabilities. The convergence of advanced computational technologies, machine learning algorithms like support vector machine (SVM), k-Nearest Neighbor (kNN), deep neural networks, convolutional neural networks [53], and sophisticated robotic infrastructure promises to revolutionize our understanding of the cosmos by enabling continuous, efficient, and precise observations that expand the boundaries of human knowledge and offer unprecedented insights into the complex and mysterious universe surrounding us.

9.7.1 Architectural Elements and Usability

AI-driven robotic telescopes use sophisticated optical and computational technologies in their technical architecture to independently monitor and study astronomical occurrences. Subsystems that are networked and intended for accurate real-time astronomical research make up these systems. High-precision optical mechanisms, sophisticated CCD imaging sensors [50], adaptive optics, and intelligent software control platforms driven by machine learning algorithms are among the essential parts. Telescope pointing mechanisms, CCD camera operations, dome and enclosure controls, focuser management, and weather condition sensing are just a few of the subsystems that are coordinated by a master control system that oversees the computing infrastructure. These telescopes have two main modes of operation: closed-loop systems, which make observations without real-time performance verification, and open-loop systems, which continuously check and modify observational parameters.

9.7.2 AI-Powered Features

Robotic telescopes are made more capable by artificial intelligence, which allows them to carry out intricate astronomical investigations effectively and independently. Its capacity to react instantly to brief astronomical occurrences, such as rapid radio bursts, gamma-ray bursts, and supernovae, is one of its primary capabilities. Advanced algorithms that evaluate incoming data streams and rank findings according to their scientific significance are used to accomplish this. Machine learning models trained on historical data, for example, can forecast the probability of an event occurring as well as its possible significance, guaranteeing the best possible use of observational time. Convolutional neural networks (CNNs) [53], a type of deep learning approach, are widely used in image processing and analysis. These networks do exceptionally well at spotting patterns in high-resolution astronomical photos, including identifying fleeting or dim objects against complex backgrounds, distinguishing between significant signals and noise, and categorizing objects according to their morphological or spectral properties. By automating these procedures, robotic telescopes drastically cut down on processing time and human intervention, allowing results to be shared with the scientific community more quickly. Combining data from robotic telescopes with inputs from other observatories, such as neutrino observatories and gravitational wave detectors, is part of multi-messenger astronomy, which is made easier by the integration of AI. These data are combined by machine learning models to produce a thorough comprehension of cosmic events. Because of this, these AI-powered systems are not only instruments for observation but also active contributors to the process of discovery, transforming our understanding of the cosmos.

9.8 Challenges in Autonomous Astronomy

While autonomous systems are revolutionizing the field of astronomy, they are also subjected to limitations and challenges. One of the major challenges is to ensure the robustness and reliability of AI algorithms in remote and harsh environments. A reliable autonomous system must operate

Table 9.3 Key challenges in autonomous astronomy.

Challenge category	Specific challenge	Description
Statistical Complexity	Nonlinear Model Fitting	Fitting astronomical spectra to highly complex astrophysical models with confidence limits
Data Management	Large Dataset Handling	Managing astronomical datasets with over 1 billion objects, exceeding standard statistical software capabilities
Observation Scheduling	Multi-objective Optimization	Balancing competing scientific goals and observational needs for different sky objects
Computational Limitations	Algorithm Performance	Developing reinforcement learning algorithms capable of optimizing telescope observation schedules
Data Interpretation	Measurement Uncertainty	Handling measurement errors, non-detections, and precise data interpretation
Technological Constraints	Autonomous System Design	Creating self-driving telescopes that can make intelligent observation decisions

seamlessly over prolonged periods, even if it comes across unforeseen challenges as mentioned in Table 9.3. Autonomous systems are transforming the field of astronomy by enabling efficient and innovative approaches to celestial observation. However, these systems face several limitations and challenges that need to be addressed for seamless and reliable operation.

9.8.1 Ensuring Robustness and Reliability

One of the primary challenges is ensuring the robustness and reliability of AI algorithms in remote and harsh environments. Autonomous systems must operate consistently over extended periods, even when encountering unforeseen circumstances such as equipment malfunctions or environmental changes. Achieving this reliability requires comprehensive testing and validation processes to ensure the system can manage unexpected obstructions. Robustness refers to the ability of these systems to maintain performance under varying and challenging conditions. For instance, autonomous telescopes must adapt to sudden weather changes, mechanical malfunctions, and power fluctuations. This necessitates the integration of advanced sensors capable of real-time environmental monitoring and intelligent algorithms that can dynamically adjust observational strategies to maintain operational efficiency.

9.8.2 Overcoming Navigation Challenges

Autonomous navigation presents another set of significant challenges, particularly in adapting to rapidly fluctuating weather conditions. The integration of multiple sensors and data streams to create precise real-time situational awareness is a complex engineering task. This requires highly sophisticated algorithms capable of synchronizing and interpreting diverse data sources to ensure uninterrupted operation and accuracy, even in unpredictable environments.

9.8.3 Addressing Data Management and Security Issues

The vast amounts of data generated by modern observatories create additional challenges in data management and security. Efficient data processing capabilities are essential for storing and analyzing these data, ensuring that insights are extracted effectively. Proper data management involves organizing information from sensors and user interactions in a way that optimizes system performance. However, this also introduces significant security risks. Autonomous systems are vulnerable to cyberattacks that could compromise their functionality or the integrity of sensitive data. To mitigate these risks, robust security measures such as encryption, secure data transmission, and regular system updates are necessary. Adhering to data privacy regulations ensures transparency and protects sensitive information from unauthorized access.

9.8.4 Ethical and Accountability Considerations

The deployment of autonomous systems raises complex ethical and accountability issues. Determining responsibility for errors or harm caused by these systems whether it lies with the developer, user, or the system itself remains a significant challenge. Additionally, the risk of job displacement due to automation highlights the need for strategies to manage workforce transitions and reduce economic inequalities. Balancing human oversight with autonomous operation is another critical consideration. Ensuring scientific integrity and compliance with ethical standards requires thoughtful coordination between human-led and autonomous activities.

9.8.5 Integration and Maintenance Requirements

Integrating autonomous systems into established workflows presents challenges in aligning them with human-led operations. These systems demand continuous maintenance and regular updates to stay aligned with technological advancements and address emerging challenges. Furthermore, the substantial investments required for research, development, and long-term sustainability pose financial constraints for institutions operating under limited resources. Addressing these challenges is crucial for the successful implementation of autonomous systems in astronomy, ensuring they achieve their full potential while maintaining reliability, security, and ethical integrity.

9.9 The Future of Autonomous Astronomy

With the use of developments in AI, machine learning, and robotics, autonomous astronomy has the potential to completely transform our understanding of the universe and push the limits of observational science [54]. These technologies have the potential to improve the capabilities of observational platforms and telescopes, allowing them to carry out intricate tasks with little assistance from humans. This section examines how these platforms can revolutionize data collecting, real-time analysis, teamwork, and the redefining of the role of human astronomers.

9.9.1 Enhanced Observational Capabilities

In order to perform advanced night sky surveys, autonomously detect and classify celestial objects, and react dynamically to environmental changes like light pollution or atmospheric turbulence, autonomous telescopes (Figure 9.14) are developing.

The efficiency of data gathering will be greatly increased by these capabilities, allowing astronomers to record extensive datasets over wider sky regions than previously possible. Even in less-than-ideal settings, consistent and trustworthy observations are guaranteed by the ability to function under a variety of situations.

9.9.2 Real-Time Data Processing and Discovery

The interpretation of enormous astronomical datasets will be completely transformed by the combination of machine learning and sophisticated data processing algorithms. Real-time patterns and abnormalities will be

Figure 9.14 A view of the 3.6-m Devasthal Optical Telescope from inside the dome. (Image credit: ARIES Nainital).

detected by autonomous systems, which will also be able to detect transitory occurrences like gamma-ray bursts or supernovae nearly instantly. In addition to speeding up discoveries, this quick analysis makes it easier to conduct follow-up observations on time, which is essential for examining the dynamics of these transient events. Such developments will greatly accelerate astronomical research by improving the efficiency and responsiveness of the observation process.

9.9.3 Collaborative Networks of Autonomous Telescopes

The possibilities for worldwide collaborative astronomy will increase with the interconnection of networks of autonomous telescopes (Figure 9.15).

The network's telescopes can instantly concentrate their observations on the same target when one detects a possible transit of an exoplanet, for example, allowing for coordinated research efforts and raising the possibility of groundbreaking findings [55]. By establishing a coordinated observational network, these systems will exchange information and insights in real time, improving our collective comprehension of astronomical occurrences [56].

Figure 9.15 T. Suchinchan comet. Observed by Devasthal Observatory. (Image credit: ARIES Nainital).

9.9.4 Transforming the Role of Human Astronomers

Human astronomers will have more time to concentrate on strategic and innovative research, such as developing hypotheses and theoretical modeling, if regular observational chores are automated. Scientists will have more time to work on complicated problems about the cosmos, such as the secrets of dark matter, dark energy, and the prerequisites for extraterrestrial life, once regular data collecting and preliminary analyses are managed autonomously. This change will reinterpret the function of astronomers, encouraging more in-depth research and creativity.

9.9.5 Paving the Way for New Discoveries

As autonomous systems develop further, they will open up hitherto unheard-of possibilities for space exploration. Together, these technologies

Table 9.4 This table showcases the future of autonomous astronomy.

Domain	Future developments	Key technologies
Telescope Technology	• 100-meter class ground telescopes • Large time-domain telescopes • Airborne stratospheric observatories	• Advanced AI/ML integration • Adaptive optics • High-resolution imaging
Space Observation	• Gravitational wave astronomy • Exoplanet detection • Deep space exploration	• Cosmic Explorer Observatory • LISA space antenna • AI-driven data processing
AI Integration	• Autonomous data analysis • Predictive cosmic event modeling • Object classification	• Neural networks • Machine learning algorithms • Digital twin technologies
Satellite Systems	• 5G-enabled networks • Proliferated satellite constellations • Interoperability	• Advanced communications • Space-based quantum communication • Autonomous operations
Research Capabilities	• Sample return missions • Infrared/X-ray telescopes • Climate monitoring	• Space exploration technologies • Advanced manufacturing • Robotic systems

(summarized in Table 9.4) have the potential to revolutionize human interaction with the cosmos and increase the effectiveness and reach of astronomical study. Astronomers will make tremendous progress in their quest to comprehend the basic nature of the cosmos by utilizing autonomy to obtain fresh thoughts and viewpoints.

Autonomous astronomy will mark a significant change in the way we investigate the universe in the future. We stand on the precipice of a new era of discovery, one that promises to reveal the secrets of the cosmos in ways that were previously unthinkable by fusing state-of-the-art technologies with collaborative frameworks and redefining the human role in research.

9.10 Conclusion

Implementation of AI in various areas of study including telescopes and observatories represents a transformative leap in our ability to explore the vast cosmos and understand the complex data collected. Here, in this chapter, we have gathered some of the details of the evolution of telescopic technologies, from the manual refracting and reflecting instruments to the highly advanced autonomous systems of today. Along with this, we have seen how advanced innovations such as AI, machine learning, and robotics have changed observational capabilities, enabling continuous, precise, and efficient data collection with minimal human assistance. The integration of AI and automation has opened a window of unseen opportunities in astronomy. Autonomous systems have created an excellent achievement in detecting transient phenomena like supernovae and gamma-ray bursts, tracking celestial objects with precision, and managing vast datasets in real time. They have made it possible for smaller organizations and independent researchers to participate in groundbreaking discoveries by enabling access to modern astronomical instruments. The field of astronomy has been even further expanded with the capacity to function without supervision in isolated or extreme conditions. With space-based telescopes and ground-based observatories in remote areas, these devices have expanded our understanding of the cosmos and offered insights into everything from radio waves to gamma rays. Moreover, the development of observatories that are linked together forming networks has made it easier for people all over the world to work together, allowing for continuous observation and multiwavelength monitoring. More thorough investigations of cosmic phenomena and improved data exchange between academics globally are the results of this networked method. Despite their huge possibilities, autonomous telescopes face significant challenges. Maintaining the adaptability and reliability of these systems is essential to maintaining the quality of the data. Their functionality and accuracy of data are at risk from cyber vulnerabilities, software bugs, and hardware malfunctions. Furthermore, complex data management techniques, cutting-edge computational skills, and reliable algorithms are required for efficient analysis due to the large and complex volume of data collected. Ethical considerations on the use of AI in scientific research also need careful attention. Biases in AI algorithms, potential misinterpretation of data, and the environmental impact of large-scale computational facilities are ongoing concerns that need to be addressed. As we look to the future, the possibilities for autonomous astronomy are both exciting and expansive. Innovations such as quantum computing, advanced AI models, and next-generation detectors promise to further enhance the capabilities of these systems. Emerging technologies like extremely large telescopes (ELTs) and gravitational wave observatories will integrate with AI to explore new frontiers in cosmic discovery. Additionally, the collaborative potential of interconnected observatories is expected to grow. Global networks of autonomous systems will enable coordinated responses to transient events, real-time data sharing, and multidisciplinary studies that break the traditional boundaries. The role of human astronomers will continue to change in this field as the role of AI advances. Researchers can concentrate on developing hypotheses, theoretical modeling, and interpreting complex occurrences when they do not have to deal with everyday observational duties. This coexistent relationship between human creativity and technological efficiency will speed up the discovery and expand our knowledge of the cosmos. Unsupervised observation facilities and telescopes represent the researcher's never-ending hard work for exploration of the universe's secrets. By pushing the limits of technology, we have created a new window into the cosmos that provides us an outstanding perspective on its composition, beginnings, and development. Although there are still few obstacles to overcome, the continuous development of these systems promises exciting developments that will have an impact on astronomy, shifting forward and inspiring future generations. One thing is certain as we approach the end of this new era, autonomous astronomy is only the beginning of many possibilities, and its potentials are as limitless as the possibilities of the cosmos.

Acknowledgement

The authors sincerely thank all the editors of this book for their valuable contributions and insightful reviews. HM acknowledges Ms. Gissmol Saji for assistance in investigation and analysis, and Mr. Kuldeep Belwal for fruitful discussions. SS & HM express gratitude to Dr. Jyoti Raghav for her valuable suggestions and help in drafting the chapter.

References

1. Dinsa, A.B., Wakjira, F.S., Teferi Demmesie, E., Negash, T.T., Indigenous knowledge of astronomical star positions and temporal patterns for seasonal weather forecasting: the case of Borana Oromo Pastoralists of Southern Ethiopia. *F1000Research*, 12, 205, 2023.
2. Rohlfs, K. and Wilson, T.L., *Tools of radio astronomy*, Springer Science & Business Media, Springer Berlin, Heidelberg, 2013.
3. Panigrahi, B.S., Artheeswari, S., Khan, W., Pavithra, G., Pathak, S.K., Bharanidharan, R., Artificial Intelligence in Astrophysics: Automated Detection of Celestial Objects and Anomaly Detection. *2024 5th International Conference on Recent Trends in Computer Science and Technology (ICRTCST)*, IEEE, pp. 210–214, 2024.
4. Gan, W.Q., Zhu, C., Deng, Y.Y., Li, H., Su, Y., Zhang, *et al.*, Advanced space-based solar observatory (ASO-S): an overview. *Res. Astron. Astrophys.*, 19, 1, 156, 2019.
5. Bhaumik, M., *The Cosmic Detective: exploring the mysteries of our universe*, Penguin Books India, 2008.
6. North, J.D., The astrolabe. *Sci. Am.*, 230, 1, 96–107, 1974.
7. Price, D.J., A collection of armillary spheres and other antique scientific instruments. *Ann. Sci.*, 10, 2, 172–187, 1954.
8. Jamieson, D., Galileo's miraculous year 1609 and the revolutionary telescope. *Aust. Phys.*, 46, 3, 72–76, 2009.
9. Mees, C.E.K., On the resolving power of photographic plates. *Proc. R. Soc. Lond. Ser. A Contain. Pap. Math. Phys. Character*, 83, 559, 10–18, 1909.
10. Van Helden, A., The invention of the telescope. *Trans. Am. Philos. Soc.*, 67, 4, 1–67, 1977.
11. Bedini, S.A., The Aerial Telescope. *Technol. Cult.*, 8, 3, 395–401, 1967.
12. Maran, S.P., Telescopes and automation. *Science*, 158, 3803, 867–871, 1967.
13. Wilson, R.N., *Reflecting telescope optics II: manufacture, testing, alignment, Modern Techniques*, Springer Science & Business Media, Springer Berlin, Heidelberg, 2013.

14. Song, M. and Yuan, Y., Autonomous navigation for deep spacecraft based on celestial objects. *2008 2nd International Symposium on Systems and Control in Aerospace and Astronautics*, IEEE, 2008.
15. University of Oregon, Refracting Telescope-Glossary, https://pages.uoregon.edu/jschombe/glossary/refracting_telescope.html.
16. Norton, R., Observing the Universe, https://physics.uwyo.edu/~ddale/research/REU/2021/ObservingTheUniverse_Norton.pdf.
17. High point scientific, Retrieved from Types of telescopes: https://www.highpointscientific.com/astronomy-hub/post/astronomy-101/types-of-telescopes.
18. Saxon, Retrieved from Different types of telescopes: https://www.saxon.com.au/different-types-of-telescopes.
19. Ozscopes, Retrieved from Types of telescopes: https://www.ozscopes.com.au/types-of-telescopes.
20. Catadioptric telescopes: https://www.abramsplanetarium.org/telescopes/index.html.
21. Cassegrain telescopes: https://outdoorandvelocity.co.za/shop/celestron-cpc-1100-gps-xlt-computerized-cassegrain-280-mm/.
22. In depth guide to catadioptric telescopes: https://www.astronomicphoto.com/exploring-the-universe-an-in-depth-guide-to-catadioptric-telescopes.
23. Klein, U., Radio astronomy: tools, applications and impacts. Course astro, p. 841, 2006.
24. Condon, J.J. and Ransom, S.M., *Essential radio astronomy*, vol. 2, Princeton University Press, Princeton, New Jersey, USA, 2016.
25. Mahoney, W. and Thanjavur, K., Artificial intelligence in autonomous telescopes. *Telescopes from Afar*, pp. 38, 2011.
26. Groenewald, E.S., Kumar, N., Groenewald, C.A., AI-Assisted Astronomical Data Analysis Unveiling Patterns and Phenomena in the Universe. *NATURALISTA CAMPANO*, 28, 1, 1916–1925, 2024.
27. Massey, P. and Hanson, M.M., Astronomical spectroscopy, arXiv preprint arXiv:1010.5270, 2010.
28. Ameen, N., Hosany, S., Tarhini, A., Consumer interaction with cutting-edge technologies: Implications for future research. *Comput. Hum. Behav.*, 120, 106761, 2021.
29. Yau, K.K.C., An investigation of some contemporary problems in astronomy and astrophysics by way of early astronomical records. Diss., Durham University, 1988.
30. Kattenborn, T., Leitloff, J., Schiefer, F., Hinz, S., Review on Convolutional Neural Networks (CNN) in vegetation remote sensing. *ISPRS J. Photogramm. Remote Sens.*, 173, 24–49, 2021.
31. Braddick, H.J.J., Photoelectric photometry. *Rep. Prog. Phys.*, 23, 1, 154, 1960.
32. Owens, A., *Semiconductor radiation detectors*, CRC Press, Boca Raton, Florida, USA, 2019.

33. Roth, M.M., A brief history of image sensors in the optical. *Astron. Nachr.*, 344, 8–9, 2023.
34. Tate, E.W., Synthesis and Characterisation of Photocatalyst Silver/Silver Halide Nanocomposites. Diss., Open Access Te Herenga Waka-Victoria University of Wellington, 2012.
35. Khaleda, M.F., Vengadaesvaran, B., Rahim, N.A., Spectral response and quantum efficiency evaluation of solar cells: A review, in: *Energy Materials*, pp. 525–566, 2021.
36. Xia, J., Qian, S., Wang, W., Ning, Z., Cheng, Y., Wang, Z., Lei, X., A performance evaluation system for photomultiplier tubes. *J. Instrum.*, 10, 03, P03023, 2015.
37. Schaber, J., Xiang, R., Gaponik, N., Review of photocathodes for electron beam sources in particle accelerators. *J. Mater. Chem. C*, 11, 9, 3162–3179, 2023.
38. Beyer, H.A., Accurate calibration of CCD-cameras. *Proceedings 1992 IEEE Computer Society Conference on Computer Vision and Pattern Recognition*, IEEE Computer Society, 1992.
39. Moomaw, B., Camera technologies for low light imaging: overview and relative advantages, in: *Methods in Cell Biology*, vol. 114, pp. 243–283, 2013.
40. Romanishin, W., *An introduction to astronomical photometry using CCDs*, vol. 17, University of Oklahoma, Norman, Oklahoma, United States, 2006.
41. Oshiro, M., Cooled CCD Versus Intensified Cameras for Low-Light Video—Applications and Relative Advantages, in: *Methods in cell biology*, vol. 56, pp. 45–62, 1998.
42. Umbaugh, S.E., *Computer imaging: digital image analysis and processing*, CRC press, Boca Raton, Florida, USA, 2005.
43. Waltham, N., CCD and CMOS sensors, in: *Observing Photons in Space: A Guide to Experimental Space Astronomy*, pp. 423–442, 2013.
44. Rane, N.L., Paramesha, M., Choudhary, S.P., Rane, J., Artificial intelligence, machine learning, and deep learning for advanced business strategies: a review. *Partn. Univers. Int. Innovation J.*, 2, 3, 147–171, 2024.
45. Christ, C.N., Montet, B.T., Fabrycky, D.C., Observations of the kepler field with tess: Predictions for planet yield and observable features. *Astron. J.*, 157, 6, 235, 2019.
46. Cunningham, P., Cord, M., Delany, S.J., Supervised learning, in: *Machine learning techniques for multimedia: case studies on organization and retrieval*, pp. 21–49, Springer Berlin Heidelberg, Berlin, Heidelberg, 2008.
47. Twidle, H., Impossible images: Radio astronomy, the square kilometre array and the art of seeing. *J. South. Afr. Stud.*, 45, 4, 767–790, 2019.
48. Snelders, M.P., Nimmo, K., Hessels, J.W.T., Bensellam, Z., Zwaan, L.P., Chawla, P., *et al.*, Detection of ultra-fast radio bursts from FRB 20121102A. *Nat. Astron.*, 7, 12, 1486–1496, 2023.
49. Sinaga, K.P. and Yang, M.-S., Unsupervised K-means clustering algorithm. *IEEE Access*, 8, 80716–80727, 2020.

50. Carlson, B.S., Comparison of modern CCD and CMOS image sensor technologies and systems for low resolution imaging. *SENSORS, IEEE*, vol. 1, pp. 171–176, 2002.
51. Hampson, K.M., Turcotte, R., Miller, D.T., Kurokawa, K., Males, J.R., Ji, N., Booth, M.J., Adaptive optics for high-resolution imaging. *Nat. Rev. Methods Primers*, 1, 1, 68, 2021.
52. Doyle, R., Kubota, T., Picard, M., Sommer, B., Ueno, H., Visentin, G., Volpe, R., Recent research and development activities on space robotics and AI. *Adv. Rob.*, 35, 21-22, 1244–1264, 2021.
53. Chauhan, R., Ghanshala, K.K., Joshi, R.C., Convolutional neural network (CNN) for image detection and recognition. *2018 first international conference on secure cyber computing and communication (ICSCCC)*, IEEE, 2018.
54. Sybilski, P.W., Pawłaszek, R., Kozłowski, S.K., Konacki, M., Ratajczak, M., Hełminiak, K.G., Software for autonomous astronomical observatories: challenges and opportunities in the age of big data. *Software and Cyberinfrastructure for Astronomy III*, vol. 9152, pp. 473–486, 2014.
55. Parker, J. and Heywood, D., The earth and beyond: Developing primary teachers' understanding of basic astronomical events. *Int. J. Sci. Educ.*, 20, 5, 503–520, 1998.
56. White, R.R., Wren, J., Davis, H.R., Galassi, M., Starr, D., Vestrand, W.T., Wozniak, P., TALON: the telescope alert operation network system: intelligent linking of distributed autonomous robotic telescopes. *Advanced Software, Control, and Communication Systems for Astronomy*, vol. 5496, SPIE, 2004.

10

Beyond Earth's Horizon: AI's Contribution to Space Exploration

Bhumika Sharma[1]*, Anil C. Mathur[2], Rama Sharma[3] and Pratibha Antil[1]

[1]*Department of Physical & Chemical Sciences, Lovely Professional University, Jalandhar, India*
[2]*Centre of Space Science and Technology (CSST), Indian Institute of Technology, Roorkee, India*
[3]*University of New South Wales, Sydney, Australia*

Abstract

As humanity pushes the boundaries of space exploration, the integration of artificial intelligence (AI) has become a transformative force, driving innovation and enabling the realization of missions that once seemed impossible. The chapter "Beyond Earth's Horizon: AI's Contribution to Space Exploration" delves into AI's pivotal role in advancing our understanding of the cosmos and enhancing the efficiency and safety of space missions. This exploration is grounded on a comprehensive overview of AI's applications in space exploration, from mission planning and autonomous navigation to data analysis and extraterrestrial research. The chapter begins by tracing the evolution of AI in space exploration, highlighting key milestones such as the use of AI in early satellite missions and its growing presence in contemporary space endeavors. As space missions become increasingly complex and distant, the need for autonomous systems that can operate independently of Earth-based control becomes paramount. AI emerges as a critical tool in this context, enabling spacecraft to make real-time decisions, optimize navigation routes, and manage unforeseen challenges in environments where communication delays are significant. One of the primary focuses of the chapter is on AI-driven navigation systems. These systems are designed to perform tasks such as trajectory optimization, hazard avoidance, and precision landing on celestial bodies. The chapter examines the sophisticated algorithms that power these systems, allowing spacecraft to navigate through the vast expanse of space with unprecedented accuracy. Case studies of missions like NASA's Mars rovers, the European Space

**Corresponding author*: bshakalya@gmail.com

Yogesh Chandra, Manjuleshwar Panda and Mahesh Chandra Mathpal (eds.) The Intelligent Universe: AI's Role in Astronomy, (359–390) © 2025 Scrivener Publishing LLC

Agency's Rosetta mission, and upcoming projects aimed at exploring asteroids and distant planets illustrate the practical applications of AI in enhancing mission success rates. In addition to navigation, AI's role in data analysis is thoroughly explored. Space missions generate vast amounts of data, much of which needs to be processed and analyzed in real time. AI techniques, including machine learning and neural networks, are employed to sift through these data, identify patterns, and extract meaningful insights. These capabilities are crucial for tasks such as analyzing planetary surfaces, detecting signs of extraterrestrial life, and understanding the complex dynamics of celestial phenomena. The chapter also discusses AI's potential in processing data from space telescopes and other observational platforms, contributing to discoveries that expand our knowledge of the universe. The chapter further explores AI's contribution to robotic exploration and the development of autonomous systems capable of performing scientific experiments, repairs, and maintenance in space. These systems are designed to operate in harsh and unpredictable environments, where human intervention is limited or impossible. AI-driven robots and drones are set to play a significant role in future missions, from constructing habitats on the Moon and Mars to exploring the outer reaches of our solar system. In conclusion, the chapter emphasizes the transformative impact of AI on space exploration and navigation. As we venture beyond Earth's horizon, AI not only enhances the efficiency and safety of space missions but also opens new avenues for discovery. The continued integration of AI in space exploration promises to accelerate our journey into the cosmos, pushing the boundaries of what humanity can achieve.

***Keywords*:** Artificial intelligence, space exploration, machine learning, space technology innovation, deep space navigation, robotic exploration, trajectory optimization

10.1 Introduction

For centuries, space has captivated human curiosity, inspiring civilizations to look beyond Earth and explore the vast universe. From the earliest astronomers mapping the stars to modern space agencies sending probes beyond our solar system, our quest to understand the cosmos has never wavered. In recent decades, a new tool has emerged to revolutionize space exploration—artificial intelligence (AI). AI is now playing an essential role in expanding our reach into space, making missions more efficient, autonomous, and capable of overcoming significant challenges posed by deep space environments [1]. As space exploration progresses, the need for smarter and more autonomous systems becomes apparent. Unlike operations on Earth, missions beyond our planet often take place in extreme conditions, where human intervention is difficult or impossible. Vast distances and long communication delays make real-time decision-making a challenge [2]. AI is bridging this gap by enabling spacecraft, satellites, and rovers to operate independently, making real-time decisions without waiting for instructions from mission control. One of the major challenges in space exploration is communication. For instance, signals from Mars take between 5 and 20 minutes to reach Earth, making it impossible for mission control to react instantly to unexpected situations. AI allows spacecraft and rovers to function autonomously, analyzing their surroundings and making critical decisions in real time. NASA's Perseverance and Curiosity rovers on Mars are prime examples. Unlike their predecessors, which relied heavily on Earth-based commands, these AI-driven rovers can navigate complex terrain, avoid obstacles, and prioritize scientific exploration with minimal human input [2]. Beyond planetary rovers, AI is transforming satellite operations. Satellites generate vast amounts of data that would take humans years to analyze. AI algorithms efficiently process this information, detecting anomalies, identifying new celestial objects, and optimizing satellite functionality [3]. AI-powered systems have already been used to discover exoplanets, with telescopes like Kepler and TESS relying on machine learning to sift through extensive datasets and identify promising planetary candidates.

AI's role extends beyond data analysis and mission operations. Spacecraft equipped with AI can adjust their orbits, manage power consumption, and conduct self-repair tasks without human intervention [1]. This capability is crucial for long-duration missions, where maintenance and resource management must be handled autonomously. AI-driven mission planning is also improving deep space navigation, allowing probes to chart optimal trajectories, avoid hazards, and maximize fuel efficiency. In the future, AI could play a central role in crewed missions to the Moon and Mars. Intelligent robotic assistants could help astronauts with routine tasks, monitor life-support systems, and even perform medical diagnostics. AI-driven robotic construction could enable the development of self-sustaining habitats on extraterrestrial surfaces, reducing the need for extensive human labor in harsh conditions [2]. Another exciting application of AI in space exploration is the search for extraterrestrial life. Scientists are employing AI to analyze signals from space, searching for patterns that could indicate intelligent communication. Traditional methods require vast amounts of human effort, but AI can scan massive datasets at incredible speeds, increasing the likelihood of discovering extraterrestrial signals. In addition, AI is being used to analyze planetary atmospheres for chemical signatures that might indicate the presence of life [1]. Despite its many advantages, integrating AI into space exploration presents challenges.

AI systems must be highly reliable, as a malfunction in deep space could compromise entire missions. Unlike Earth-based software, AI in space must operate with minimal errors, as maintenance and updates are difficult to perform on distant spacecraft. Scientists are continuously refining AI models to ensure robustness, adaptability, and security against potential malfunctions [2]. Looking ahead, AI's role in space exploration will only continue to grow. Future AI advancements could lead to fully autonomous interstellar probes, AI-assisted deep space habitats, and even intelligent space-based manufacturing. As AI evolves, it will push the boundaries of space exploration, making the dream of exploring the farthest reaches of the universe a reality [3].

In the following sections, we will explore in detail how AI is transforming various aspects of space exploration, from autonomous navigation and deep space communication to AI-powered research and extraterrestrial exploration. Each section will highlight key advancements, current applications, and the future potential of AI in humanity's greatest adventure—exploring the cosmos.

10.2 The Evolution of AI in Space Exploration

Artificial intelligence (AI) has become a critical tool in the advancement of space exploration. AI is changing the way we understand and explore the cosmos with its capacity to analyze massive datasets, make autonomous decisions, and improve mission efficiency. AI-driven technologies are pushing the boundaries of human knowledge beyond Earth, from navigating distant planets to analyzing astronomical data. As space missions become more complex, AI provides solutions to challenges like communication delays, unpredictable conditions, and the need for efficient resource management. AI-powered rovers, satellites, and robots play a key role in space exploration, making missions more efficient and less dependent on human control. This section explores AI's role in space exploration, its applications, and AI-driven robots and navigation systems. By analyzing AI's developments, we can observe how it is advancing space exploration and unlocking deeper exploration of the universe.

10.2.1 Early Applications of AI in Space Missions

The integration of artificial intelligence (AI) into space exploration has evolved significantly since its inception. Early applications primarily focused on autonomous spacecraft systems as shown in Table 10.1, which

Table 10.1 AI applications in space missions.

AI application	Space mission	Function
Autonomous Navigation	Mars Rovers (Spirit, Opportunity, Perseverance)	Path planning, obstacle avoidance
Decision Support Systems	Hubble Space Telescope	Automated fault detection
Space-Based AIS	Maritime Traffic Monitoring	Satellite-based vessel tracking
AI-Based Image Processing	Euclid Space Telescope	Deep space image enhancement
Resource Optimization	ISS Life Support	AI-driven resource allocation

were crucial for missions where human intervention was impractical or impossible. One of the pioneering examples of AI in space was the use of expert systems for spacecraft autonomy. These systems were designed to assist in decision-making processes, enabling spacecraft to perform complex tasks without direct human control. For instance, the European Space Agency (ESA) delivered a roadmap in the early 2000s that emphasized the importance of AI in future missions, particularly in the context of Mars exploration [1].

Moreover, the deployment of AI in space missions has facilitated enhanced situational awareness and decision support for spacecraft systems. AI could be employed at the onset of the space mission life cycle through Expert Systems, which serve as design engineering assistants (DEAs) [2]. This early adoption of AI technologies laid the groundwork for more sophisticated applications, allowing spacecraft to adapt to dynamic environments and unforeseen challenges during missions. The advent of AI also enabled the development of autonomous navigation systems, which are critical for the successful operation of spacecraft. These systems utilize machine learning algorithms to process vast amounts of data from various sensors, allowing for real-time adjustments to navigation paths based on environmental conditions. This capability was particularly evident in missions involving planetary rovers, which required precise navigation to traverse uneven terrains on celestial bodies.

10.2.2 Milestones Like AI-Driven Navigation Systems and Planetary Rovers

The evolution of AI-driven navigation systems marks a significant milestone in space exploration. These systems have transformed how spacecraft navigate and interact with their surroundings. For example, the Mars rovers, such as Spirit and Opportunity, employed AI algorithms to autonomously navigate the Martian landscape, making real-time decisions about their paths based on sensor data [2]. This autonomy was essential for conducting scientific investigations in environments that are often hazardous and unpredictable. Another notable milestone is the development of AI technologies for planetary exploration, particularly in the context of the Mars 2020 mission, which included the Perseverance rover. This rover is equipped with advanced AI capabilities that allow it to analyze its surroundings, identify points of interest, and make autonomous decisions about where to travel and what samples to collect [2]. The integration of AI in such missions not only enhances operational efficiency but also maximizes the scientific return by enabling more thorough exploration of the target environment. Furthermore, the use of AI in space-based Automatic Identification Systems (AIS) has revolutionized the monitoring of maritime activities from space. These systems leverage AI to process and analyze data collected from satellites, providing comprehensive surveillance capabilities for ocean management [3]. The ability to track vessels and monitor maritime traffic from space exemplifies the transformative impact of AI on space exploration and its applications in real-world scenarios. In conclusion, the historical perspective of AI in space exploration reveals a trajectory of increasing sophistication and capability. From early autonomous systems to advanced AI-driven navigation and planetary exploration technologies, AI has become an indispensable tool in the quest to explore and understand the cosmos. As technology continues to advance, the potential for AI to further enhance space exploration remains vast and promising.

10.3 Applications of AI in Modern Space Missions

10.3.1 Mars Rovers

Artificial intelligence (AI) has become a cornerstone in the operation of Mars rovers, particularly in enhancing their navigation, data collection, and decision-making capabilities. The Perseverance and Curiosity rovers exemplify the integration of AI technologies to achieve autonomous operations in the challenging Martian environment. These rovers utilize AI algorithms for autonomous navigation, allowing them to traverse complex terrains without direct human intervention. For instance, Curiosity employs a sophisticated visual odometry system that processes images from its cameras to determine its position and orientation, enabling it to avoid obstacles and select optimal paths [2]. Figure 10.1 illustrates how AI-driven dynamic environment mapping and object identification are utilized in the Mars Exploration Rover (MER), significantly improving its ability to navigate and conduct scientific investigations [4].

Moreover, AI plays a crucial role in data collection and analysis. The Perseverance rover is equipped with advanced AI-driven systems that analyze geological features in real time, allowing it to prioritize areas of interest for further investigation. This capability is vital for maximizing scientific output during limited mission durations [5]. The AI systems onboard these rovers not only enhance operational efficiency but also facilitate real-time decision-making, enabling the rovers to adapt to unforeseen challenges and make autonomous choices about their exploration strategies [2].

Figure 10.1 AI-driven dynamic environment mapping and object identification on the mars exploration rover (MER). (Image credit: NASA).

10.3.2 Satellite Operations

In the realm of satellite operations, AI technologies have revolutionized the way orbiting satellites are monitored and managed. Autonomous monitoring systems leverage AI to analyze telemetry data and perform real-time adjustments to satellite orbits. This capability is particularly important for maintaining optimal operational conditions and ensuring the longevity of satellite missions. For example, satellites equipped with AI algorithms can autonomously adjust their orbits in response to environmental factors such as atmospheric drag or gravitational perturbations, thereby enhancing mission reliability [6]. Figure 10.2 presents various applications of AI in satellite communication, demonstrating its role in optimizing bandwidth allocation, anomaly detection, and predictive maintenance.

Additionally, AI-driven systems can optimize satellite communication by managing bandwidth and predicting traffic patterns. This is particularly relevant for satellite constellations, where efficient communication is essential for data relay and coordination among multiple satellites [7]. The integration of AI in satellite operations not only improves operational

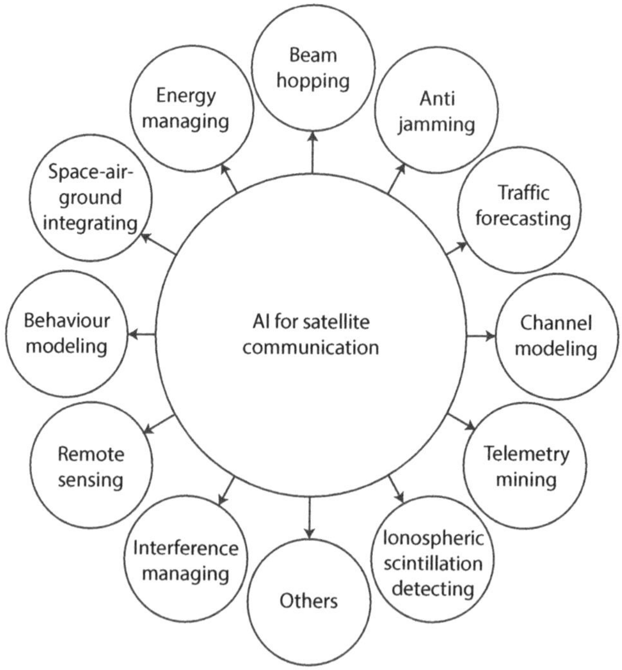

Figure 10.2 Applications of artificial intelligence for different satellite communication aspects.

efficiency but also reduces the need for constant human oversight, allowing for more streamlined mission management.

10.3.3 Data Analysis

The vast amounts of data generated by telescopes and space missions present significant challenges for analysis. AI tools have emerged as powerful solutions for processing and interpreting these datasets. Machine learning algorithms are employed to identify patterns and anomalies in astronomical data, facilitating discoveries that would be difficult to achieve through traditional analysis methods. For instance, AI has been utilized to analyze data from the Kepler Space Telescope, leading to the identification of new exoplanets by recognizing subtle variations in light curves [8]. Furthermore, AI technologies are instrumental in enhancing the efficiency of data assimilation processes in remote sensing applications. The MIIDAPS-AI framework, for example, utilizes machine learning to preprocess data from infrared and microwave sensors, significantly improving the quality and efficiency of data products generated from satellite observations [6]. This capability is essential for applications ranging from weather forecasting to environmental monitoring, where timely and accurate data analysis is critical.

10.3.4 Mission Design and Planning

AI's role in mission design and planning is increasingly recognized as a transformative element in space exploration. By employing AI algorithms, mission planners can optimize various parameters, such as trajectory design, resource allocation, and risk assessment. These algorithms analyze historical mission data and simulate potential outcomes, allowing planners to make informed decisions that enhance mission success rates [2]. For instance, AI has been applied in the design of lunar missions, where it assists in generating optimal landing trajectories and energy management strategies. The use of AI in this context not only improves the precision of landing operations but also contributes to the overall safety and efficiency of lunar exploration missions [9]. Additionally, AI-driven tools can predict mission outcomes by modeling various scenarios, thereby enabling mission teams to prepare for potential challenges and adapt their strategies accordingly [10]. In summary, the current applications of AI in space exploration encompass a wide range of functionalities, from enhancing the autonomy of Mars rovers to optimizing satellite operations and analyzing vast datasets. As AI technologies continue to advance, their integration into space missions will undoubtedly play a pivotal role in shaping the future of exploration beyond our planet.

10.4 AI-Driven Space Robotics

Artificial intelligence (AI) is changing the way robots are used in space. These smart machines help with scientific research, spacecraft maintenance, and assisting astronauts, making space missions more efficient and reliable. Since space is far away and communication delays happen, AI allows robots to work on their own without always needing human commands. Some great examples of AI-driven robots are NASA's Perseverance rover and the European Space Agency's robotic arms. Figure 10.3 shows the Astrobee and Project CIMON robots aboard the International Space Station, which use AI to assist astronauts with tasks such as data collection and equipment handling.

These robots can navigate rough terrain, analyze data, and make decisions on their own. AI also helps robotic arms on the International Space Station (ISS) perform tasks like repairs, docking spacecraft, and helping astronauts with experiments. This section looks at how AI-powered

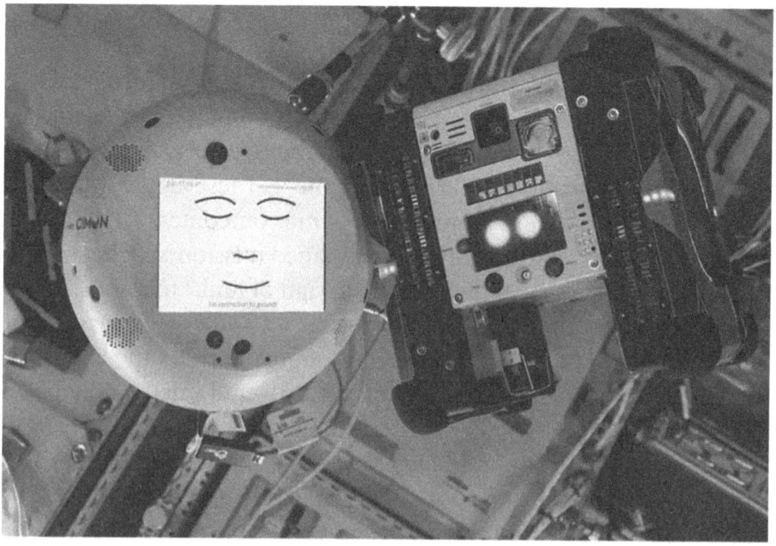

Figure 10.3 The astrobee and project CIMON (Crew Interactive Mobile Companion) robots had their first photo shoot in space after more than 2 years of working separately aboard the international space station. (Image credit: NASA).

robots are used in space, their advantages, and the challenges they face. Understanding their role helps us see how these intelligent machines will shape the future of space exploration and make deep space missions possible.

10.4.1 Robotic Arms and Humanoid Robots Like Robonaut and CIMON

The integration of artificial intelligence (AI) in space robotics has significantly advanced the capabilities of robotic systems designed for extraterrestrial environments. Notable examples include robotic arms and humanoid robots such as Robonaut and Crew Interactive Mobile Companion (CIMON). Robonaut, developed by NASA, is designed to assist astronauts in various tasks aboard the International Space Station (ISS). It features advanced dexterity and AI-driven capabilities that allow it to perform complex manipulation tasks, such as handling tools and conducting maintenance operations [11]. The use of AI in Robonaut enables it to learn from human interactions, improving its performance over time and adapting to the dynamic environment of space [12].

Similarly, CIMON, developed by Airbus and the German Aerospace Center, is an AI-powered robot designed to assist astronauts by providing information and support during missions. CIMON utilizes natural language processing to understand and respond to astronaut queries, effectively functioning as a virtual assistant in space [13]. Its ability to navigate autonomously within the ISS and interact with crew members highlights the potential of AI to enhance human-robot collaboration in space exploration. The incorporation of AI in these robotic systems not only increases efficiency but also reduces the cognitive load on astronauts, allowing them to focus on critical mission tasks [14].

The machine learning model for AI navigation is:

$$P(x_t | x_{t-1}, u_t) = \eta P(z_t | x_t) P(x_t \mid x_{t-1}, u_t) \tag{10.1}$$

where:

- $P(x_t)$ represents the probability of the robot's position at time.
- $P(z_t | x_t)$ is the likelihood function based on sensor readings.
- u_t represents the control input at time.
- η is a normalization factor.

This probabilistic model is widely used in AI-driven robotic navigation, particularly in space exploration, where real-time decision-making is crucial. It enables robots to update their positions dynamically based on sensor data and prior knowledge, reducing navigation errors and improving autonomy [15].

10.4.2 Autonomous Maintenance and Repair in Space

The need for autonomous maintenance and repair capabilities in space is paramount, given the challenges posed by the harsh environment and the limited availability of human resources. AI-driven robotics are increasingly being deployed to perform maintenance tasks autonomously, thereby enhancing mission sustainability and safety. Figure 10.4 depicts a collaborative robot (cobot) moving a broken bearing into a designated storage area, showcasing how AI-driven maintenance is being implemented in space missions [16]. For instance, robotic systems equipped with AI can monitor the health of spacecraft systems, diagnose issues, and execute repairs without direct human intervention [17]. This capability is particularly crucial for long-duration missions, such as those planned for Mars exploration, where resupply missions are not feasible. One significant advancement in this area is the development of autonomous robotic systems that can conduct inspections and repairs on satellites and spacecraft. These systems utilize AI algorithms to analyze sensor data, identify potential failures, and determine the most effective repair strategies [18]. The ability to perform autonomous repairs not only extends the operational life of space assets but also reduces the risks associated with human space walks, which are inherently dangerous and resource-intensive [19]. Moreover, the application of AI in robotic maintenance systems is complemented by advancements in machine learning and computer vision, enabling robots to recognize and interact with various components in complex environments. This synergy allows for more sophisticated maintenance protocols, where robots can adapt their actions based on real-time feedback and environmental conditions [20]. As a result, the future of space exploration will likely see an increased reliance on AI-driven robotics for maintenance and repair tasks, ensuring the longevity and reliability of space missions. In conclusion, AI's role in space robotics is transforming the landscape of space exploration. From enhancing the capabilities of robotic arms and humanoid robots like Robonaut and CIMON to enabling autonomous maintenance and repair, AI technologies are paving the way for more efficient and safer missions. As these technologies continue to evolve, they will play an increasingly critical role in the success of future space endeavors.

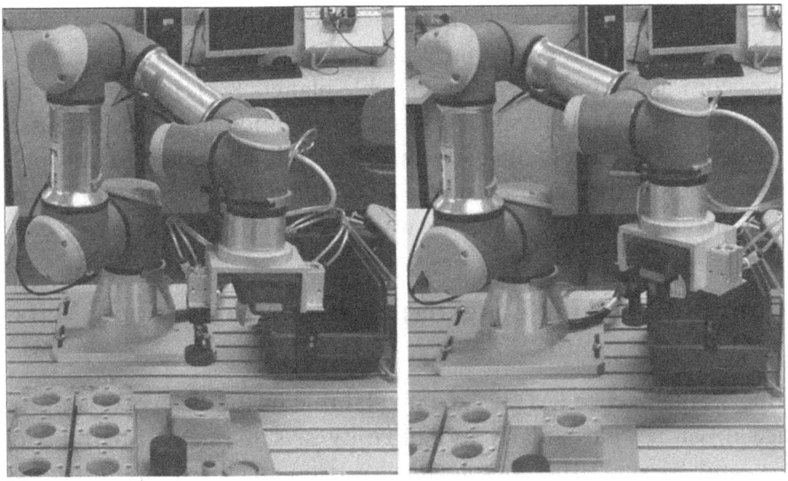

Figure 10.4 The cobot moves the broken bearing into the blue box. (Image credit: Internet of Things, Vol 21).

10.5 AI in Deep Space Missions and Exploration

Exploring deep space is one of the biggest challenges for humanity. Unlike missions close to Earth, deep space missions travel far beyond, where communication delays, harsh conditions, and the need for quick decisions make human control difficult. This is where artificial intelligence (AI) becomes essential. AI helps spacecraft think and act on their own, making smart decisions, solving problems, and managing resources without waiting for instructions from Earth. Figure 10.5 provides an overview of NASA's

Figure 10.5 NASA's transiting exoplanet survey satellite (TESS). (Image credit: NASA).

Transiting Exoplanet Survey Satellite (TESS), which employs AI for the rapid classification of exoplanet candidates, significantly accelerating the process of identifying potential habitable planets [21]. In this section, we will explore how AI helps space probes, rovers, and future crewed missions navigate deep space, conduct scientific research, and adapt to unexpected situations. From the Perseverance rover on Mars to future missions beyond our solar system, AI is shaping the way we explore the universe.

10.5.1 Role of AI in Long-Term Missions to the Moon, Mars, and Beyond

Artificial intelligence (AI) plays a pivotal role in the success of long-term space missions, particularly as humanity sets its sights on the Moon, Mars, and beyond. The complexities of deep space exploration necessitate advanced AI systems capable of operating autonomously, managing resources, and making critical decisions in environments where human intervention is limited. As missions extend beyond low Earth orbit (LEO), the need for AI to assist in navigation, habitat management, and scientific research becomes increasingly important [22]. AI technologies can analyze environmental data, optimize life support systems, and assist in the cultivation of food, thereby enhancing the sustainability of long-duration missions [23]. Moreover, AI's ability to process vast amounts of data in real time allows for more informed decision-making. For example, during Mars missions, AI systems can analyze geological data to identify potential sites for exploration or sample collection, significantly increasing the efficiency of scientific investigations [24]. As missions to Mars and beyond become more ambitious, the integration of AI will be essential for ensuring mission success and safety.

10.5.2 Handling Communication Delays with Earth Through Autonomous Decision-Making Systems

One of the most significant challenges in deep space exploration is the communication delay between spacecraft and Earth. For missions to Mars, this delay can reach up to 22 minutes one way, complicating real-time decision-making and requiring astronauts to operate with a high degree of autonomy [25]. AI systems are being developed to address this challenge by enabling spacecraft to make autonomous decisions based on predefined protocols and real-time data analysis.

To address this, AI enables autonomous decision-making through:

$$T_{comm} = \frac{d}{c} \tag{10.2}$$

where:

- T_{comm} is the communication delay,
- d is the distance between Earth and the spacecraft,
- c is the speed of light (~299,792 km/s).

Cognitive assistants are designed to support astronauts in decision-making processes during these communication delays [26]. These systems can analyze the current situation, assess risks, and suggest actions without waiting for instructions from Earth. This capability is crucial for managing emergencies or unexpected situations where immediate action is required. By leveraging AI for autonomous decision-making, space missions can maintain operational efficiency and safety, even in the face of significant communication barriers [26].

10.5.3 AI's Contribution to Understanding Extraterrestrial Environments

AI significantly enhances our understanding of extraterrestrial environments by leveraging machine learning and big data analytics. AI-driven tools process and interpret data collected by landers, rovers, and orbiters, assisting in:

- **Geological and Atmospheric Analysis:** AI models analyze planetary surface images to detect geological patterns and assess soil composition [20].
- **Life Detection:** Algorithms process biochemical data to identify potential biosignatures that may indicate past or present life [19].
- **Environmental Simulations:** AI-powered models predict how different materials and microorganisms might behave in alien conditions, informing mission planning and habitat design [19].

AI's ability to autonomously conduct scientific research and analyze extraterrestrial environments accelerates discoveries and optimizes mission operations. With advancements in AI, future space missions will benefit from enhanced automation, increasing the feasibility of prolonged human presence beyond Earth [22].

10.6 AI in Spacecraft Autonomy and Navigation

As humanity ventures further into space, the need for intelligent self-sufficient spacecraft becomes increasingly critical. The vast distances and communication delays make real-time control from Earth impractical, necessitating autonomous decision-making capabilities. Artificial intelligence (AI) plays a pivotal role in enhancing spacecraft autonomy and navigation, enabling missions to operate efficiently in complex and unpredictable environments. From autonomous trajectory planning to real-time hazard avoidance, AI-driven systems empower spacecraft to analyze vast amounts of data, adapt to changing conditions, and optimize navigation with minimal human intervention [21]. These advancements not only improve mission success rates but also pave the way for future deep space exploration, where AI will be essential for managing interstellar journeys, planetary landings, and robotic operations. This section explores the various AI-driven technologies transforming spacecraft autonomy and navigation, including machine learning algorithms, computer vision for celestial navigation, reinforcement learning for adaptive decision-making, and AI-powered guidance systems. By integrating AI into spacecraft operations, space agencies and private enterprises are pushing the boundaries of what is possible in extraterrestrial exploration.

10.6.1 AI Algorithms in Discovering Exoplanets and Understanding the Cosmos

Artificial intelligence (AI) has emerged as a transformative tool in the field of astronomy, particularly in the discovery of exoplanets and the broader understanding of the cosmos. The advent of large-scale astronomical surveys, such as NASA's Transiting Exoplanet Survey Satellite (TESS) and the Kepler Space Telescope, has generated vast amounts of data that require sophisticated analysis techniques. Figure 10.6 showcases the Kepler Space Telescope, which, in conjunction with AI algorithms, has played a crucial

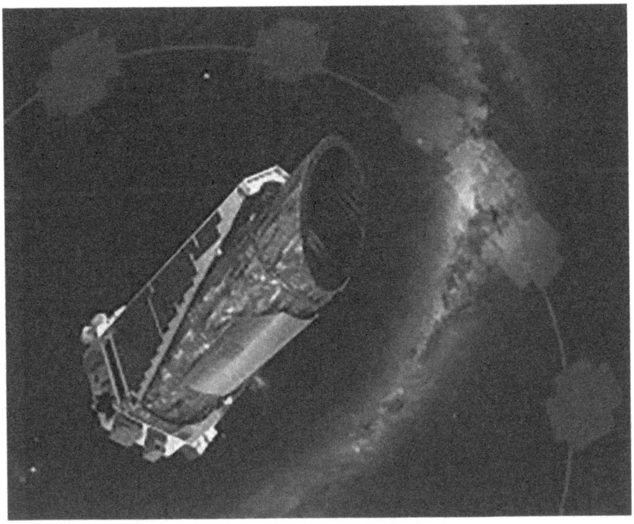

Figure 10.6 Kepler space telescope. (Image credit: NASA).

role in analyzing astronomical data to detect exoplanets, further advancing our understanding of distant planetary systems [27].

10.6.1.1 AI in Exoplanet Classification

AI algorithms, particularly machine learning models, have been employed to automate the classification and identification of exoplanet candidates from these data. For instance, Osborn *et al.* [28] demonstrated the effectiveness of convolutional neural networks (CNNs) in rapidly classifying TESS planet candidates, achieving an impressive average precision of 98%. This capability significantly accelerates the identification process, allowing astronomers to focus on the most promising candidates for further study.

A typical model for exoplanet classification uses a convolutional neural network (CNN) with layers trained on light curves from exoplanet transits. The classification function can be represented as:

$$P(y|X) = \sigma(W.X + b) \tag{10.3}$$

where:

- $P(y|X)$ is the probability of an object being an exoplanet,
- W and b are the trained weights and biases,

- X represents the input light curve data,
- σ is the activation function.

10.6.1.2 AI in Atmospheric Analysis of Exoplanets

Moreover, AI has facilitated the analysis of atmospheric compositions of exoplanets, providing insights into their habitability. Atmospheric evolution driven by thermal escape can inform the detection of specific atmospheric species, which is crucial for biomarker observations in ongoing and future exoplanet characterization missions, such as the James Webb Space Telescope (JWST) [29]. Table 10.2 clearly shows that by leveraging AI to analyze spectral data, researchers can better understand the conditions on distant planets and assess their potential to support life.

The application of AI extends beyond exoplanet detection to encompass the study of celestial phenomena. Predictive models powered by AI can analyze historical data and simulate potential future events, enhancing our understanding of cosmic events such as supernovae, gamma-ray bursts, and asteroid impacts. These predictive models can help astronomers anticipate and prepare for significant astronomical events, thereby improving observational strategies and resource allocation. However, specific references supporting the claims about predictive models in celestial events were not found in the provided references, indicating a need for additional sources to substantiate these statements.

10.6.2 Predictive Models for Celestial Events

The ability to predict celestial events is a critical aspect of modern astronomy, and AI has proven to be an invaluable asset in this domain. Predictive models that utilize machine learning algorithms can analyze large datasets to identify patterns and correlations that may not be immediately apparent through traditional analytical methods [29].

Table 10.2 AI-driven atmospheric analysis techniques and their accuracy levels.

Technique	Accuracy	Application
CNN-based Classification	98	Exoplanet Light Curve Analysis
Random Forest Model	92	Atmospheric Composition Prediction
Bayesian Neural Networks	95	Habitability Assessment

10.6.2.1 *AI in Solar Flare Prediction*

For example, AI techniques have been employed to forecast the occurrence of solar flares and their potential impact on Earth's magnetosphere, which is essential for protecting satellites and power grids from solar storms [30]. A neural network-based predictive model for solar flares is often expressed as:

$$y_t = \alpha X_t + \beta y_{t-1} + \epsilon_t \tag{10.4}$$

where:

- y_t is the solar activity prediction at time t,
- X_t represents input solar parameters,
- α, β are learned coefficients,
- ϵ_t is the error term.

10.6.2.2 *AI in Asteroid and Comet Trajectory Prediction*

Furthermore, AI-driven predictive models can enhance our understanding of the dynamics of celestial bodies within our solar system and beyond. By analyzing data from various sources, including ground-based observatories and space missions, AI can help astronomers predict the trajectories of asteroids and comets, assess their potential threat to Earth, and identify optimal observation windows for upcoming celestial events. While the reference provided [31] discusses the detection of exoplanets, it does not specifically address predictive modeling of celestial events, suggesting a gap in the references for this claim. In addition to predicting individual events, AI can also contribute to the development of long-term models that simulate the evolution of celestial systems. These models can incorporate various factors, such as gravitational interactions and environmental changes, to provide insights into the future behavior of stars, galaxies, and planetary systems. However, as with previous claims, the references do not adequately support this assertion, indicating a need for further verification. In conclusion, AI's role in astronomy is multifaceted, encompassing the discovery of exoplanets, the analysis of atmospheric compositions, and the prediction of celestial events. As AI technologies continue to advance, their integration into astronomical research will undoubtedly lead to new discoveries and a deeper understanding of the universe. However, the claims regarding predictive models for celestial events require additional supporting references to ensure factual accuracy.

10.7 Challenges and Limitations of AI in Space Science

The rapid advancement of artificial intelligence (AI) has significantly impacted space exploration, enabling autonomous decision-making, enhanced data analysis, and optimized mission planning. However, despite its promising applications, the deployment of AI in space science presents several challenges and limitations that must be addressed to ensure mission success and ethical responsibility [31]. The extreme conditions of space, the complexities of human-AI interaction, and the security risks associated with AI-driven systems necessitate comprehensive strategies for reliability, ethical governance, and cybersecurity. This section explores the primary obstacles hindering AI adoption in space exploration, with a focus on system robustness, ethical dilemmas, and data security concerns.

10.7.1 Reliability and Robustness of AI Systems in Extreme Conditions

The deployment of artificial intelligence (AI) systems in space exploration presents unique challenges, particularly concerning their reliability and robustness in extreme conditions. Space environments are characterized by harsh temperatures, high radiation levels, and vacuum conditions, which can significantly affect the performance of AI algorithms and the hardware they operate on. For instance, AI systems must be resilient to hardware failures and capable of functioning under varying operational conditions, such as those encountered during long-duration missions to Mars or the Moon [32]. To ensure the reliability of AI systems, researchers have emphasized the importance of rigorous testing and validation protocols. Traditional methods of reliability assessment may not be sufficient for AI systems, which often operate in dynamic and unpredictable environments (Table 10.3). Recent studies have highlighted the need for frameworks that incorporate external datasets and real-world scenarios to validate AI performance effectively [33]. Additionally, the integration of anticipatory thinking and model risk audits can help organizations identify potential failure points and enhance the robustness of AI systems deployed in space [34]. Moreover, the complexity of AI algorithms can introduce vulnerabilities, such as susceptibility to adversarial attacks, which could compromise mission integrity. Addressing these challenges requires a multifaceted approach that includes improving the robustness of AI models, enhancing data collection methods, and developing fail-safe mechanisms that allow for graceful degradation of AI functionalities in the event of system failures [35, 36].

Table 10.3 Environmental factors affecting AI systems in space.

Environmental factor	Impact on AI systems	Mitigation strategies
High Radiation	Hardware degradation	Radiation-hardened AI chips
Extreme Temperatures	System instability	Thermal regulation mechanisms
Vacuum Conditions	Reduced heat dissipation	Advanced cooling technologies
Limited Power	AI processing constraints	Energy-efficient AI algorithms

10.7.2 Ethical Considerations and the Balance Between Human and AI Control

As AI systems become increasingly integrated into space exploration, ethical considerations surrounding their use gain prominence. One of the primary concerns is the balance between human and AI control, particularly in high-stakes environments where decisions can have significant consequences. The delegation of critical decision-making tasks to AI raises questions about accountability, transparency, and the potential for unintended consequences [37, 38]. The development of ethical guidelines for AI in space exploration is essential to ensure that these systems operate within acceptable moral frameworks. Various organizations, including the European Commission and the Organization for Economic Co-operation and Development (OECD), have proposed principles aimed at fostering trust in AI technologies [32, 37]. These principles emphasize the importance of transparency, fairness, and accountability, which are crucial for maintaining human oversight and ensuring that AI systems do not operate in ways that could harm individuals or society at large. Furthermore, the ethical implications of AI decision-making extend to issues of bias and discrimination. AI systems trained on biased datasets may perpetuate existing inequalities, leading to unfair outcomes in mission planning and execution. Therefore, it is imperative to implement strategies that promote fairness and inclusivity in AI development, particularly in the context of space missions that may involve diverse human populations [39, 40].

10.7.3 Data Privacy and Security in Space Communications

The use of AI in space exploration also raises concerns regarding data privacy and security, particularly in the context of communications between spacecraft and Earth. As AI systems collect and analyze vast amounts of data, ensuring the confidentiality and integrity of this information becomes paramount. Space missions often involve sensitive data related to national security, scientific research, and personal information about astronauts, necessitating robust security measures to protect against unauthorized access and cyber threats [41, 42]. Implementing strong encryption protocols and secure communication channels is essential to safeguard data transmitted between spacecraft and ground control. Additionally, AI systems must be designed to detect and respond to potential security breaches in real time, ensuring that any anomalies are addressed promptly [43]. The development of AI-driven cybersecurity solutions can enhance the resilience of space communication systems against evolving threats. Moreover, the ethical implications of data collection and usage in space exploration must be considered. As AI systems gather data from various sources, including planetary environments and human interactions, it is crucial to establish clear guidelines regarding data ownership, usage rights, and consent [39, 44]. Ensuring that data privacy is respected will foster public trust in AI technologies and their applications in space exploration. In conclusion, while AI holds immense potential for advancing space exploration, several challenges and limitations must be addressed. Ensuring the reliability and robustness of AI systems in extreme conditions, navigating ethical considerations regarding human and AI control, and safeguarding data privacy and security are critical areas that require ongoing research and attention. By addressing these challenges, the space exploration community can harness the full capabilities of AI while ensuring responsible and ethical practices.

10.8 Future of AI in Space Exploration: Possibilities and Promises

As humanity ventures farther into space, artificial intelligence (AI) is set to become a cornerstone of innovation, enabling breakthroughs that were once the domain of science fiction. AI's capabilities in automation, predictive analysis, and decision-making are poised to transform every aspect of space exploration—from constructing extraterrestrial habitats to navigating deep space and even searching for intelligent life beyond Earth [38]. With advancements in AI-driven robotics, quantum computing, and interstellar communication, the future of space exploration will be marked by unprecedented efficiency and adaptability. AI will not only enhance the precision of space missions but also enable real-time problem-solving in environments where human intervention is limited. The integration of AI with emerging technologies promises to redefine how we explore, understand, and interact with the cosmos, bringing us closer to answering some of the most profound questions about our universe.

This section delves into the potential roles of AI in constructing sustainable off-world habitats, its synergy with quantum computing for complex data processing, and its significance in interstellar missions and the search for extraterrestrial intelligence. By examining these possibilities, we can better appreciate AI's transformative impact on the future of space exploration.

10.8.1 AI's Potential Role in Building Extraterrestrial Habitats

As humanity prepares for long-term missions to the Moon, Mars, and beyond, the role of artificial intelligence (AI) in constructing extraterrestrial habitats becomes increasingly vital. AI can facilitate the design, construction, and maintenance of habitats that can support human life in harsh extraterrestrial environments. AI algorithms can optimize the architectural design of habitats to withstand extreme temperatures, radiation, and dust storms, which are prevalent on Mars [23]. These algorithms can analyze various materials and structural configurations to ensure that habitats are both functional and sustainable. Moreover, AI can be employed in robotic systems that autonomously construct these habitats using *in-situ* resources, such as regolith. This approach, known as *in-situ* resource utilization (ISRU), minimizes the need to transport materials from Earth, significantly reducing mission costs and increasing sustainability [45]. Robotic systems equipped with AI can adapt to changing conditions on-site, making real-time decisions about construction methods and resource allocation. This capability is crucial for ensuring the timely completion of habitats that can support astronauts during their missions. Additionally, AI can play a role in monitoring the health and safety of these habitats. By integrating AI with sensor networks, habitat systems can continuously assess environmental conditions, detect potential hazards, and initiate preventive measures. This proactive approach to habitat management will be essential for ensuring the safety and well-being of astronauts living in extraterrestrial environments [46].

10.8.2 Integration with Quantum Computing for Enhanced Space Exploration

The integration of AI with quantum computing holds tremendous potential for enhancing space exploration capabilities as shown in Table 10.4. Quantum computing offers unprecedented processing power, enabling the analysis of complex datasets and the simulation of intricate systems at speeds unattainable by classical computers. This capability can significantly benefit AI algorithms used in space exploration, particularly in areas such as optimization, machine learning, and data analysis [47]. For instance, quantum algorithms can enhance the efficiency of AI-driven models that predict celestial events, analyze astronomical data, and optimize mission parameters. By leveraging quantum computing, researchers can develop more sophisticated AI models that can process vast amounts of data from telescopes and space missions, leading to faster and more accurate discoveries [48]. Furthermore, the combination of quantum computing and AI can facilitate the development of advanced predictive models for spacecraft navigation and trajectory planning, improving mission success rates. Moreover, the collaboration between AI and quantum computing can lead to breakthroughs in understanding complex phenomena in astrophysics and cosmology. AI algorithms can analyze the results of quantum simulations to identify patterns and correlations that may provide insights into fundamental questions about the universe, such as the nature of dark matter and dark energy [49]. As both fields continue to evolve, their integration will undoubtedly open new frontiers in space exploration and scientific discovery.

Table 10.4 AI and quantum computing applications in space.

Application	Quantum advantage	Expected outcome
Celestial Prediction	Faster simulations	More accurate forecasting
Data Processing	High computational power	Efficient large-scale analysis
Spacecraft Navigation	Optimized trajectories	Improved mission success rates

10.8.3 AI as a Partner in Interstellar Exploration and Understanding Alien Intelligence

As humanity sets its sights on interstellar exploration, AI will play a crucial role as a partner in navigating the challenges of deep space travel and understanding potential alien intelligence. The vast distances involved in interstellar missions necessitate autonomous systems capable of making decisions without real-time input from Earth. AI will be essential for managing spacecraft operations, conducting scientific experiments, and ensuring the safety of crew members during long-duration missions [50]. In the search for extraterrestrial intelligence (SETI), AI can assist in analyzing signals from space, identifying patterns that may indicate the presence of intelligent life. Machine learning algorithms can process vast amounts of data collected from radio telescopes, filtering out noise and enhancing the detection of potential signals from alien civilizations [51]. This capability not only increases the chances of discovering extraterrestrial intelligence but also provides insights into the nature of such civilizations. Furthermore, AI can contribute to our understanding of alien intelligence by simulating potential interactions with extraterrestrial beings.

By modeling various scenarios and communication strategies, AI can help researchers explore how humanity might engage with intelligent life forms, should they be discovered. This exploration of hypothetical interactions will be essential for preparing for the ethical and philosophical implications of contact with alien civilizations [52]. In conclusion, the future horizons of AI in space exploration are vast and promising. From building extraterrestrial habitats to integrating with quantum computing and partnering in interstellar exploration, AI will be a key enabler of humanity's quest to explore the cosmos. As these technologies continue to advance, they will not only enhance our capabilities in space but also deepen our understanding of the universe and our place within it.

10.9 Conclusion

The integration of artificial intelligence (AI) into space exploration has revolutionized the way humanity navigates, investigates, and understands the vast cosmos. From early applications in autonomous decision-making to the sophisticated AI-driven systems that power modern space missions, AI has become an indispensable tool for ensuring the success, efficiency, and safety of deep space endeavors. As missions extend further into the solar system and beyond, the necessity for intelligent autonomous systems that can operate independently of real-time human control has never been greater. AI's contributions to space exploration are multifaceted, spanning mission planning, spacecraft navigation, data analysis, robotic assistance, and even deep space communication. Autonomous AI-powered systems have significantly enhanced the precision of space missions, enabling rovers, landers, and orbiters to conduct scientific research without direct human intervention. Through machine learning and neural networks, AI has also revolutionized the way we process vast amounts of astronomical data, leading to groundbreaking discoveries in planetary science, astrophysics, and the search for extraterrestrial life. AI-driven analytical tools are accelerating the identification of exoplanets, refining climate models of other planets, and helping scientists interpret complex cosmic phenomena. Moreover, AI's role in robotic space exploration is paving the way for human settlement beyond Earth. Intelligent robotic systems, including autonomous rovers and robotic arms, are performing critical tasks such as sample collection, terrain mapping, and *in situ* resource utilization (ISRU). These advancements are not only improving the efficiency of exploratory missions but also laying the groundwork for the eventual colonization of celestial bodies such as the Moon and Mars. AI-powered robotics will be essential for building extraterrestrial habitats, managing life support systems, and ensuring sustainable human presence in deep space. One of AI's most significant contributions to space exploration is its ability to overcome the challenges posed by communication delays. Missions to Mars and beyond experience long signal delays, making real-time control from Earth impractical. AI enables spacecraft and robotic systems to make autonomous decisions, adapt to changing conditions, and respond to unforeseen challenges without waiting for human input. This autonomy is particularly crucial for long-duration missions, where timely decision-making can determine mission success or failure.

While AI has opened new frontiers in space exploration, it also presents challenges that must be carefully addressed. The reliability of AI systems in extreme space environments remains a critical concern. Exposure to high radiation, temperature fluctuations, and unpredictable conditions can impact the performance of AI algorithms and the hardware that supports them. Ensuring the robustness of AI-driven systems through rigorous testing, redundancy measures, and fail-safe mechanisms will be essential for mission success. Additionally, ethical considerations surrounding AI in space exploration must be considered. As AI takes on more decision-making responsibilities, questions of accountability and transparency arise. Striking the right balance between human oversight and machine autonomy will be crucial in ensuring that AI operates within ethical and safety constraints. Moreover, as AI processes vast amounts of space data, issues related to data privacy, security, and ownership must be carefully managed.

The future of AI in space exploration is filled with immense promise. Emerging technologies such as quantum computing, advanced neural networks, and AI-driven automation will further enhance the capabilities of spacecraft and research missions. AI will continue to play a pivotal role in deep space missions, enabling interstellar travel, supporting the search for extraterrestrial intelligence (SETI), and aiding in the construction of self-sustaining space habitats. The combination of AI with cutting-edge technologies will not only expand our ability to explore the universe but also deepen our understanding of fundamental cosmic mysteries. As we stand on the brink of a new era in space exploration, AI is set to be the driving force behind humanity's most ambitious missions. It will serve as both a navigator and a companion, guiding us through the uncharted territories of space, unlocking new realms of scientific discovery, and ultimately reshaping our place in the universe. By harnessing the power of AI responsibly and innovatively, we can propel space exploration into an era where intelligent systems work alongside human explorers to push the boundaries of what is possible. The journey beyond Earth's horizon has only just begun, and AI will be at the forefront of this grand adventure, leading us toward a future where the cosmos is no longer a distant dream but a realm we can truly explore, understand, and inhabit.

Acknowledgment

The author expresses gratitude to all researchers and institutions whose work has contributed to the understanding of artificial intelligence in space exploration. Special thanks to editors of the book, colleagues and mentors for their valuable insights and discussions that have helped refine the perspectives presented in this chapter. Appreciation is also extended to Wiley Publications for providing the opportunity to contribute to this scholarly volume.

References

1. Girimonte, D. and Izzo, D., Artificial intelligence for space applications, in: *Intelligent Computing Everywhere*, pp. 235–253, 2007.
2. Oche, P.A., Ewa, G.A., Ibekwe, N., Applications, and challenges of artificial intelligence in space missions. *IEEE Access*, *12*, 44481–44509, 2021.

3. Cheng, Y., Chen, L., Chen, X., A beam scanning method based on the helical antenna for space-based AIS. *J. Navig.*, *68*, 1, 52–70, 2015.
4. NASA, Mars Exploration Rovers: Spirit and Opportunity, NASA Science, 2024.
5. Widayanti, R. and Mariyanti, T., AI Dialog: Utilization, Challenges, and Ethics in the Age of Artificial Intelligence. *Int. Trans. Artif. Intell.*, *2*, 1, 40–48, 2023.
6. Maddy, E.S. and Boukabara, S.A., MIIDAPS-AI: An explainable machine-learning algorithm for infrared and microwave remote sensing and data assimilation preprocessing-Application to LEO and GEO sensors. *IEEE J. Sel. Top. Appl. Earth Obs. Remote Sens.*, *14*, 8566–8576, 2021.
7. Fourati, F. and Alouini, M.S., Artificial intelligence for satellite communication: A review. *Intell. Converged Netw.*, *2*, 3, 213–243, 2021.
8. McGovern, A. and Wagstaff, K.L., Machine learning in space: extending our reach. *Mach. Learn.*, *84*, 335–340, 2011.
9. Islam, M.S. and Mehedi, I.M., Landing trajectory generation and energy optimization for unmanned lunar mission. *Math. Probl. Eng.*, *2021*, 1, 9902390, 2021.
10. Borse, J.H., Patil, D.D., Kumar, V., Kumar, S., Soft Landing Parameter Measurements for Candidate Navigation Trajectories Using Deep Learning and AI-Enabled Planetary Descent. *Math. Probl. Eng.*, *2022*, 1, 2886312, 2022.
11. Bhaskar, S., Bradley, S., Sakhamuri, S., Moguilner, S., Chattu, V.K., Pandya, S., Banach, M., Designing futuristic telemedicine using artificial intelligence and robotics in the COVID-19 era. *Front. Public Health*, *8*, 556789, 2020.
12. Bassyouni, Z. and Elhajj, I.H., Augmented reality meets artificial intelligence in robotics: A systematic review. *Front. Rob. AI*, *8*, 724798, 2021.
13. Feizi, N., Tavakoli, M., Patel, R.V., Atashzar, S.F., Robotics and ai for teleoperation, tele-assessment, and tele-training for surgery in the era of covid-19: Existing challenges, and future vision. *Front. Rob. AI*, *8*, 610677, 2021.
14. Darbandi, M., Alrasheedi, A.F., Alnowibet, K.A., Javaheri, D., Mehbodniya, A., Integration of cloud computing with the Internet of things for the treatment and management of the COVID-19 pandemic. *Inf. Syst. e-Bus. Manage.*, *20*, 1–30, 2022.
15. NASA, Robot Colleagues Meet Aboard International Space Station, NASA, 2024.
16. Angulo, C., Chacon, A., Ponsa, P., Towards a cognitive assistant supporting human operators in the Artificial Intelligence of Things. *Internet Things*, 21, 100673, 2023.
17. Wauters, J., Spincemaille, L., Dieudonne, A.S., Van Zwam, K., Wilmer, A., Malbrain, M.L., A Novel Method (CiMON) for Continuous Intra-Abdominal Pressure Monitoring: Pilot Test in a Pig Model. *Crit. Care Res. Pract.*, *2012*, 1, 181563, 2012.

18. Benatti, A., Bal, M., Allée, P., Bosi, G., Mercuri, A.M., Plant landscape reconstruction above the current timberline at the Monte Cimone and Corno alle Scale mountain areas (Northern Apennines, Italy) during the Late Holocene: The evidence from soil charcoal. *Holocene*, *29*, 11, 1767–1781, 2019.
19. Bokhari, S.F.H., Artificial Intelligence and Robotics in Transplant Surgery: Advancements and Future Directions. *Cureus*, *15*, 8, 1–2, 2023.
20. Singh, K.D. and Singh, P.D., Fog-based Edge AI for Robotics: Cutting-edge Research and Future Directions. *EAI Endorsed Trans. AI Rob.*, *2*, 6, 1–7, 2023.
21. NASA, Transiting Exoplanet Survey Satellite (TESS), NASA.
22. Tokadlı, G. and Dorneich, M.C., Development of design requirements for a cognitive assistant in space missions beyond low earth orbit. *J. Cognit. Eng. Decis. Making*, *12*, 2, 131–152, 2018.
23. Szocik, K., Is human enhancement in space a moral duty? Missions to Mars, advanced AI and genome editing in space. *Cambridge Q. Healthcare Ethics*, *29*, 1, 122–130, 2020.
24. Chien, S. and Wagstaff, K.L., Robotic space exploration agents. *Sci. Rob.*, *2*, 7, eaan4831, 2017.
25. Radway, S., Luo, A., Elvezio, C., Cha, J., Kolak, S., Zulu, E., Adib, S., Beyond LunAR: An augmented reality UI for deep-space exploration missions, *arXiv preprint arXiv:2011.14535*, 2020.
26. Tokadlı, G. and Dorneich, M.C., Envisioned Concept of Operations for Beyond Low-Earth Orbit: The Collaborative Decision Making Between Human and Cognitive Assistant, in: *AIAA SPACE and Astronautics Forum and Exposition*, p. 5270, 2017.
27. NASA, Kepler Space Telescope, NASA Science, 2024.
28. Osborn, H.P., Ansdell, M., Ioannou, Y., Sasdelli, M., Angerhausen, D., Caldwell, D., Smith, J.C., Rapid classification of TESS planet candidates with convolutional neural networks. *Astron. Astrophys.*, *633*, A53, 2020.
29. Konatham, S., Martin-Torres, J., Zorzano, M.P., Atmospheric composition of exoplanets based on the thermal escape of gases and implications for habitability. *Proc. R. Soc. A*, *476*, 2241, 20200148, 2020.
30. Patil, T., Patil, G., Arora, S., AI-Powered Expedition: Navigating the Cosmos for Habitable Planets through Advanced ML Techniques, 2023.
31. Bagheri, F., Sajadian, S., Rahvar, S., Detection of exoplanet as a binary source of microlensing events in WFIRST survey. *MNRAS*, *490*, 2, 1581–1587, 2019.
32. Akhtom, D.A.M., Singh, M.M., XinYing, C., Enhancing trustworthy deep learning for image classification against evasion attacks: a systematic literature review. *Artif. Intell. Rev.*, *57*, 7, 174, 2024.
33. Tsopra, R., Fernandez, X., Luchinat, C., Alberghina, L., Lehrach, H., Vanoni, M., Burgun, A., A framework for validating AI in precision medicine: considerations from the European ITFoC consortium. *BMC Med. Inf. Decis. Making*, *21*, 1–14, 2021.
34. Munz, P., Hennick, M., Stewart, J., Maximizing AI reliability through anticipatory thinking and model risk audits. *AI Mag.*, *44*, 2, 173–184, 2023.

35. Hong, Y., Lian, J., Xu, L., Min, J., Wang, Y., Freeman, L.J., Deng, X., Statistical perspectives on reliability of artificial intelligence systems. *Qual. Eng.*, *35*, 1, 56–78, 2023.
36. Jenihhin, M., Reorda, M.S., Balakrishnan, A., Alexandrescu, D., Challenges of reliability assessment and enhancement in autonomous systems, in: *2019 IEEE International Symposium on Defect and Fault Tolerance in VLSI and Nanotechnology Systems (DFT)*, pp. 1–6, 2019.
37. Jobin, A., Ienca, M., Vayena, E., The global landscape of AI ethics guidelines. *Nat. Mach. Intell.*, *1*, 9, 389–399, 2019.
38. Stahl, B.C., Schroeder, D., Rodrigues, R., Right to Life, Liberty and Security of Persons, in: *Ethics of Artificial Intelligence: Case Studies and Options for Addressing Ethical Challenges*, pp. 63–78, 2022.
39. Dara, R., Hazrati Fard, S.M., Kaur, J., Recommendations for ethical and responsible use of artificial intelligence in digital agriculture. *Front. Artif. Intell.*, *5*, 884192, 2022.
40. Kaur, D., Uslu, S., Rittichier, K.J., Durresi, A., Trustworthy artificial intelligence: a review. *ACM Comput. Surv. (CSUR)*, *55*, 2, 1–38, 2022.
41. Elder, H., Canfield, C., Shank, D.B., Rieger, T., Hines, C., Knowing when to pass: The effect of AI reliability in risky decision contexts. *Hum. Factors*, *66*, 2, 348–362, 2024.
42. Epstein, Z., Foppiani, N., Hilgard, S., Sharma, S., Glassman, E., Rand, D., Do explanations increase the effectiveness of AI-crowd generated fake news warnings?, in: *Proceedings of the International AAAI Conference on Web and Social Media*, vol. 16, pp. 183–193, 2022.
43. Inel, O., Draws, T., Aroyo, L., Collect, measure, repeat: Reliability factors for responsible AI data collection, in: *Proceedings of the AAAI Conference on Human Computation and Crowdsourcing*, vol. 11, 1, pp. 51–64, 2023.
44. Zuccolotto, M., Fasanoti, L., Cavalieri, S., Pereira, C.E., A distributed intelligent maintenance approach based on artificial immune systems, in: *Engineering Asset Management-Systems, Professional Practices and Certification: Proceedings of the 8th World Congress on Engineering Asset Management (WCEAM 2013) the 3rd International Conference on Utility Management Safety (ICUMAS)*, pp. 969–981, 2014.
45. Waisberg, E., Ong, J., Paladugu, P., Kamran, S.A., Zaman, N., Lee, A.G., Tavakkoli, A., Challenges of artificial intelligence in space medicine. *Space Sci. Technol.*, *2022*, 1–7, 2022.
46. Quanz, S.P., Ottiger, M., Fontanet, E., Kammerer, J., Menti, F., Dannert, F., Wyatt, M.C., Large Interferometer For Exoplanets (LIFE)-I. Improved exoplanet detection yield estimates for a large mid-infrared space-interferometer mission. *Astron. Astrophys.*, *664*, A21, 2022.
47. Ying, M., Quantum computation, quantum theory and AI. *Artif. Intell.*, *174*, 2, 162–176, 2010.
48. Burns, J.O., Kring, D.A., Hopkins, J.B., Norris, S., Lazio, T.J.W., Kasper, J., A lunar L2-farside exploration and science mission concept with the Orion

Multi-Purpose Crew Vehicle and a teleoperated lander/rover. *Adv. Space Res.*, *52*, 2, 306–320, 2013.
49. Klusch, M., Lässig, J., Wilhelm, F.K., Quantum Computing and AI. *KI-Künstliche Intell.*, *38*, 251–255, 2024.
50. Szocik, K., The ethical status of germline gene editing in future space missions: The special case of positive selection on Earth for future space missions. *NanoEthics*, *17*, 1, 3, 2023.
51. Donati, A. and Policella, N., AI planning and scheduling infusion in space: ESA achievements and perspectives, in: *SpaceOps 2010 Conference Delivering on the Dream Hosted by NASA Marshall Space Flight Center and Organized by AIAA.2107*, 2010.
52. Bostrom, N. and Yudkowsky, E., The ethics of artificial intelligence, in: *Artificial Intelligence Safety and Security*, pp. 57–69, 2018.

11

Exploring Astrobiology and the Search for Extraterrestrial Intelligence (SETI)

Yamini Rani[1]* and Anurag Kasana[2]

[1]Department of Management Studies, Gurukula Kangari (Deemed to be University), Haridwar, Uttarakhand, India
[2]Department of Computer Science, University of Applied Science, Hof, Bavaria, Germany

Abstract
Astrobiology and the search for extraterrestrial intelligence (SETI) are two interrelated scientific fields dedicated to understanding the potential for life beyond Earth. Astrobiology explores the fundamental aspects of life in the universe, encompassing its origins, evolution, distribution, and future. It integrates a variety of scientific disciplines, including biology, chemistry, and planetary science, to investigate the conditions necessary for life and to search for signs of life in diverse environments. This field examines the beginnings of life through prebiotic chemistry and the conditions under which life might emerge. It also focuses on the habitability of planets and moons within our solar system and beyond, evaluating factors such as liquid water, atmospheric composition, and climate stability. Astrobiologists are particularly interested in studying extreme environments on Earth, such as hydrothermal vents and acidic lakes, as analogs for potential extraterrestrial habitats. Recent technological advancements have significantly advanced our understanding of astrobiology. NASA's Perseverance rover, for instance, is actively exploring Mars's surface to uncover signs of past microbial life and assess the planet's habitability. Meanwhile, the James Webb Space Telescope (JWST) provides unprecedented observational capabilities, allowing scientists to study the atmospheres of exoplanets in detail and search for potential biosignatures, such as specific atmospheric gases that could be indicative of biological processes. SETI, conversely, is dedicated to the search for intelligent extraterrestrial civilizations. Traditionally, SETI's primary approach has involved scanning the cosmos for radio signals that might be broadcast by technologically advanced civilizations.

**Corresponding author*: raniyamini0616@gmail.com

Yogesh Chandra, Manjuleshwar Panda and Mahesh Chandra Mathpal (eds.) The Intelligent Universe: AI's Role in Astronomy, (391–430) © 2025 Scrivener Publishing LLC

This search has expanded beyond radio waves to include optical signals, such as laser pulses, and other potential technosignatures. The goal of SETI is not just to detect extraterrestrial intelligence but to understand the nature and extent of such civilizations if they exist. The field of SETI also delves into the philosophical and ethical implications of discovering extraterrestrial life. The potential contact with an advanced civilization could profoundly impact humanity's self-perception, societal structures, and even our philosophical and ethical frameworks. While astrobiology aims to study the conditions that enable life and where it might exist, on the other hand, SETI focuses on detecting intelligent life using advanced technology. Ultimately, the quest to understand whether life exists beyond Earth encompasses not only scientific exploration but also profound philosophical questions. They reflect humanity's constant effort to ascertain whether we are alone in the universe, which has far-reaching implications for science, philosophy, and our view of life. The search for extraterrestrial life and intelligence not only pushes the boundaries of science and technology but also invites profound existential reflections. These fields underscore a deep-rooted curiosity and the enduring quest to comprehend the broader universe and our role within it.

Keywords: Astrobiology, SETI (search for extraterrestrial intelligence), biosignatures, technosignature, habitability, Perseverance rover, James Webb Space Telescope (JWST)

11.1 Introduction to Astrobiology and Search for Extraterrestrial Intelligence

Astrobiology is the interdisciplinary scientific study of life in space, encompassing its origin, evolution, distribution, and potential future beyond Earth. It integrates knowledge from biology, chemistry, physics, astronomy, geology, and planetary science to address fundamental questions about life's potential elsewhere. The study of life's origins is a core aspect of astrobiology, focusing on how life first appeared on Earth and the conditions necessary for its development. Researchers examine extreme environments on Earth, such as deep-sea vents and acidic lakes, to gain insights into the habitability of other planets and moons, such as Mars or Europa, one of Jupiter's frozen moons [1].

The search for biosignatures—indicators of life such as unique chemical patterns or atmospheric gases like oxygen and methane—is also a key component of astrobiology. Astrobiologists use telescopes and space missions to detect these signs, particularly focusing on exoplanets within their stars' habitable zones where conditions might support liquid water. Additionally, astrobiology explores how life might adapt to various environmental

Figure 11.1 Astrobiology exploration overview (Image Credit: NASA Astrobiology Institute).

conditions, including the possibility of life forms differing from those on Earth. This research informs the design of future space missions and the search for extraterrestrial intelligence. Ultimately, astrobiology seeks to answer profound questions about humanity's place in the cosmos, including the existence of life beyond Earth and the implications of such discoveries for science, philosophy, and society. Figure 11.1 visually represents the astrobiology exploration overview.

"Astrobiology is the bridge between planetary science and the search for life beyond Earth."

NASA Astrobiology Institute

11.1.1 The Key Components of Astrobiology

Astrobiology is a multidisciplinary field that investigates the possibility of life beyond Earth using knowledge from numerous scientific disciplines to better understand the origins, evolution, and future of life in the cosmos. This text describes the major aspects of astrobiology, such as the study of severe environments, the search for extraterrestrial life, planetary habitability, and the ramifications of discovering life beyond our planet [2]. Figure 11.2 illustrates the origin of astrobiology.

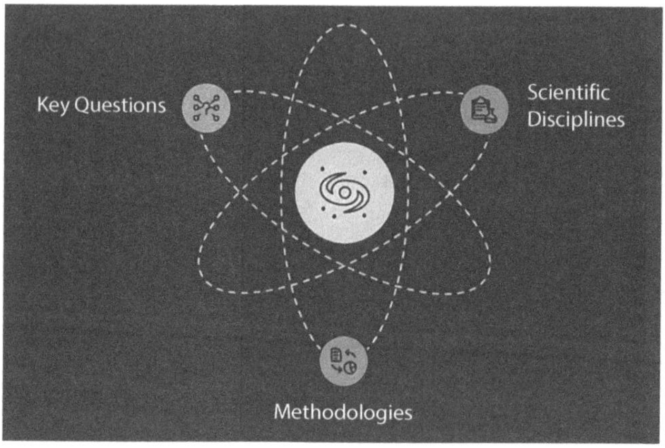

Figure 11.2 The origin of astrobiology (Image Credit: NASA).

- **Origins of Life:** The origins of life is the study of how life initially appeared on Earth, focusing on the chemical and environmental factors that facilitated this process.
- **Habitability:** Habitability studies examine the potential for life to exist in space environments, such as Mars and the icy moons of Jupiter and Saturn. This research involves investigating factors like the availability of water, essential chemicals, energy sources, and environmental conditions that could support microbial or more complex life forms.
- **Biosignature:** Biosignatures are detectable signals or markers of life, such as specific molecules, isotopic ratios, or patterns in atmospheric gases, which indicate the presence of biological activity. These markers can include gases like oxygen or methane, chemical compounds, or even surface features, and they are crucial targets in the search for life in other worlds.
- **Search for Extraterrestrial Life:** The hunt for extraterrestrial life employs telescopes, space missions, and cutting-edge technologies to identify indications of life beyond Earth. This includes examining exoplanets located in their stars' habitable zones, where conditions might support liquid water, as well as evaluating signals or anomalies that could suggest the presence of alien life forms.

- **Life's Adaptation:** Life's adaptations entail investigating how organisms may survive and thrive in harsh conditions such as deep-sea vents, polar ice, and acidic lakes on Earth. This research extends to hypothetical life forms in space, exploring how they might adapt to extreme conditions such as intense radiation, extreme temperatures, and low gravity on other planets or moons.

11.1.2 The Quest for Extraterrestrial Life

The search for alien life is a significant scientific activity that investigates whether we are alone in the universe. This quest spans numerous fields, including astronomy, planetary science, and biology, and employs cutting-edge technology and approaches. Central to this endeavor is locating habitable environments beyond Earth, focusing on planets and moons within their stars' habitable zones—regions where conditions might allow liquid water to exist. Mars, which shows evidence of ancient water activity, and frozen moons such as Europa and Enceladus are prime candidates for exploration. Researchers examine these celestial bodies for evidence of water, critical molecules, and energy sources that could support life.

Another essential component of the quest is discovering biosignatures or evidence of life. On Earth, biosignatures include specific chemicals and gases like oxygen and methane, which are frequently associated with life activity. Scientists use these indicators to examine the atmospheres of exoplanets, looking for similarities that could suggest the possibility of life. Technological developments have greatly improved our search capabilities. Space telescopes, such as the James Webb Space Telescope, and missions like the Mars rovers provide critical information about distant planets. These instruments enable scientists to investigate planetary atmospheres, surface conditions, and potential direct evidence of life, such as microbial fossils [3].

11.1.3 Future Prospects in Astrobiology

Astrobiology advances with new technologies and collaborative research, increasing the likelihood of discovering extraterrestrial life. Future exploration will focus on space missions, biosignature detection, synthetic biology, and collaborative endeavors. Upcoming missions such as NASA's Europa Clipper and ESA's JUICE will investigate the deep oceans of ice worlds, while the Artemis program sets the framework for future Mars exploration. Proposed interstellar probes could let us go beyond the solar system. Advances in spectroscopy and remote sensing, enabled by telescopes like JWST and upcoming initiatives like the Habitable Worlds Observatory, improve our ability to detect molecular indicators of life in exoplanetary atmospheres. Experiments that simulate harsh planetary conditions help scientists better grasp hypothetical alien biochemistry. Synthetic biology could potentially help with space colonization by creating microbes that produce resources and adapt to their surroundings. AI and big data analytics improve astrobiological research, while collaborations between space agencies, academia, and private industry speed up exploration efforts. Astrobiology is getting closer to answering the question of whether life exists beyond Earth, which will reshape our knowledge of the universe.

11.2 The Role of SETI in the Search for Extraterrestrial Intelligence

The search for extraterrestrial intelligence (SETI) is an important step in the search for intelligent life beyond Earth. SETI employs advanced technology and procedures to identify potential signals from extraterrestrial civilizations particularly *via* radio telescopes. SETI tries to detect communications that could indicate the presence of intelligent life by scanning the universe for anomalous electromagnetic signals. One of SETI's core strategies is to monitor specific radio frequencies thought to be suitable for interstellar communication. This strategy is predicated on the idea that sophisticated civilizations may use these frequencies to convey information over long distances. SETI's use of massive radio arrays, such as the Arecibo Observatory and the Green Bank Telescope, allows researchers to explore a broad area of the sky for potential signals [4].

11.2.1 Exploring Astrobiology and SETI

This chapter presents a broad review of astrobiology, an interdisciplinary science that studies life in space. Astrobiology combines biology, chemistry, physics, astronomy, geology, and planetary science to study the origins, evolution, distribution, and potential future of life beyond Earth. The study of life's origins is central to this area, concentrating on how life first appeared on Earth and determining the conditions required for its growth. Understanding whether alien environments are habitable is an important component of astrobiology. Researchers study harsh habitats on Earth, such as deep-sea vents and acidic lakes, to determine the possibility of life on other planets and moons, such as Mars or Europa. These studies aim to evaluate whether similar settings in the solar system or beyond could support life.

11.2.2 Technological Advancement in SETI

SETI has benefited from significant technological breakthroughs, which have increased our ability to identify and analyze potential extraterrestrial transmissions. Modern radio telescopes, such as the Allen Telescope Array (ATA) and the MeerKAT observatory, offer more sensitivity and frequency coverage, increasing the likelihood of detecting alien signals. AI and machine learning techniques are currently used to filter massive volumes of data, discovering anomalies that may imply extraterrestrial sources. Furthermore, optical SETI has evolved as a complementary strategy, looking for laser pulses that could represent messages from advanced civilizations. These advancements broaden SETI's search capability, raising the prospect of detecting sentient species beyond Earth.

11.2.3 Challenges and Future Prospects of SETI

Despite technological breakthroughs, SETI poses enormous hurdles, including the vastness of space, limited financing, and the potential of alien civilizations using unknown communication technologies. Signal interference from man-made sources further complicates data processing, necessitating advanced filtering approaches. Future SETI initiatives may include multinational cooperation, larger telescope arrays, and broader search criteria, such as new wavelength ranges and deep-space investigations. With continuous breakthroughs and interdisciplinary backing, SETI remains a viable route for addressing one of humanity's most profound questions: Are we alone in the universe?

11.3 The Origin of Astrobiology

This graphic portrays the evolution of humanity's awareness and study of alien life, starting with ancient philosophical questions that represent early ideas and interest in life beyond Earth. It then examines the historical evolution of astrobiology, tracking how our understanding of the universe and the possibility of extraterrestrial life has changed over time.

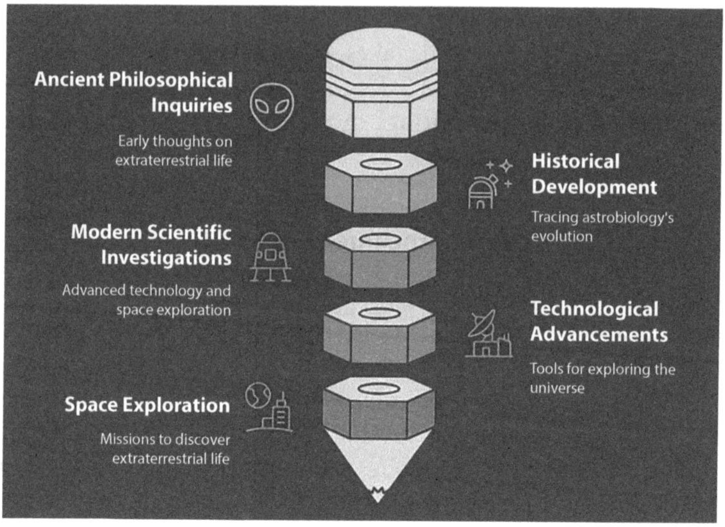

Figure 11.3 The formation and evolution of astrobiology (Image Credit: Planck Collaboration, 2018).

Modern scientific investigations, fueled by enhanced technology and space exploration, represent the next step, focused on methodical ways of looking for indications of life [5]. Figure 11.3 depicts the formation and evolution of astrobiology.

11.3.1 Historical Context and Early Speculations

The concept of life beyond Earth has ancient roots, as evidenced by numerous societies' beliefs and cosmologies. Philosophers like Anaximander and Epicurus in ancient Greece hypothesized about the existence of other realms inhabited by life. The Copernican revolution of the 16th century significantly altered humanity's perception of the universe, challenging the geocentric paradigm and opening the door to the possibility of additional livable planets.

11.3.2 The Formation of Astrobiology as a Scientific Discipline

Astrobiology emerged as a formal scientific study in the middle of the 20th century. The discovery of extremophiles, or species that flourish in harsh circumstances on Earth, called into question traditional assumptions of the conditions required for life and broadened the universe's potential habitable zones. These discoveries, combined with advances in space exploration, fueled the creation of astrobiology as scientists tried to understand how life might persist in places previously assumed to be hostile [6]. NASA had a key role in establishing astrobiology as a recognized science, particularly with the foundation of the NASA Astrobiology Institute (NAI) in 1998. This interdisciplinary research institute has since been at the vanguard of exploring life's potential in the universe, encouraging experts from diverse professions to collaborate on the most fundamental astrobiology topics.

11.3.3 Key Milestones in the Development of Astrobiology

Several significant milestones have marked the advancement of astrobiology. The Viking missions to Mars in the 1970s were among the first to look for direct evidence of life on another planet, but their findings were inconclusive. The discovery of extremophiles in the 1980s and 1990s transformed our view of life's resilience, proving that life may exist in conditions previously considered too extreme. Another watershed moment came in 1995 when the first exoplanet orbiting a sun-like star was discovered, proving the existence of planets outside our solar system and boosting the prospect of inhabited worlds. Subsequent missions, such as the Kepler Space Telescope, have discovered thousands of exoplanets, some of which are near their parent stars' habitable zones, where conditions may be suitable for life.

11.4 Understanding the Universe: A Foundation for Astrobiology

Astrobiology, or the study of life beyond Earth, is based on an understanding of the universe's origins, evolution, and the physical circumstances that allow life to exist. The hunt for extraterrestrial life is guided by three key aspects: celestial body formation, chemical building blocks, and habitable zones. The formation of planets and moons influences their ability to support life, whereas the presence of critical chemical components supplies the raw ingredients required for biological activity. Identifying habitable zones around stars helps to focus the search for settings in which life could thrive. Together, these components form a framework for investigating life's possibilities beyond Earth [7]. Figure 11.4 shows the concept of the habitable zone.

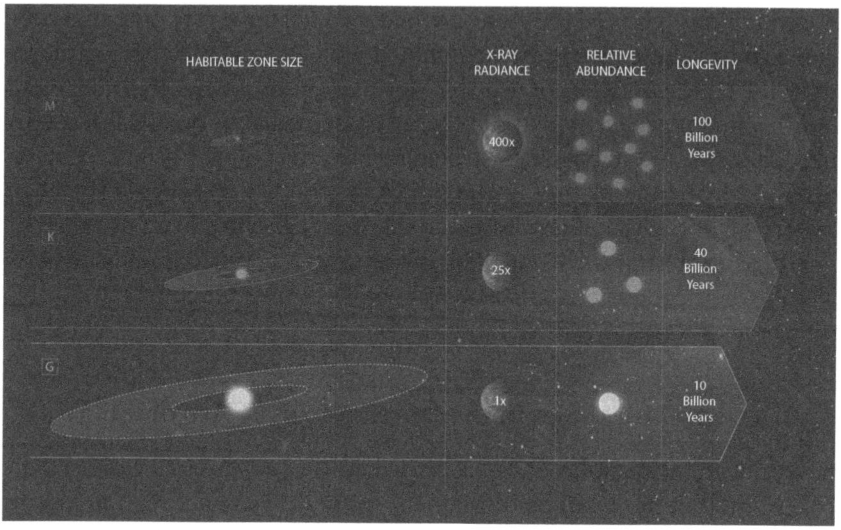

Figure 11.4 The habitable zone (Image Credit: NASA).

11.4.1 The Formation and Evolution of the Universe

The universe began around 13.8 billion years ago with the Big Bang, which formed the fundamental particles, forces, and structures that later gave rise to stars, galaxies, and planets. Initially, the cosmos was a hot dense plasma made primarily of hydrogen and helium. As the cosmos expanded and cooled, these elements started to form the first stars and galaxies.

Star Formation and Heavy Element:

- Nebulae, or clouds of gas and dust, are where stars form.
- The initial generation of stars, known as Population III stars, was huge but short-lived. Its nuclear fusion produced heavier elements including carbon, oxygen, and nitrogen, which were then disseminated by supernova explosions.
- This process, which occurred over billions of years, endowed the cosmos with the components required for life.

Planetary System Formation:

- After star formation, the residual gas and dust create proto-planetary disks, the birthplaces of planets and moons.

- Dust grains stay together during accretion to form planetesimals, which collide and merge to produce full-sized planets.
- Some of these planets have atmospheres, surface water, and conditions suitable for life.
- Understanding cosmic evolution assists astrobiologists in determining where to hunt for habitable planets and how life-sustaining settings may evolve.

11.4.2 The Conditions Necessary for Life

Life, as we know it, requires specific circumstances to thrive. These include the presence of liquid water, a consistent energy supply, and a variety of chemical components and compounds required to construct complex organic molecules. The study of extremophiles on Earth has demonstrated that life may flourish in a wide range of settings, from deep ocean vents to acidic hot springs, implying that life could potentially thrive in a variety of environments elsewhere in the cosmos. Astrobiologists are very interested in planets and moons that may meet these parameters. The habitable zone, sometimes known as the "Goldilocks zone," is an area surrounding a star where conditions are ideal for liquid water to exist on a planet's surface. However, as our understanding of life's adaptability has expanded, so has the number of conditions considered potentially habitable [8].

11.4.3 The Habitable Zone and Exoplanets

The habitable zone, often known as the "Goldilocks Zone," is the region around a star where circumstances are ideal for liquid water to exist on a planet's surface—not too hot or too cold. This concept is essential to the quest for life beyond Earth, as liquid water is thought to be necessary for life as we know it. The habitable zone's distance changes with the size and temperature of the star; for example, it is closer to a cooler star and distant from a hotter one. Exoplanets are planets orbiting stars outside our solar system. Since the first confirmed finding of an exoplanet in 1995, many have been discovered, with some positioned within their star's habitable zone. These planets are ideal possibilities for the hunt for extraterrestrial life. However, simply existing in the habitable zone does not guarantee that a planet possesses life-supporting circumstances; other characteristics such as atmosphere, composition, and geologic activity all play important roles.

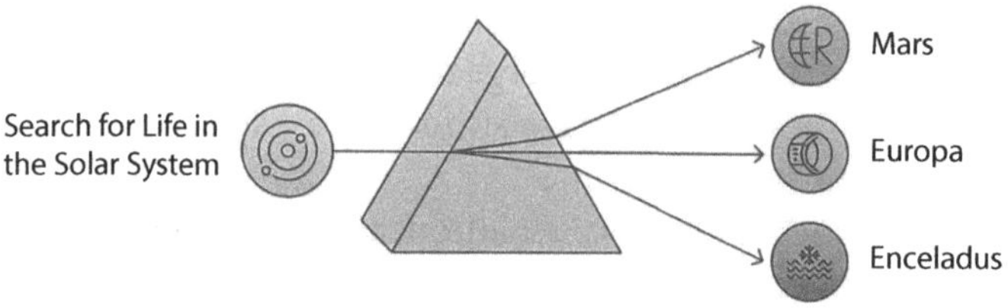

Figure 11.5 Mars: The red planet and its potential for life (Image credit: NASA/JPL-Caltech).

11.5 The Search for Life in the Solar System

The hunt for life in the solar system is centered on locating settings that potentially support microbial or sophisticated creatures, aided by the availability of water, organic molecules, and energy sources. Scientists are looking for planets and moons like Mars, Europa (a Jupiter moon), and Enceladus (a Saturn moon) that have subterranean seas and or liquid water and are potentially habitable. Robotic missions, such as Mars' Perseverance rover and the planned Europa Clipper, examine surface and atmospheric conditions for biosignatures—chemical or geological indicators of life. Astrobiology and space exploration advancements are further refining our understanding of where and how life might exist beyond Earth [9]. Figure 11.5 highlights key features of Mars and its potential for life.

11.5.1 Mars: The Red Planet and Its Potential for Life

Mars, nicknamed the "Red Planet" because of its reddish hue, has long attracted scientists and the general public. Its distinctive color is caused by iron oxide, sometimes known as rust, on the surface. Mars is the fourth planet from the Sun and almost half the size of Earth. Despite its hostile environment, Mars has been a top contender in the quest for extraterrestrial life and a possible future human settlement.

11.5.2 The Surface and Atmosphere

Mars has a thin atmosphere made up largely of carbon dioxide, with traces of nitrogen and argon. This atmosphere is too thin to maintain liquid water for long on the surface, although evidence implies that Mars originally had a thicker atmosphere and more water. Today, water exists primarily as ice, while some liquid water may exist underneath. The planet's surface features include the solar system's greatest volcano, Olympus Mons, and a canyon system called Valles Marineris that dwarfs Earth's Grand Canyon. The surface is littered with impact craters, dried-up riverbeds, and polar ice caps composed of water and dry ice.

For decades, scientists have been intrigued by the potential of life on Mars. While there is no evidence of contemporary life on Mars, it is possible that life existed there in the past. The discovery of ancient river valleys, lake basins, and minerals that grow in the presence of water indicates that liquid water was previously plentiful on Mars. The hunt for life on Mars is centered on discovering biosignatures or evidence of past or present life. Scientists search for organic compounds, isotopic patterns, and microfossils in Martian rocks. Several missions, including NASA's Curiosity and Perseverance rovers, have been launched to investigate Mars and collect data on its geology, climate, and possible habitability [10].

11.5.3 Human Exploration and Colonization

Mars is also viewed as a prospective site for human exploration and habitation. The idea of sending humans to Mars has been the subject of extensive scientific and engineering research. Challenges include the planet's thin atmosphere, severe temperatures, radiation, and the need to bring or generate water, oxygen, and food. NASA's Artemis program and SpaceX's plans for the Starship rocket are part of continuous attempts to make human flights to Mars a reality. If successful, humanity could establish outposts on Mars, with the ultimate goal of creating a self-sustaining colony. Mars remains a primary focus in the hunt for extraterrestrial life and future space travel. Whether or not there is life on Mars, the planet continues to teach us about the history of our solar system and the potential for life in severe conditions. The exploration of Mars may one day lead to people residing on another planet, ushering in a new era of space exploration.

11.5.4 Europa, Enceladus, and Other Icy Moons

Europa, Enceladus, and other ice worlds in the outer solar system are thought to be among the most promising areas for finding extraterrestrial life. These moons, which circle the gas giants Jupiter and Saturn, have sparked substantial scientific interest due to their subterranean seas and the possibility of livable conditions underneath their frozen crusts. Europa, one of Jupiter's largest moons, is particularly intriguing due to substantial evidence indicating the presence of a massive global ocean beneath its dense ice shell.

This ocean may come into contact with the moon's rocky mantle, allowing hydrothermal activity to occur and offer the chemical elements required for life. Europa's surface is characterized by fissures and ridges, indicating that the ice is moving and interacting with the liquid water beneath, making it a prime target for future missions such as NASA's Europa Clipper. Enceladus, Saturn's smaller moon, has also emerged as a promising possibility in the quest for life. The Cassini spacecraft observed plumes of water vapor and ice particles erupting from its south pole, indicating a subterranean ocean beneath the icy crust [11].

11.5.5 Europa

Europa, one of Jupiter's largest moons, is particularly intriguing due to substantial evidence indicating the presence of a massive global ocean beneath its dense ice shell. This ocean is thought to be in contact with the moon's rocky mantle, which may give the chemical elements for life *via* hydrothermal activity. Much of the early evidence for this ocean came from the Galileo spacecraft, which revealed a surface defined by ice fractures and ridges indicating movement and interaction with liquid water underneath. NASA's Europa Clipper is one of several planned missions to examine Europa's possible habitability.

11.5.6 Enceladus and Other Icy Moons

Saturn's moon Enceladus has also emerged as a leading contender in the quest for life. The Cassini spacecraft observed plumes of water vapor and ice particles erupting from the moon's South Pole, indicating the presence of a subterranean ocean beneath the frozen crust. The plumes contain organic chemicals, raising the prospect that Enceladus's ocean could support microbial life, analogous to Earth's deep-sea hydrothermal vents. Other ice moons, including Ganymede, Callisto, and Titan, provide intriguing prospects. Ganymede, the solar system's largest moon, has a magnetic field and may contain a deep ocean, albeit it is less accessible than Europa's. Callisto, another of Jupiter's moons, may potentially have an ocean, although its surface is extensively cratered and old, indicating less geological activity. Titan, Saturn's largest moon, stands out for its thick atmosphere and surface lakes of liquid methane and ethane. While the chemistry of life on Titan would be significantly different from that on Earth, the availability of liquid on its surface makes it a fascinating astrobiological target.

11.6 Venus and the Possibility of Aerial Biospheres

Venus, sometimes known as Earth's "sister planet" because of its comparable size and composition, has long piqued scientists' interest in the quest for extraterrestrial life. While Venus's surface circumstances are very hostile—with temperatures hot enough to melt lead, crushing atmospheric pressure, and clouds of sulfuric acid—new research has revived interest in the possibility of life in its upper atmosphere. This zone, located 50 to 60 kilometers above the surface, has a more temperate climate, with temperatures and pressures comparable to those on Earth's surface. The concept of an "aerial biosphere" has evolved, implying that life may exist in the form of microscopic creatures suspended in Venus's atmosphere.

The discovery of phosphine, a chemical associated with biological activities on Earth, in Venus's clouds in 2020 sparked renewed interest in the possibility of life in the planet's atmosphere. This study, while contentious and still being investigated, suggests that Venus's clouds may contain microbial life or, at the very least, unexplored chemical reactions. The dense cloud layers, which are primarily formed of sulfuric acid, provide obstacles for life as we know it, yet extremophiles on Earth—organisms that flourish in extreme environments—set the standard for life in such conditions. The concept of airborne biospheres on Venus has reignited interest in future missions to investigate the planet's atmosphere in greater detail [12].

11.6.1 Astrobiological Mission and Technological Advances

Astrobiological missions are critical to the hunt for alien life and the study of potentially livable habitats beyond Earth. These missions use advanced technologies to investigate, study, and collect data from distant planets, moons, and celestial bodies, therefore expanding our understanding of the universe's potential for life. Technological developments have greatly accelerated astrobiological research. Robotic missions, including NASA's Curiosity and Perseverance rovers, use advanced instruments to investigate soil, rock samples, and the atmosphere on Mars. These rovers are outfitted with capabilities such as the Sample Analysis at Mars (SAM) suite, which can identify organic substances and evaluate their chemical signatures. Similarly, the Europa Clipper mission, due in the 2020s, will carry a variety of scientific instruments to delve beneath Europa's icy crust and assess its habitability. Space observatories, including the James Webb Space Telescope, are expected to transform our search for exoplanets and their atmospheres. These telescopes can detect chemical signals of life in extraterrestrial atmospheres. Advances in spectroscopy, robotic technology, and deep-space communication are constantly improving our abilities to investigate other worlds and look for indications of life.

11.6.2 Exoplanets and the Potential for Life Beyond the Solar System

Exoplanets, or planets that orbit stars outside our solar system, are one of the most fascinating frontiers in astronomy and the quest for extraterrestrial life. The discovery of exoplanets has significantly increased our understanding of the cosmos and raised the tantalizing possibility of life beyond Earth. The first confirmed exoplanet, orbiting a sun-like star, was discovered in 1995, marking a watershed moment in astronomy. Since then, many more exoplanets have been discovered, many of which are positioned in their stars' habitable zones—the region where conditions are just ideal for liquid water to exist, a key component of life as we know it. Scientists have made tremendous progress in exoplanet discovery, thanks to advancements in space telescopes and detecting methods.

One of the most important missions in this area was NASA's Kepler Space Telescope, which was launched in 2009. Kepler detected exoplanets using the transit method, which involved viewing stars dimming when planets passed in front of them. These discoveries have not only increased the number of known exoplanets but have also helped scientists improve their understanding of the frequency of planets in habitable zones and the conditions that may allow life to develop [13].

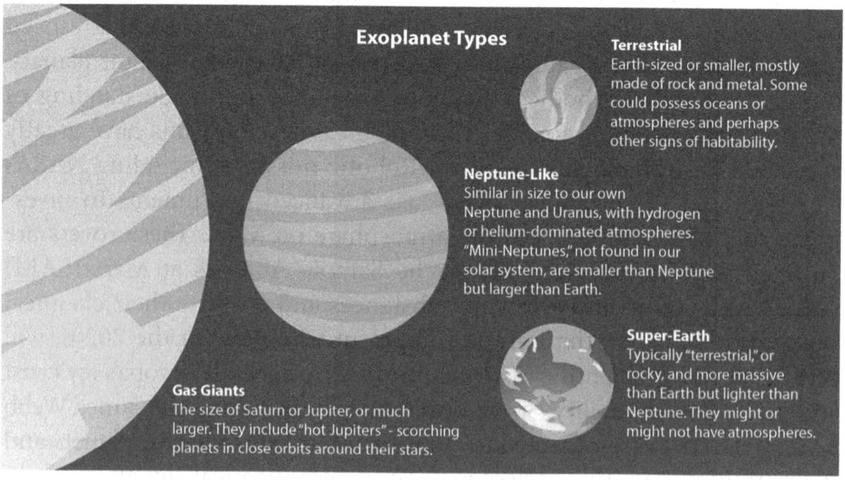

Figure 11.6 Types of exoplanets (Image Credit: NASA).

The study of exoplanet atmospheres has emerged as another crucial aspect of understanding their potential to host life. The upcoming launch of the James Webb Space Telescope (JWST) in the 2020s will offer unprecedented infrared capabilities to investigate the composition of exoplanet atmospheres and search for potential biosignatures—chemicals like oxygen, methane, and water vapor that could indicate biological activity. These advancements in technology and research bring us closer to answering one of the most profound questions in science: Is there life beyond our solar system? Figure 11.6 illustrates types of exoplanets.

11.6.3 Discovering Exoplanet Methods and Technologies

Several methods have been devised to discover and analyze exoplanets. Each method has advantages and limitations, and by combining multiple methodologies, scientists have been able to improve the accuracy and scope of exoplanet identification [14]. Figure 11.7 outlines the fundamental components of exoplanet research. The discovery of exoplanets leads to two primary areas of investigation.

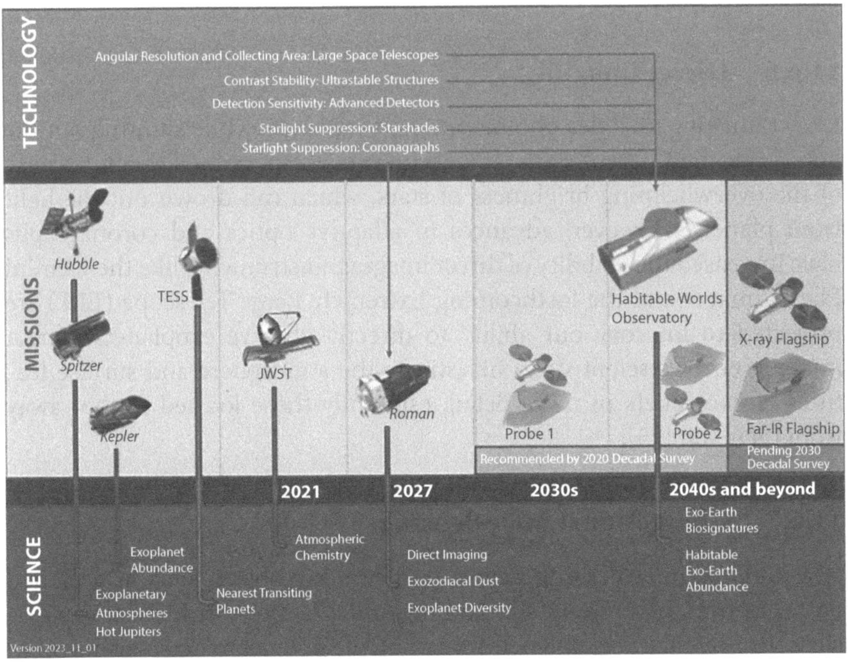

Figure 11.7 Key aspects of exoplanet discovery and habitability. (Image Credit: NASA).

11.6.4 Transit Method

The transit method includes monitoring a star's light and detecting periodic brightness drops when a planet passes in front of it. This approach is quite good at detecting smaller planets, particularly Earth-sized exoplanets. When a planet transits its host star, the star's light dims slightly, and the depth and duration of this dimming can reveal important information about the planet's size, orbital period, and distance from the star. The Kepler Space Telescope, which used this technology, has helped discover thousands of exoplanets and identified probable Earth-like worlds.

11.6.5 Radial Velocity Method

The radial velocity approach discovers exoplanets by measuring their gravitational pull on their parent stars. This gravitational force causes the star to wobble slightly, causing shifts in the star's light spectrum to the red or blue end. Scientists can determine an exoplanet's presence, mass, and orbit by monitoring these shifts. Instruments such as the HARPS spectrograph have enabled exact measurements of these Doppler shifts, making this method very useful for detecting bigger exoplanets, particularly those close to their sun.

11.6.6 Direct Imaging

Direct imaging includes photographing exoplanets while shutting out the light from their parent stars. This technique is especially difficult because of the overwhelming brightness of stars, which can drown out the light from planets. However, advances in adaptive optics and coronagraphs have increased the viability of direct imaging. Instruments like the Gemini Planet Imager and the forthcoming Extremely Large Telescope (ELT) are projected to improve our ability to directly observe exoplanets. Direct imaging enables scientists to investigate the atmosphere and surface features of exoplanets in more detail, especially those located farther away from their sun [15].

11.6.7 Gravitational Microlensing

Gravitational microlensing happens when a big object, such as a planet, passes in front of a distant star and magnifies its light. This effect can disclose the presence of exoplanets that would be difficult to detect using other approaches. Gravitational microlensing is particularly useful for finding exoplanets in areas of the sky where other methods, such as transit and radial velocity, are ineffective. This method can also provide useful information regarding the size and distance of exoplanets, even those found in distant star systems.

11.6.8 Astrometry

Astrometry is the exact measurement of star positions and the tracking of minor shifts produced by orbiting exoplanets' gravitational attraction. This method needs exceedingly accurate measurements and, when combined with other methods, can confirm the presence of exoplanets. Astrometry has been essential in validating exoplanets discovered using other approaches, such as the transit and radial velocity methods.

11.7 The Role of Space Telescopes (Kepler, TESS, JWST)

Space telescopes have transformed our understanding of exoplanets by giving precise studies of faraway worlds outside the purview of Earth's atmosphere. Key missions like Kepler, TESS, and the JWST have made substantial contributions to exoplanet research. NASA's Kepler Space Telescope, which launched in 2009, was a pioneering project in the search for exoplanets. Kepler used the transit method to observe over 150,000 stars to discover hundreds of exoplanets, many of which are Earth-sized and potentially habitable. Kepler's discoveries have contributed to our understanding of the frequency of planets in habitable zones and the conditions that may allow life to exist on other planets. Much of our understanding of exoplanet features and distributions is based on Kepler's findings. Launched in 2018, the Transiting Exoplanet Survey Satellite (TESS) continues Kepler's work by exploring the entire sky for exoplanets orbiting the brightest and nearest stars.

TESS focuses on stars within 300 light-years of Earth, enabling more detailed follow-up investigations of exoplanets, particularly those in the habitable zone. TESS data will be used to identify which exoplanets are most promising for future research, particularly those that may show evidence of life. The James Webb Space Telescope (JWST), which is planned to launch in December 2021, will revolutionize exoplanet research because of its improved infrared capabilities. JWST will investigate starlight traveling through exoplanet atmospheres in search of chemical clues of possible life.

With its enormous mirror and high-resolution imaging, JWST is expected to reveal unparalleled data on exoplanet chemical compositions and climatic conditions, providing fresh insights into the habitability of distant worlds [16].

11.7.1 Classification of Exoplanets: Super-Earths, Hot Jupiters, and More

Exoplanets differ widely in size, composition, and circumstances. Scientists have created a classification system for these planets based on parameters such as mass, radius, and proximity to their parent stars. This classification allows researchers to better understand the processes of planetary formation and evolution, as well as the possibility of habitability. Table 11.1 categorizes different types of exoplanets based on their characteristics.

11.7.2 Earth-Like Planets and Rogue Planets

Earth-like exoplanets are those that are similar to Earth in size and composition, and they are usually found in their star's habitable zone. These planets are particularly interesting in the hunt for extraterrestrial life because they may have the necessary circumstances for liquid water and a stable atmosphere, two essential components of life as we know it. Rogue planets are those that do not orbit a star and instead travel freely through space. These planets may have formed around a star before being expelled, or they may have formed independently in interstellar space. Rogue planet research helps scientists better comprehend planetary formation and migration, as well as the dynamics of planet-star interactions.

11.7.3 Assessing Habitability: Atmospheres, Biosignatures, and Remote Sensing

The habitability of an exoplanet is determined by analyzing various parameters, including its atmospheric composition, the presence of biosignatures, and the use of remote sensing technology to collect data from a distance. These technologies enable scientists to determine whether a planet has the necessary circumstances for life to exist [17]. Table 11.2 highlights key factors in assessing exoplanet habitability.

Table 11.1 Exoplanets and their classification.

Type of exoplanet	Characteristics	Key features	Potential for habitability
Super-Earths	• Larger than Earth but smaller than Neptune (1.2 - 2× Earth's radius).	• Can have rocky surfaces or thick atmospheres. Some may have liquid water.	• High potential if in the habitable zone with a stable atmosphere.
Mini-Neptunes	• Intermediate between Earth and Neptune (2 - 4× Earth's radius).	• Thick hydrogen-helium atmospheres, possibly with deep oceans.	• Uncertain; thick atmospheres may create a greenhouse effect.
Hot Jupiters	• Gas giants orbiting extremely close to their stars.	• Extremely high temperatures, fast orbits (days-long years).	• Very low; extreme heat prevents habitability.
Cold Gas Giants	• Similar to Jupiter and Saturn but farther from their stars.	• Thick atmospheres, potentially with large moons that could be habitable.	• The planets themselves are inhospitable, but their moons may support life.
Ocean Worlds	• Planets covered in deep global oceans.	• May have ice-covered surfaces with liquid water beneath.	• High, especially if hydrothermal vents exist like those on Europa or Enceladus.
Rogue Planets	• Free-floating planets that do not orbit a star.	• No solar energy but could have internal heat sources.	• Extremely low, but subsurface oceans could exist if geothermal heating is sufficient.

Table 11.2 Key factors in assessing exoplanet habitability.

Category	Key aspects	Description
Atmospheres	Temperature and Pressure	• The atmospheric conditions must support liquid water. Instruments like the JWST analyze thermal emissions to determine surface conditions.
	Composition	• The presence of gases such as oxygen, carbon dioxide, and methane provides insights into habitability. A balanced oxygen-methane ratio may indicate biological activity.
Biosignatures	Chemical Markers	• Certain gases, such as ozone, methane, and oxygen, are potential indicators of biological processes on an exoplanet.
	Surface Features	• The detection of specific minerals, complex organic compounds, or isotopic ratios may suggest biological activity.
Remote Sensing	Spectroscopy	• The study of light passing through an exoplanet's atmosphere helps determine its composition. Instruments like Hubble and JWST use this method.
	Transit Photometry	• Observing light variations when an exoplanet transits in front of its star provides insights into atmospheric properties.

11.8 The Search for Extraterrestrial Intelligence

The search for extraterrestrial intelligence (SETI) is a scientific effort to identify evidence of intelligent life beyond Earth. It entails using various methods and technology to search the universe for signs that could indicate the presence of technologically advanced civilizations. Scientists such as Frank Drake and Carl Sagan spearheaded the hunt for extraterrestrial intelligence (SETI) in the mid 20th century. Drake launched the first contemporary SETI experiment, Project Ozma, in 1960, using a radio telescope to listen for signals from neighboring stars, ushering in the era of dedicated SETI research. A fundamental method in SETI is to detect radio waves, which may travel long distances and are relatively straightforward to identify. The SETI Institute uses enormous radio telescopes to hunt for narrowband signals that may imply an artificial origin, separating them from the natural background noise of space. In addition to radio waves, researchers are looking into optical communication methods like laser pulses, which advanced civilizations could utilize to send messages across galaxies. Optical SETI, which looks for brief flashes of light that could represent beacons or messages from extraterrestrial sources, supplements radio SETI by scanning a different region of the electromagnetic spectrum [18]. Figure 11.8 represents the methodologies used in the search for extraterrestrial intelligence.

The Fermi Paradox asks why, despite the number of stars and possibly habitable planets, humanity has not discovered intelligent life. This contradiction drives much of the discussion in SETI research. The Drake Equation, developed by Frank Drake during a 1961 meeting at Green Bank Observatory, seeks to estimate the number of communicative extraterrestrial civilizations in the Milky Way. It considers the pace of star formation, the fraction of stars with planetary systems, the number of planets capable of supporting life, and the likelihood of intelligent life developing. Although the equation remains hypothetical due to the ambiguity of its variables, it serves as the cornerstone for SETI research and has become a

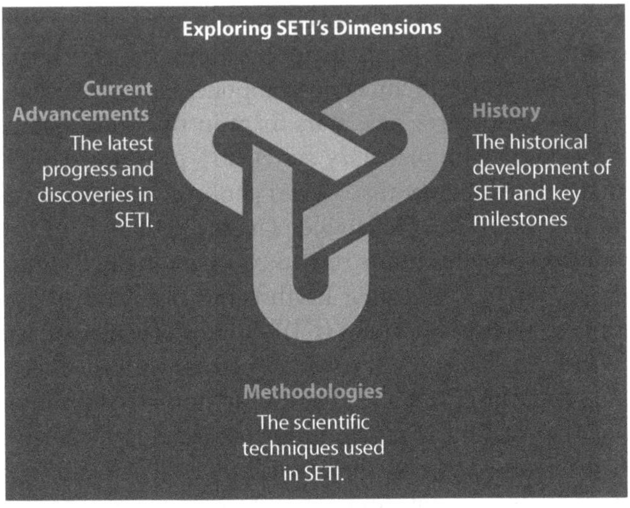

Figure 11.8 SETI and the search for extraterrestrial intelligence (Image Credit: SETI Institute).

symbol of humanity's search for extraterrestrial intelligence. SETI's capabilities have grown dramatically in recent years because of technological and data analysis improvements. Breakthrough Listen, which debuted in 2015, uses the most sensitive radio telescopes to search for signals of intelligence among the million stars closest to Earth. Furthermore, the use of artificial intelligence and machine learning has improved the ability to efficiently handle large datasets, potentially improving the possibility of discovering extraterrestrial signals [19].

11.8.1 The Drake Equation and Estimating the Probability of Extraterrestrial Civilization

The Drake Equation is a probabilistic model for estimating the number of active communicative extraterrestrial civilizations in the Milky Way galaxy. The equation, proposed by American astronomer Frank Drake in 1961, provides a framework for understanding the circumstances that lead to the existence of such civilizations and has become a cornerstone of the Search for Extraterrestrial Intelligence (SETI). Equ. (11.1) illustrates the variables of the Drake Equation for estimating extraterrestrial life.

$$N=R*\cdot fp\cdot ne\cdot fl\cdot fi\cdot fc\cdot L \tag{11.1}$$

The Drake Equation provides a framework for quantifying our uncertainty about the existence of extraterrestrial civilizations rather than predicting their exact number. Each variable in the equation represents a huge unknown, with increasing levels of supposition. The star formation rate (R*) is one of the more predictable components, with recent estimates indicating that one to three new stars form in the Milky Way each year. The fraction of stars with planetary systems (f_p) is presently thought to be close to 1, with exoplanet discoveries confirming that most stars contain planets. The number of habitable planets per star (n_e) is unknown, but conservative estimates imply one to two potentially livable planets in each planetary system. Given that Earth is our only known example, the fraction of planets where life arises (f_l) is highly speculative, with a range of almost 0 to 1. Similarly, the fraction of planets with intelligent life (f_i) is unknown, given the conditions that lead to human civilization could be uncommon. The fraction of civilizations that develop detectable technology (f_c) is determined by how many societies establish and use technologies such as radio waves that can be monitored over interstellar distances. Finally, civilizational longevity (L) is an important consideration, as the survival of technologically advanced societies is dependent on their ability to endure self-destruction and other existential challenges, which ultimately determines how long they can be detected [20].

11.8.2 Implications and Applications

The Drake Equation has important implications for our understanding of the universe. It gives a framework for SETI researchers to plan where and how to look for extraterrestrial life. The equation also emphasizes the possible rarity of communicative civilizations, as even small fractional values can significantly lower the number of detectable civilizations. Despite its hypothetical character, the Drake Equation has sparked scientific investigation and debate, spurring interdisciplinary research in astronomy, biology, and even sociology to better understand the elements influencing the origin of intelligent life. It also stresses the importance of ongoing exploration, as each discovery—such as the identification of habitable exoplanets or the detection of biosignatures—helps to revise our estimations for each term in the equation.

11.8.3 Challenges and Criticisms

The Drake Equation has been questioned for its speculative character, especially regarding unclear parameters like L, which refers to the lifetime of communicative civilizations. Many say that the equation is more of a thought experiment than a practical tool because the variables are unpredictable, making meaningful predictions difficult. Concerns include the speculative nature of inputs such as the probability of life emergence ($f\ l$ f l) and intelligence development ($f\ i$ f i), the lack of empirical data beyond Earth, and the unresolved Fermi Paradox, which questions why we have yet to detect extraterrestrial signals if civilizations are common. Despite these shortcomings, the equation remains a critical intellectual framework in the hunt for extraterrestrial life because it shapes scientific inquiry, fosters debate, and drives research in astrobiology and SETI.

11.8.4 SETI Techniques and Technologies: Radio Signals, Optical SETI, and More

The search for extraterrestrial intelligence (SETI) employs a wide range of techniques and technology to identify potential signals or evidence of intelligent life beyond Earth. These technologies have changed over time, incorporating breakthroughs in radio astronomy, optical technology, and processing capacity. The following is an overview of the key techniques and technologies utilized in SETI, such as radio signal detection, optical SETI, and new approaches.

Radio signal detection is the most classic and well-known method in SETI. This concept is predicated on the assumption that advanced civilizations may employ radio waves, particularly those in the microwave region of the spectrum, to communicate across great distances in space. Radio waves are preferred because they may cover large distances without being severely absorbed or dispersed by cosmic dust. Optical SETI investigates whether extraterrestrial civilizations could communicate using light rather than radio frequencies. This approach is designed to detect quick strong flashes of light, such as laser pulses, which could be transferred across interstellar distances [21].

- **Advantages of Optical SETI:** Light, particularly laser pulses, may transport a large amount of information and be finely focused, allowing for precise communication over long distances. Laser pulses can be as brief as a millisecond, making them easily distinct from natural light sources.
- **Techniques:** Optical SETI typically employs big optical telescopes outfitted with sensitive photodetectors capable of capturing brief flashes of light. The Harvard Optical SETI program's detectors are capable of searching the sky for these nanosecond pulses.
- One of the challenges of optical SETI is dealing with the massive volume of data generated, as well as detecting potential signals from background noise like cosmic rays or satellite reflections.

11.8.5 Emerging SETI Technologies and Techniques

As technology has advanced, SETI has evolved beyond classic radio and optical technologies to encompass new approaches and instruments.

- **Search for Technosignatures:** SETI researchers are increasingly interested in finding technosignatures, which are signs of advanced technology. This could involve detecting large-scale energy use, artificial illumination of planetary surfaces, or unique chemical compositions in planetary atmospheres.

- **Machine Learning and AI:** With the massive amounts of data created by SETI projects, machine learning and artificial intelligence are becoming increasingly important for evaluating and detecting potential signals. These technologies can aid in sifting through noise to discover patterns that may imply intelligent beginnings.
- **Quantum Communication:** Although highly theoretical, some academics believe that advanced civilizations may use quantum communication methods, such as quantum entanglement, to enable instantaneous communication across huge distances.
- In addition to radio and optical signals, some SETI initiatives are looking into the infrared spectrum, specifically for waste heat from advanced civilizations or to detect interstellar spacecraft utilizing infrared emissions.

11.8.6 Collaborative Efforts, Challenges, and Future Prospects in SETI

SETI is gradually becoming a worldwide collaborative undertaking. Initiatives such as SETI, which allows individuals to contribute computing power for analyzing radio signals, and the Allen Telescope Array, which uses multiple telescopes to conduct simultaneous observations, demonstrate the value of collaborative efforts in the search for extraterrestrial intelligence. Crowdsourcing programs and worldwide cooperation have considerably increased SETI's reach, allowing for larger searches and improving the efficiency of analyzing large datasets. Despite these advances, the quest for extraterrestrial intelligence remains a daunting task. The vastness of space, the uncertainty about the presence of intelligent civilizations, and the limitations of modern technology all provide enormous challenges. However, as computing, artificial intelligence, and observational techniques progress, SETI methodologies will become more advanced. As our understanding of the cosmos grows, the possibility of making revolutionary discoveries in the search for sentient life increases [22].

11.8.7 Major SETI Projects and Initiatives

Several important projects and activities have contributed to the ongoing search for evidence of intelligent life beyond Earth. Project Ozma, the

Table 11.3 Key SETI projects and their contributions to the search for extraterrestrial intelligence.

Project	Overview	Goal/Purpose	Method/ Technology	Outcome/ Significance
Project Ozma (1960)	• The first modern SETI experiment was conducted by Frank Drake at the Green Bank Observatory, West Virginia.	• Detect radio signals from intelligent extraterrestrial civilizations.	• Used a 25-meter radio telescope to observe Tau Ceti and Epsilon Eridani at 1.42 GHz (natural hydrogen frequency).	• No signals were detected but laid the foundation for future SETI studies and inspired the Drake Equation.
Arecibo Message (1974)	• A binary-coded radio transmission was sent from the Arecibo Observatory in Puerto Rico to M13, a globular star cluster.	• Symbolic demonstration of humanity's technological advancement and potential for interstellar communication.	• Designed by Frank Drake, Carl Sagan, and others; included numbers, atomic elements, DNA structure, human figure, solar system, and telescope schematic. Transmitted as binary digits.	• Unlikely to be received due to M13's movement but showcased the power of Arecibo Observatory and SETI efforts.
Breakthrough Listen (2015 - Ongoing)	• One of the largest and most ambitious SETI initiatives, backed by $100 million in funding from Yuri Milner and Stephen Hawking.	• Conduct a large-scale search for extraterrestrial signals using radio and laser transmissions.	• Uses Green Bank Telescope (USA), Parkes Observatory (Australia), and Automated Planet Finder (California). Employs AI to analyze large datasets across a wide frequency range.	• Ongoing; represents a significant expansion of SETI research, leveraging advanced technology to analyze vast amounts of data.

Arecibo Message, and the Breakthrough Listen program are among the most noteworthy in SETI history. Here is a summary of the major search for extraterrestrial intelligence (SETI) projects and initiatives, including Project Ozma, the Arecibo Message, and Breakthrough Listen. Table 11.3 shows the key SETI projects and their contribution to the search for extraterrestrial intelligence.

11.9 The Fermi Paradox and the Great Silence

The hunt for extraterrestrial intelligence presents a deep contradiction: while the universe appears to be wide enough to support many civilizations, we have yet to discover any evidence of intelligent alien life. This conundrum, known as the Fermi conundrum, poses an important question: Where is everyone? Despite the high likelihood of habitable planets and the probable lifespan of advanced civilizations, no conclusive signs or remains of their presence have been discovered. This section delves into the fundamentals of the Fermi Paradox, the different potential explanations, and the ramifications for our understanding of intelligence, civilization, and cosmic evolution [23]. Figure 11.9 depicts possible explanations for the lack of extraterrestrial contact.

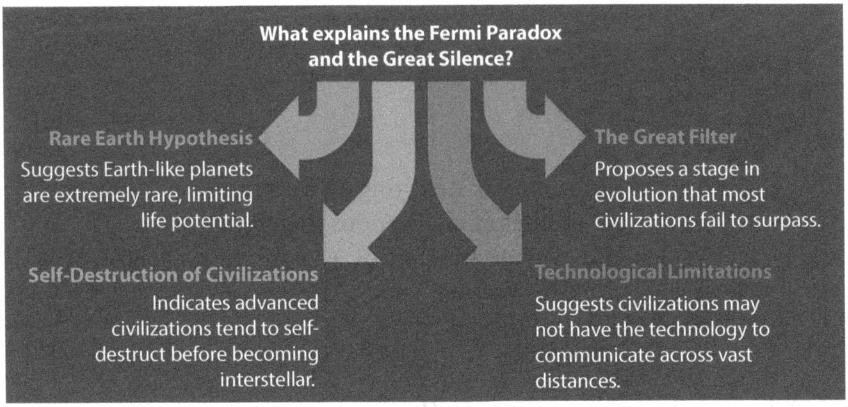

Figure 11.9 The fermi paradox and the great silence (Image credit: NASA/ESA).

11.9.1 Core Aspects of the Fermi Paradox

The Fermi Paradox is based on three major observations:

- **The vastness of the cosmos:** The observable universe is made up of billions of galaxies, each with billions of stars, many of which have planetary systems. Given the vast number of potentially habitable exoplanets, intelligent life should have appeared several times.
- **Technological possibilities:** A sufficiently sophisticated society may develop interstellar travel or communication technologies, allowing it to explore or broadcast signals throughout the cosmos.
- **The absence of evidence:** Despite decades of scientific research, including radio signal searches (such as SETI), no conclusive evidence of extraterrestrial intelligence has been discovered. The absence of detectable signals or visible mega structures indicates a fundamental mismatch between expectation and reality.

The Fermi Paradox asks why, despite the vastness of the universe and the high probability of extraterrestrial civilizations, we have yet to detect any evidence of them. This can be framed mathematically by examining the number of stars and how long it takes for civilizations to develop:

$$P = TsNc \cdot Tc \tag{11.2}$$

where:

- (P) = the probability of detecting a civilization
- (N_c) = the number of civilizations that have existed
- (T_c) = the average lifespan of a civilization capable of interstellar communication
- (T_s) = the age of the galaxy

11.9.2 Proposed Solutions to the Fermi Paradox

There have been various proposed solutions to the Fermi Paradox, ranging from scientific and technical explanations to philosophical and speculative ideas. These proposals seek to balance the high likelihood of intelligent alien life with the scarcity of visible evidence [24].

- **Rare Earth Hypothesis:** Intelligent life may be exceptionally scarce in the universe. While microbiological or primitive life forms may be ubiquitous, the evolution of complex intelligent life may necessitate an unusual mix of circumstances.
- **Self-Destruction Hypothesis:** Advanced civilizations may annihilate themselves before reaching the point of interstellar communication or exploration. This could be due to nuclear war, natural disasters, or technological failures.
- **Zoo Hypothesis:** Advanced extraterrestrial civilizations may be aware of humans but opt not to contact us. They may be watching us like we would animals in a zoo, enabling us to develop organically without meddling.
- **Technological Incompatibility:** Advanced extraterrestrial civilizations may use technologies or communication methods that humans cannot detect or comprehend. For example, they could communicate using unusual forms of matter or dimensions that are beyond our current scientific understanding.

11.9.3 The Great Filter and the Fragility of Civilization

- The Great Filter Hypothesis: One explanation for the Great Silence is the concept of a Great Filter, a severe barrier in the evolution of life that most civilizations fail to overcome. This filter could exist in the past (for example, the beginning of life, the evolution of intelligence) or in the future (for example, self-destruction or environmental collapse).
- Potential threats: If the Great Filter exists in the future, it implies that advanced civilizations may confront unavoidable problems that will lead to their extinction, such as nuclear war, environmental damage, or technological failure. This raises questions about humanity's long-term viability [25].

11.9.4 Anthropocentric Assumptions

- Assumption Reevaluation: The Great Silence calls into doubt many of our assumptions regarding extraterrestrial life, such as the premise that other civilizations will employ radio waves or have expansionist tendencies. It challenges our anthropocentric perspective on intellect, technology, and communication.

- Different Types of Intelligence: Extraterrestrial life may be so different from ours that their modes of communication, technology, or even existence are unintelligible to humans. The Great Silence could indicate that we are unprepared to identify life or intelligence that acts in ways that are completely foreign to us.

11.10 Ethical and Philosophical Implications of Contacting Extraterrestrial Life

Contact with extraterrestrial life raises fundamental ethical and philosophical issues that necessitate serious examination of humanity's obligations, potential perils, and broader consequences for both civilizations. Ethically, humans must handle the risks of cultural contamination and exploitation, especially when encountering less technologically advanced species. The philosophy of noninterference, derived from concepts such as the Zoo Hypothesis and Star Trek's Prime Directive, implies that we should let such civilizations evolve freely. Furthermore, the finding of intelligent extraterrestrial life would require us to reconsider human uniqueness, challenging our long-held assumptions about intelligence and consciousness [26].

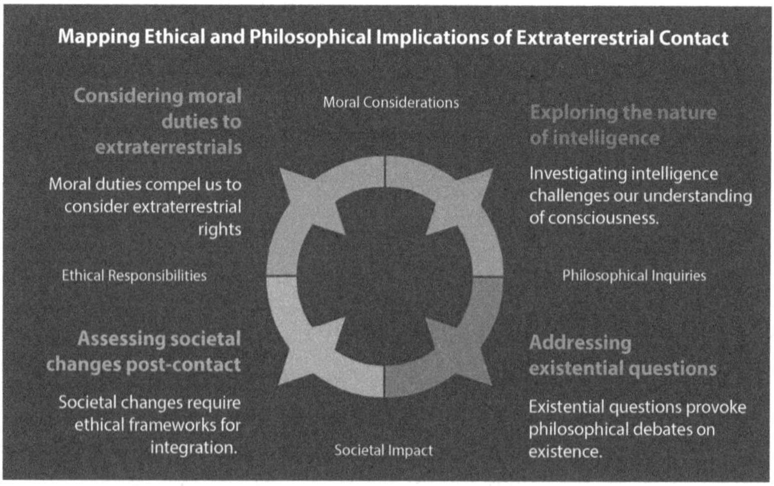

Figure 11.10 Ethical and philosophical implications of extraterrestrial contact.

Figure 11.10 depicts the ethical and philosophical implications for extraterrestrial contact.

The risk of hostility or conflict further complicates the ethical landscape, raising questions about self-defense, diplomacy, and the possibility of cooperation to ensure survival. There are also concerns about speciesism and the moral rights of alien beings, which could require the development of a cosmic ethics framework.

11.10.1 The Impact on Human Society and Civilization

Contact with extraterrestrial life would have far-reaching consequences for human culture and civilization, causing broad changes in cultural, social, political, and scientific spheres. One of the most immediate consequences would be a fundamental upheaval in humanity's worldview, altering long-held assumptions about our place in the universe. Religions based on human exceptionalism or divine creation may be reinterpreted, and some people may experience existential distress, whereas others may find a sense of unity or purpose in knowing that we are not alone. Politically, states might either band together in the face of such a historic occurrence or battle for control over how contact is handled, potentially leading to new global alliances or conflicts. Economically, the discovery of advanced extraterrestrial technologies has the potential to change industries, shifting the balance of global power, as well as disrupting present scientific paradigms by introducing knowledge that questions our understanding of biology, physics, and engineering [27].

11.10.2 Ethical Considerations: Messaging to Extraterrestrial Intelligence

Messaging to extraterrestrial intelligence (METI) raises serious ethical concerns that must be thoroughly addressed before mankind transmits intentional signals into space. One of the most pressing worries is the threat to humanity's safety. Some ethicists and scientists warn that actively reaching out to unknown civilizations risks exposing ourselves to potentially dangerous or hostile alien creatures. According to the "Lurking Predator" hypothesis, an evolved extraterrestrial society may perceive humanity as a threat or a resource, resulting in a scenario in which contact encourages danger. As a result, METI proponents must balance the potential benefits of contact with the existential risks. Another important ethical issue is consent. Because METI includes representing the entire human species, there are concerns regarding who has the authority to determine whether or not to transmit signals. Currently, METI programs are mainly driven by small groups of scientists or private corporations, raising questions about whether these acts represent humanity's collective desire. Some say that before transmitting messages, there should be worldwide consensus or at least a larger international debate to guarantee that such a significant decision is made democratically and with proper caution.

Finally, there are long-term implications for the species involved. METI may establish relationships that affect future generations, and there is an ethical need to assess the possible results of these interactions—whether they result in cooperation, knowledge exchange, or conflict. Given the profound and irreversible consequences that METI might have, careful consideration and broad ethical discussions are essential before humanity reaches out to communicate with other intelligent beings in the universe [28].

11.10.3 The Potential for Interstellar Communication

The possibility of interplanetary communication is one of humanity's most significant prospects and difficulties. Given the huge distances between stars, existing communication means like radio waves and lasers, which we now employ, are severely limited owing to time delay. Messages sent to even the nearest stars could take years, decades, or centuries to arrive, and receiving a response would take just as long. This delay calls into doubt the viability of real-time communication, pushing us to consider long-term communication methods, such as the exchange of knowledge or cultural artifacts, rather than real-time discourse. The language barrier is one of the most significant hurdles to interstellar communication. It is highly unlikely that extraterrestrial civilizations share human languages, iconography, or cognitive frameworks. This needs the creation of a universal language—likely based on mathematics, physics, or universal scientific principles—that may serve as a foundation for mutual understanding. Projects such as the Arecibo Message and Voyager Golden Record have attempted to encode human knowledge in ways that alien intelligence may understand, but these attempts are not guaranteed to be successful.

Furthermore, the ethical implications of interplanetary communication must be addressed. Communication with another civilization carries the possibility of miscommunication or misunderstanding, particularly given cultural, biological, and psychological differences. Even the best-intentioned message may be viewed as unfriendly or misunderstood by another species. This could have unforeseen effects, including violence if the extraterrestrial culture perceives our message as a threat or an infringement of privacy [29].

11.10.4 Future Prospects in Astrobiology and SETI

Astrobiology and the search for extraterrestrial intelligence (SETI) have strong future possibilities, thanks to technological improvements, rising scientific knowledge, and a growing commitment to exploring life beyond Earth. Astrobiology, the study of life in the universe, will benefit from missions to investigate the habitats of planets, moons, and other celestial bodies in our solar system, including Mars, Europa (a Jupiter moon), and Enceladus (a Saturn moon). These missions, combined with the discovery of organic compounds in deep space, present great opportunities for discovering microbial life or chemical precursors to life. SETI, which has traditionally searched for signs of intelligent extraterrestrial civilizations by detecting radio transmissions or laser emissions, is set for a technological renaissance. The development of more sensitive observatories, such as the James Webb Space Telescope (JWST) and the Square Kilometre Array (SKA), will improve our ability to search distant stars and galaxies for indications of life. With these tools, SETI can examine a broader span of the electromagnetic spectrum and seek deeper into space than ever before. New approaches, such as looking for technosignatures—observable signs of alien technology, such as megastructures or industrial pollutants—are also gaining popularity. These efforts may broaden our understanding of how we detect signals of intelligent life beyond Earth.

Furthermore, AI and machine learning will transform the quest for extraterrestrial life. These tools can process massive volumes of data from telescopes, spacecraft, and laboratory tests, detecting patterns and anomalies that may imply life more efficiently than human researchers. AI is already being used to sift through massive datasets created by SETI and planetary surveys, increasing the likelihood of identifying small signals or biosignatures that would otherwise go unnoticed. Looking ahead, international collaboration will be critical to the success of astrobiology and SETI. As space exploration becomes more accessible, efforts to detect extraterrestrial life will necessitate greater collaboration among government space organizations (such as NASA, ESA, and CNSA), commercial firms (such as SpaceX), and international scientific institutes. Such collaboration can result in pooled resources, knowledge, and plans for robotic and crewed trips to distant planets. While we have yet to unearth solid proof of extraterrestrial life, the future of astrobiology and SETI seems promising. With continued technological advancement, the finding of microscopic life on a nearby moon or evidence of an intelligent civilization in a distant galaxy could revolutionize our concept of life and our place in the universe [30].

11.11 Conclusion

Astrobiology and the search for extraterrestrial intelligence (SETI) are interdisciplinary studies that investigate the possibility of life beyond Earth. Astrobiology investigates the origins, evolution, distribution, and future of life in the cosmos by combining biology, chemistry, and planetary science to look for livable habitats and biosignatures—chemical signs of life on exoplanets such as oxygen and methane. SETI tries to identify signs of intelligent extraterrestrial civilizations usually through radio transmissions but is now looking into technosignatures such as industrial pollutants or constructed constructions. Advances in technology, such as NASA's James Webb Space Telescope (JWST) and the Square Kilometre Array (SKA), are increasing our ability to identify life, including examining exoplanet atmospheres. AI and machine learning are transforming data processing, making it easier to discover patterns that could signal life. International collaboration among space agencies, private firms, and scientific institutions is critical to the success of these initiatives. The Fermi Paradox tackles the discrepancy between the high probability of extraterrestrial civilizations and the lack of proof, while the Great Silence discusses the absence of detectible signals. Proposed solutions include the Rare Earth Hypothesis, which proposes that intelligent life is uncommon, and the Self-Destruction Hypothesis, which proposes that civilizations destroy themselves before making contact.

The possibility of interstellar communication presents difficulties due to time delays, language hurdles, and technology gaps. Overall, both astrobiology and SETI have exciting prospects, with the possibility of startling findings that could revolutionize our knowledge of life and the universe. Upcoming space missions will investigate planets and moons in our solar system that have the potential for habitability. Mars, Europa (a Jupiter moon), and Enceladus (a Saturn moon) are particularly interesting because they may contain liquid water or underground oceans. These missions, which include NASA's Perseverance rover and the future Europa Clipper mission, seek to find biosignatures or indications of microbial life. The discovery of thousands of exoplanets, or planets outside our solar system, has pushed the field of astrobiology forward. Many of these exoplanets are positioned in their stars' habitable zones, where conditions may be favorable for liquid water and, maybe, life. Future missions, such as those employing the James Webb Space Telescope (JWST) and the Square Kilometre Array (SKA), will investigate the atmospheres of these exoplanets for biosignatures such as oxygen and methane, which could indicate the presence of life. Advances in spectroscopy will enable more detailed examinations of planetary atmospheres, surface conditions, and potential evidence of life. New methods, such as artificial intelligence and machine learning, will improve data analysis by detecting small signals or anomalies that could indicate biological activity.

Continued research into extremophiles—organisms that thrive in extreme circumstances on Earth—will help astrobiologists predict how life can adapt to hostile environments on other planets or moons. The study of creatures that survive deep-sea vents, polar ice, and acidic lakes allows scientists to conceive life forms in severe extraterrestrial environments. The search for technosignatures—observable evidence of alien technology—is progressing beyond ordinary radio signal detection. The researchers will hunt for large-scale structures (such as Dyson spheres), industrial pollution in planetary atmospheres, and artificial lights on distant worlds. These technological signatures may suggest the presence of advanced civilizations. The search for alien life is one of humanity's most profound and persistent undertakings, stemming from our curiosity about the universe and our role in it.

While we have yet to discover solid proof of life beyond Earth, the search itself continues to expand our understanding and inspire us. Whether or not we find extraterrestrial life, the hunt encourages us to push the boundaries of science and our imagination. It represents our intrinsic yearning to discover our origins, explore the unknown, and connect to a larger cosmos. Observing the stars helps us better comprehend ourselves and our place in the universe. The study of life beyond Earth forces us to reconsider the essence of life itself. Exploring planets, moons, and exoplanets helps us better grasp what makes a world livable and how life might emerge and persist in harsh conditions. The tremendous speed of technological improvements in telescopes, spectroscopy, artificial intelligence, and machine learning gives hope that we are on the verge of making important discoveries, maybe within the next decade. A revelation, whether it is the discovery of life on a neighboring moon or the reception of a signal from an intelligent civilization, has the potential to unite humanity in a common sense of purpose.

Declaration Section

Data Availability

The data referenced and discussed in this chapter are derived from publicly available astronomical datasets and survey archives, including but not limited to those from NASA, ESA, and other collaborative research observatories.

Acknowledgement

The authors would like to express their sincere gratitude to their respective institutions for providing the necessary resources and support throughout this research. A special note of thanks to **Dr. Yogesh Chandra**, whose editorial guidance has been instrumental in refining this manuscript. We are also deeply grateful to our publisher and institute for their unwavering support in facilitating this research. Furthermore, the authors would like to acknowledge the invaluable mentorship of **Prof. Bindu Arora**, whose guidance and encouragement have been pivotal in shaping this study. Her continuous support and scholarly advice have been a source of inspiration. Finally, we extend our appreciation to all those who have contributed, directly or indirectly, to the successful completion of this work.

References

1. Bains, W., Many chemistries could be used to build living systems. *Astrobiology*, 4, 2, 137–167, 2004.
2. Barrett, A.J., Brown, G.D., Blackwell, J., Extreme environments and the search for extraterrestrial life. *Annu. Rev. Microbiol.*, 60, 149–173, 2006.
3. Borucki, W.J., *et al.*, Kepler planet-detection mission: Introduction and first results. *Science*, 327, 5968, 977–980, 2010.
4. Borucki, W.J., *et al.*, Characteristics of planetary candidates observed by Kepler. *Astrophys. J.*, 736, 1, 19, 2011.
5. Bostrom, N., Where are they? Why I hope the search for extraterrestrial life finds nothing. *Technol. Rev.*, 111, 3, 72–77, 2008.
6. Byrne, S., The polar deposits of Mars. *Annu. Rev. Earth Planet. Sci.*, 37, 535–560, 2009.
7. Carr, M.H., *Water on Mars*, Oxford University Press, New York, NY, 2006.
8. Carr, M.H. and Head, J.W., Geologic history of Mars. *Earth Planet. Sci. Lett.*, 294, 3-4, 185–203, 2010.

9. Charbonneau, D., *et al.*, Detection of planetary transits across a sun-like star. *Astrophys. J.*, 529, 2, L45–L48, 2000.
10. Chyba, C.F. and Hand, K.P., Astrobiology: The study of the living universe. *Annu. Rev. Astron. Astrophys.*, 43, 31–74, 2005.
11. Clayton, D.D., *Principles of stellar evolution and nucleosynthesis*, University of Chicago Press, Chicago, IL, 1983.
12. Cockell, C.S., *et al.*, The nature and extent of life in extreme environments. *Philos. Trans. R. Soc. B Biol. Sci.*, 361, 1474, 1835–1845, 2009.
13. Cocconi, G. and Morrison, P., Searching for interstellar communications. *Nature*, 184, 4690, 844–846, 1959.
14. Crowe, M.J., *The extraterrestrial life debate, 1750-1900: The idea of a plurality of worlds from Kant to Lowell*, Cambridge University Press, Cambridge, UK, 1986.
15. Des Marais, D.J. and Walter, M.R., Astrobiology: Exploring the origins, evolution, and distribution of life in the universe. *Annu. Rev. Ecol. Syst.*, 30, 397–420, 1999.
16. Des Marais, D.J., *et al.*, The NASA Astrobiology Roadmap. *Astrobiology*, 2, 1, 1–48, 2002.
17. Dick, S.J., *Plurality of worlds: The origins of the extraterrestrial life debate from Democritus to Kant*, Cambridge University Press, Cambridge, UK, 1982.
18. Drake, F. and Sagan, C., *Is anyone out there?*, Delacorte Press, New York, NY, 1995.
19. Dyson, F.J., Search for artificial stellar sources of infrared radiation. *Science*, 131, 3414, 1667–1668, 1960.
20. Ehrenreich, D., *et al.*, A giant comet-like cloud of hydrogen escaping the warm Neptune-mass exoplanet GJ 436b. *Nature*, 529, 7584, 59–62, 2016.
21. Farley, K.A., *et al.*, A summary of the Perseverance rover mission. *J. Geophys. Res.: Planets*, 125, 3, e2020JE006523, 2020.
22. Gardner, J.P., *et al.*, The James Webb Space Telescope. *Space Sci. Rev.*, 123, 4, 485–606, 2006.
23. Greaves, J.S., *et al.*, Phosphine gas in the cloud decks of Venus. *Nat. Astron.*, 5, 7, 655–665, 2020.
24. Greene, T.P., *et al.*, characterizing exoplanet atmospheres with JWST. *Nat. Astron.*, 5, 737–747, 2021.
25. Grotzinger, J.P., *et al.*, Mars Science Laboratory mission and science investigation. *Space Sci. Rev.*, 170, 1-4, 5–56, 2012.
26. Haldane, J.B.S., The origin of life. *Rationalist Annu.*, 148, 3–10, 1929.
27. Harp, G.R., The Allen Telescope Array: The first wideband, wide-field, partially steerable telescope. *Publ. Astron. Soc. Pac.*, 124, 921, 329–341, 2012.
28. Haughton, G., *et al.*, Astrobiological implications of subsurface liquid water on Mars. *Icarus*, 157, 2, 498–502, 2002.
29. Hoffman, S.J. and Kaplan, D.I., The Mars reference mission: A vision for the first human mission to Mars. NASA Special Report, 1997.
30. Howard, A.W., *et al.*, A search for optical pulses from extrasolar planets. *Astrophys. J.*, 613, 2, 1270–1282, 2004.

12

Anticipating the Unseen: AI's Promise in Illuminating Astronomy's Future

Ritika Joshi* and Pratibha Fuloria

Department of Physics, Soban Singh Jeena University, Almora, Uttarakhand, India

Abstract

With the dawn of the new age of astronomy, AI is the perfect prospect to change how we perceive space. AI refers to a combination of software applications that facilitate solving problems that would naturally be solved by human input. In the case of AI, ML is a type of AI where the algorithms help in determining the results from data without being programmed to do so. DL is a subcategory of ML and uses neural networks with many layers to map the data into desired patterns. That is why DL has adapted to the requirements prevailing in various tasks such as image and voice recognition, among others. The remaining chapter is devoted to the prospects of AI, ML, and DL's application in astronomy and discusses how data-centric approaches revolutionize the studies of the cosmos. To start with, we present an overview of some of the existing issues in astronomy nowadays: the overwhelming amount of data provided by the advanced telescopes and the problem of quantity that exceeds the abilities of qualitative analysis. Example: AI's contribution is detailed for particular domains, for example, AI seems to be great in exoplanet detection, as its algorithms help to find out potential exoplanets from large datasets. Furthermore, due to AI's real-time observational data processing capability, it has played a crucial role in the identification and study of transient sources such as supernovae and fast radio bursts. The chapter also presents a discussion on the feasibility of implementing AI with quantum computing as a way to look into the future of astronomical research. In the next section, we give a very brief description of how quantum computing could extend AI's abilities to perform modeling and prediction, which in turn could lead to improved conclusions about the nature of the universe's greatest enigmas. It is worth stressing that this chapter also describes existing opportunities within the field of astronomical research and the threats and disadvantages of using AI technology. We consider

**Corresponding author*: Joshi.r1312@gmail.com

Yogesh Chandra, Manjuleshwar Panda and Mahesh Chandra Mathpal (eds.) The Intelligent Universe: AI's Role in Astronomy, (431–474) © 2025 Scrivener Publishing LLC

their limitations, for example, possibilities of making an algorithm biased, requirements to the base that should be large and selected carefully, and ethical problems of artificial intelligence application in research. Possible approaches to managing these issues are also discussed, for instance, creating explainable artificial intelligence systems. In conclusion, this chapter offers reasonable positive expectations for the future development of AI in astronomy. Overall, we hope to encourage researchers, astronomers, and practitioners who work with AI to take full advantage of AI in exploring other facets of the world beyond our limited ability to see. This collaborative work shows that the combination of increasingly embedded AI and astronomy will lead to discoveries and a better understanding of the universe.

Keywords: Artificial intelligence, astronomy, exoplanets, quantum computing, transient phenomena, data-driven research, ethical AI

12.1 Introduction

At the start of the new millennium, the world continues to witness rapid changes in the growth of technology and science. One of the most outstanding of these changes includes artificial intelligence (AI), the realization of which is however embryonic as it starts to alter many industries including the healthcare and finance sectors. AI's future role appears to be more promising than ever, especially given the fact that, as we discuss this, the world looks likely to enter a new era in identifying space bodies, which therefore means that, in the world of astronomy, which is often said to be the voice of science, AI could play the voice of transformation than we can imagine [1]. The tremendous extent of space added to the amount of data resulting from the new-generation telescopes calls for other techniques and tools. Thus, AI as an algorithm that analyzes huge quantities of input data, recognizes elaborate patterns, makes immediate choices, and seems to offer a perfect solution to the difficulties of contemporary astronomers [2]. In its simplest sense, AI is a field that studies how to give machines the capacity to act intelligently. Such activities include as simple as vision and speech to as complex as decision and planning [3]. Despite being an umbrella term for disparate methods, its most useful paradigms in astronomy are machine learning (ML) and deep learning (DL). ML is a collection of computational procedures in which a system can learn from data and refine its performance without being reprogrammed. In ML, DL continues this idea by using neural networks with more than one layer to model nonlinear dependencies that may characterize various data [4]. Of all these neural networks, these are better suited for the kinds of data that modern astronomical instruments output: high-dimensional data. In the case of astronomical sciences, the standard approaches to the analysis of gathered data and information are, although efficient for the relatively small amounts of data, incapable of handling the current inflow of information. For instance, the upcoming Square Kilometre Array radio telescope will produce more than an exabyte of data daily. They are enormous datasets that cannot be analyzed by simply employing more human astronomers. Herein lies the importance of AI: Its capability of enabling automation and enhancement of these data, into patterns, as well as classification of objects and detection of new phenomena in the universe is significant [5]. This kind of methodology results in the ability to perform tasks that, if done manually, would have taken months at best, and in some cases, minutes. So, if we are to give our best shot at understanding the role of AI in modern astronomy, we must analyze the data pipeline where AI steps in and how useful it is at that stage.

AI is an integral part of the scientific data life cycle, encompassing data acquisition, preparation, analysis, and discovery. A common example of AI utilization in astronomy is illustrated in Figure 12.1, which depicts the overall workflow from raw observational data to actionable scientific insights. Adapted from NASA's Earth data framework, this figure highlights how

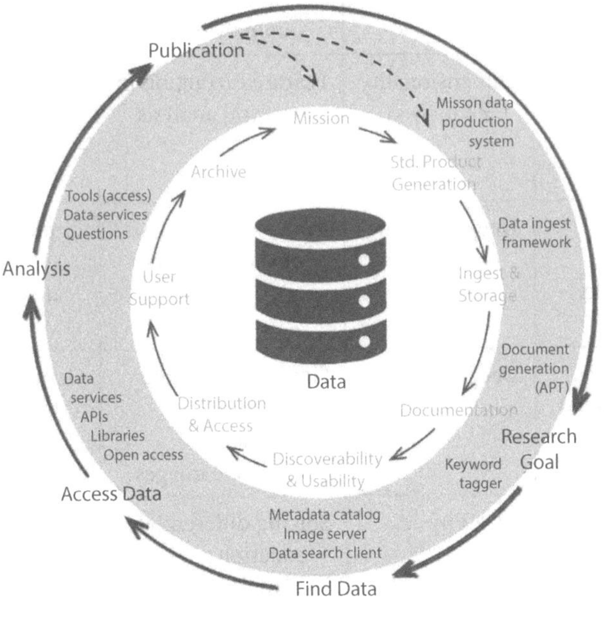

Figure 12.1 AI's role in the scientific data life cycle, illustrating the progression from raw observational data to actionable insights. (Image Source: NASA Earthdata).

AI-driven models facilitate data processing, automate feature extraction, and enhance predictive analysis, ultimately transforming vast observational datasets into meaningful discoveries.

12.1.1 Optimized Data Collection, Preprocessing, and Cutting-Edge Algorithms

Depending on the specific area of astronomy, the primary data can come from any observing instrument, from radio telescopes to optical telescopes, to space observatories [6]. Some of these instruments provide raw images, spectra, and time-series measurements of the universe. These are original facts upon which all subsequent analyses are built as indicated in the following sections. However, any analysis can be performed only after the

Table 12.1 Machine learning and deep learning algorithms for optimized data processing in astronomy.

Algorithm	Category	Application in astronomy	Advantages
Random Forest	Machine Learning	Classification of celestial objects, anomaly detection	High accuracy, robust to noise
Principal Component Analysis (PCA)	Dimensionality Reduction	Feature extraction in spectral analysis	Reduces computational complexity
Recurrent Neural Networks (RNNs)	Deep Learning	Time-series analysis of variable stars	Captures temporal dependencies
Generative Adversarial Networks (GANs)	Deep Learning	Simulating cosmic structures, enhancing telescope images	Generates high-fidelity data
Physics-Informed Neural Networks	AI + Physics	Solving differential equations in modeling astrophysical phenomena	Ensures physical consistency

raw data have been preprocessed. This step includes the deletion of additional unwanted signals, synchronizing various observations, and making preliminary calibration. Data preprocessing [7] helps feed pure data to AI systems without biases or noise that are likely to cause the system to give wrong results. After data preprocessing, it is the turn of AI algorithms in the process. These algorithms, which might be machine learning (ML) models or deep learning (DL) networks, are expected to learn patterns, outliers, as well as other forms of attributes in the data. For instance, in exoplanet [8] identification, the AI algorithms can also be used to study "light curves" in a bid to search for slight dips that suggest there is a planet. Likewise, in the search for transient phenomena, including supernovae or fast radio bursts, for instance, AI can quickly survey the big datasets and determine their nature in the nearest real time. Various machine learning and deep learning algorithms have been employed for data collection, preprocessing, and analysis in astronomical studies. The above Table 12.1 summarizes key algorithms and their applications in astronomy.

12.1.2 Real-Time Intelligence and Simulation

The strength of AI mostly lies in its ability to work in the context of the actual data processing performed in real time [9]. This is important in identifying the transient point events where detection and response must occur swiftly. One of the most important applications is the possibility of constant surveillance of the hemisphere, with automatic marking of potentially interesting phenomena as they unfold, freeing up the attention of astronomers to encompass only the most exciting events. Subsequently, after identifying and analyzing data in real time, an AI system can make predictions and classifications of astronomical objects or events [10]. For example, by making a set of rules about the patterns of the data, the AI can sort between different sorts of stars, galaxies, or additional staking objects and even predict their future behavior. This capability is important for the determination of the life cycles of stars, the evolution of galaxies, and other essential astrophysical phenomena. Once predictions and classifications are made, the results have to be visualized and simulated to enhance the understanding [10]. Civilian uses of AI have been prominent in explaining large trends and patterns in data, in helping astrophysicists create models of astronomical events, and in assisting scientists in understanding the implications of their results. These visualizations not only serve the needs of researchers but also allow for the expression of complex astronomy to the public.

12.1.3 Scientific Discovery

Last but not least, applying AI at each stage of data processing allows making scientific discoveries. These capabilities of recognizing, analyzing, and classifying objects allow AI to open new insights about the universe. Since the identification of new exoplanets to the unexplained concept of dark matter and energy, AI is showing how much can be done in the field of astronomy.

Probably one of the most transformative ways AI has taken a part in the field of astronomy is the discovery of exoplanets, planets that exist around stars beyond the solar system. Conventional techniques of exoplanet hunting like the analysis of fluctuations in light intensity due to transiting exoplanets are time-consuming and often involve many false alarms. Still, especially with the NASA Kepler mission data, AI algorithms have been able to sort out much larger data volumes and find new exoplanets with much higher precision and much less time [11]. The utilization of AI in the said discipline has not only enhanced the finding of new exoplanets but also introduced other planets that could be unnoticed. AI is also transforming how transient astronomical events can be detected and analyzed—phenomena, which occur in the sky and do not last very long. These are supernovae, gamma-ray bursts, and fast radio bursts (FRBs). Due to their ephemeral character, such events must be identified and monitored continuously—a task that certainly belongs to the high-priority assignments within the sphere of AI utilization. Data analysis involves employing machine learning models such that a stream of observational data can be analyzed as it comes in and alerts important events for follow-up analysis. This capability has been especially useful in extending our study of these events, as many of them are crucial to the explanation of such phenomena as the formation of black holes and neutron stars. However, the fact that AI can be integrated with the next-generation techniques like the quantum computing system will open up further possibilities in the field of astronomy. Quantum computing [12] is still an emerging field, but it utilizes exponential computational capacity as well as addresses problems that non-quantum computers cannot solve. With quantum computing deep learning, it would be possible to model higher complex systems, for instance, the way galaxies are formed, the characteristics of dark matter and dark energy, or even developing computation for black holes. It might also put light into some of the greatest philosophical questions of astronomy and possibly revolutionize understanding of the universe. There are ample opportunities for the application of AI in astronomy, but using it is not free from difficulties. Algorithmic biases that can be dangerous are again arrived at by making mistakes like arriving at wrong conclusions due to restricted/imbalanced training data. Additionally, AI models heavily depend on the quality and quantity of the training dataset, and although current telescopes provide extensive amounts of data, their high-quality processing and variety are not an easy task. Technical or method-related issues of interest are raised here also concerning ethical matters such as transparency or interpretability. For the methods and tools based on AI to be acceptable to astronomers, they must be developed in a manner that reveals how the decisions are being made instead of the components being shrouded in mystery or "black box." However, its potential to revolutionize our perception of the universe can be scarcely contested. In particular, it helps to save the most time-consuming tasks for data analysis and provides astronomers with data that could not be obtained before. From finding new planets, analyzing eruptions of new stars, short-lived astronomical phenomena, to writing computer-based cosmological simulations, the progression of AI is set to be the next big discovery in astronomy. One of the most significant contributions of AI to modern astronomy is its ability to drastically enhance both the speed of computations and the precision of predictions, which are crucial for advancing our understanding of the cosmos. The equation below encapsulates this impact:

$$AI\ impact = Speed\ of\ computation + Precision\ of\ predictions \tag{12.1}$$

Equ. (12.1) reflects two core aspects of AI's role in astronomy: accelerating the pace of computational processes and refining the accuracy of astronomical predictions. In this chapter, various instances where AI, ML, and DL are being used in astronomy have been presented. We will provide glimpses of particular usages—justifications including detecting exoplanets, analyzing transient phenomena, and simulating cosmological phenomena—explaining how the technologies we present provide solutions throughout astronomy and astrophysics. Also, proposals for future perspectives of AI usage in astronomy will be discussed, especially taking into consideration the effect of such drivers as quantum computing on the field. Finally, the presented problems and risks of the growing application of AI in astronomical research will be discussed along with potential solutions to these emerging issues. At this bright intersection of the future AI integration and astronomy, it is becoming rather obvious that the assistance of those two sciences can open the gate to a new era of space exploration and get rid of the lack of knowledge about the existence of the universe.

12.2 Modern Issues in Astronomy

In view of the advancements in the field of astronomy, a wide range of discoveries has been made, from exoplanets to distant galaxy clusters. Nevertheless, the development of technologies has been rapid and astronomers are confronted with many obstacles [13] that are likely to slow down discovery. These challenges emanate from the increasing rate at which data are growing, the complexity of the universe, and constraints in available technologies and human capital. It is important to be aware of these challenges to creatively solve them—essential for growing the field. The data have grown exponentially. The field of astronomy has made remarkable strides in recent decades, uncovering phenomena ranging from distant exoplanets to the vast structures of galaxy clusters. However, despite these advances, astronomers face a host of challenges that threaten to slow the pace of discovery. These challenges arise from the rapid growth of data, the complexity of the universe, and limitations in technology and human resources. Understanding these hurdles is crucial to developing innovative solutions that can propel the field forward.

12.2.1 The Exponential Growth of Data

Astronomy has now become a big data science owing to the vast amounts of information that observational instruments produce [14]. Until now, telescopes recorded small portions of data that were manageable to be processed by the scientists. However, modern telescopes and their observational complexes receive terabytes of information at a rate approaching real time. For example, the Large Synoptic Survey Telescope (LSST) is expected to generate about 30 terabytes of data per night when at its full capacity; over a decade, it will deliver a sample of 500 petabytes. Gigantic products such as the next-generation radio telescope (Square Kilometre Array or SKA) will produce exabytes per year. These figures are well beyond the manageable capacity of traditional data analyzing tools, an aspect that makes it extremely difficult for researchers to engage in such volumes of data analysis manually. It is not just the amount of big data, but its very nature: big data is not simple, it is complex. The instruments of the modern age gather data in various wavelengths, at different time ranges and spatial scales. Handling, storing, and analyzing such high-dimensional data remain a challenging task due to the availability of only limited computational facilities and non-robust and computationally complex algorithms.

Table 12.2 Exponential growth in astronomical data over time.

Year	Data collected (TB)	Major telescope/Misson
2000	~1	Sloan Digital Sky Survey (SDSS)
2010	~50	Kepler Space Telescope
2020	~1000+	Gaia, LSST (Vera C. Rubin Obs.)

Astronomical data have been growing exponentially due to advances in observational technology. The above Table 12.2 illustrates the rapid increase in data collected by major telescopes and missions over time.

12.2.2 Real-Time Data Processing and Complexity of Data Interpretation

As real-time observational infrastructure has become possible, a demand for on-the-spot analysis has emerged. Such applications concern transient astrophysical phenomena for which near-real-time processing is necessary to identify sources such as gravitational wave observatories or gamma-ray burst detectors. For instance, gravitational waves are the ripples in space-time created by chains of massive happenings such as black hole collisions and emitting a signal that is just microseconds long. Astronomers require systems that can easily detect such signatures, characterize them from noise, and initiate further observations. However, the quick efficient processing of large datasets is a difficult task. Due to the amount of information produced in the field, if there are delays in analyzing specific data, it may cause the inability to document and explore key celestial phenomena. Modern computational environments may not be designed to process these streams of data in a time frame that is fast enough for real-time analysis, which results in analysis gaps and lost opportunities [9].

Astronomy is deploying its ways of knowing, operating in a data-intensive and intricate knowledge environment. In addition to the amount of data, the greater complexity of the data collected constitutes a major challenge [15]. Many instruments today offer more dimensions, and they include spectra, images, time-series data, or geographic information. For example, spectroscopic surveys are responsible for data that cover all wavelengths ranging from details concerning the composition, motion, and temperature of certain astronomical objects. Dealing with multidimensional data is computationally and mathematically intensive and cannot ordinarily be solved using simple methods. Astronomy data scientists are in a unique position where they need to adapt computations that help them conduct noise elimination, symbolize discovery, and signify signal unmasking across huge volumes of high-dimensional data. Additional complexity arises from the combination of different types of data within the same analysis; for example, spectroscopy is often combined with photometry. Human and computer constraints in space-time caused by massive cosmic events like black hole mergers produce signals that last mere fractions of a second. Astronomers need systems capable of immediately identifying these signals, distinguishing them from noise, and triggering follow-up observations. However, processing vast amounts of data quickly is a significant challenge. The sheer volume of information means that delays in analysis can result in missed opportunities to capture and study important cosmic events.

12.2.3 Computational Limits and Data Archiving

Today's observatories and telescopes are delivering over previous years the amount of data being produced is, however, insufficient in addressing by astronomers and data analysts. The role of human resources is stretched to the limit as researchers struggle to manage the large amount of data generated daily and the amount of work required to analyze, interpret, and report these data. However, it is recognized that even with these great supercomputers, computation-based searches are somewhat bounded. Most complex calculations, such as challenging simulations of galaxy formation [16] or black hole coalescence, can take days, weeks, or even months and involve parallel HPC clusters expensive to support. These computational resources, even though are efficient in terms of the amount of work they can do for each task, are not always sufficient for the increasing need for astronomy computation. Therefore, maintaining the rate of data acquisition has become an increasing challenge and research is slowly lagging.

Data storage in astronomy [17] has become a big challenge as data collected through observation increase in volume. Big datasets require storage that allows such information to be easily retrievable for current and future analysis. However, the storage of data in the long term is costly and requires more resources in terms of storage size capacity and power source. Storage for institutions and observatories has to be advanced, a lot of the time cloud-based, for data to be archived safely. There is also a big issue in data organization and annotation of the data collected in the process.

Inefficient structuring of datasets can compromise productive investigation by missing exceptional information or taking a long time to locate data. Today's data archival systems depend on compatibility across institutions, yet the establishment of such systems is expensive in terms of money and technology.

12.2.4 Theoretical Gaps and Interpretative Challenges

One of the deepest issues in astronomy is that of the space between observations and theory. Nevertheless, there is a great amount of information; still, there are numerous questions concerning the essence of the universe that remain unanswered. Dark matter and dark energy, the two constituents of most of the total mass-energy of the universe, remain mysterious despite a multidimensional approach to the study and experimental investigation that has been launched over the years. This is true even with advanced observations assuming data, and making conclusions mostly involves assumptions arising out of partial models at best. These models have to be modified as observations are made, but anomalies between theory and observation still constrain astronomers from gaining further insights into the universe. To sum up, contemporary astronomy is presented with several issues as regard to the availability of data gathered by new telescopes and the lack of time and space both in terms of human and computational analysis. Interpretation of large amounts of data, challenges with real-time analysis, as well as prospective archive preservation, are the main challenges that astronomers face. Solving these issues will require fresh approaches, and this will be where the introduction of AI, ML, and DL into the sphere of astronomy in the future will change. AI with its transformative capabilities in the field of astronomy must be further developed, but gaps between theory and observation continue to limit astronomers' ability to fully understand the cosmos.

In summary, astronomy today faces a range of challenges, from the sheer volume of data generated by new telescopes to the limitations of human and computational resources. The complexity of data interpretation, issues with real-time processing, and difficulties in long-term data storage all contribute to the obstacles that astronomers must navigate. Addressing these challenges will require innovative solutions, and this is where the integration of AI, ML, and DL can play a transformative role in the future of astronomy.

12.3 AI's Transformative Role in Astronomy

This has provided an unprecedented shift of paradigm in several disciplines, and astronomy is one of them that has benefited immensely from the development of AI. Historically, the study of outer space belonged to the category of research where observers picked out crucial information and interpreted findings with simple tools and intuition. Though the history of the use of such methods produced impressive results, modern astronomy has produced an enormous amount of data that exceeds the ability to use these techniques. Now, with the help of technology that can consume and analyze vast amounts of data and understand complex patterns, AI has become one of the most revolutionary tools that change the process of astronomical research.

12.3.1 The Big Data Challenge in Astronomy

Undoubtedly, one of the most significant challenges in modern astronomy is handling the vast volumes of data generated by advanced observatories and telescopes. Cutting-edge facilities like the Square Kilometre Array (SKA) and the Vera C. Rubin Observatory, formerly known as the Large Synoptic Survey Telescope (LSST), are revolutionizing astronomical data collection. As depicted in Figure 12.2, the Rubin Observatory alone, once operational, will gather approximately 30 terabytes of data each night. These data are collected in various forms, including images, spectra, and time-series measurements. However, traditional astronomical data processing techniques were designed for much smaller datasets, making it essential to develop new AI-driven methodologies for managing, analyzing, and extracting meaningful insights from this vast and complex data stream.

AI provides an effective solution to this problem because it can handle all challenges that arise when collecting, processing, and making sense of data. These are huge datasets that would take ages if analyzed manually; this is where and when the machine learning (ML) algorithms play their part by analyzing the data and pointing out such issues as patterns and other things. A subfield of ML, DL or multilayered neural networks are especially fitting for managing the intricacies of astronomy data. These neural networks can then approximate the high-dimensional nonlinear relationship as exhibited by the large noisy data generated by today's telescopes. Because details are processed in real time by the use of artificial intelligence, astronomers are achieving remarkable discoveries that have

Figure 12.2 The Vera C. Rubin Observatory, an advanced telescope designed for large-scale sky surveys, expected to generate 30 terabytes of data per night. (Image Source: LSST Project/NSF/AURA).

deepened the understanding of the universe. A revolution in spying out exoplanets is being driven by the challenge of automating the entire data pipeline—from collection to interpretation. Machine learning (ML) algorithms can quickly analyze these enormous datasets, detecting patterns and anomalies that would otherwise go unnoticed. Deep learning (DL) [18], a subset of ML that uses neural networks with multiple layers, is particularly well-suited to handling the complexities of astronomical data. These neural networks can model the nonlinear high-dimensional relationships that are characteristic of the vast noisy datasets produced by modern telescopes. AI's ability to process data in real time is enabling astronomers to make discoveries at an unprecedented pace, leading to a deeper understanding of the universe.

12.3.2 Revolutionizing Exoplanet Discovery

The use of AI in astronomy is perhaps most impactful, especially in the discovery of so-called exoplanets—celestial bodies, which orbit stars other than the Sun. Within exoplanet detection, previous methodologies were to use graphs for light intensities where the subtle variation in brightness due to transit could only be plotted manually, and this was time-consuming, erroneous, and inefficient. This process entailed human researchers spending time wading through the massive amount of data from telescopes such as Kepler that could lead them to wrong observations or fail to observe certain events. AI has democratized the process of discovery of exoplanets and enhanced the speed as well as the relevancy of their discoveries. With the use of advanced algorithms, what took researchers days and even months to sieve through can now be analyzed within a few microseconds to detect exoplanets with immense precision. For instance, AI systems can identify faint recurrent patterns of light curves from stars, meaning that there is a planet out there. These algorithms can also be trained to differentiate the real planetary system transits from other effects that cause similar excursions in luminosity, namely, star, or binary systems. Using AI in space research has already yielded new exoplanets that had not been discovered through conventional means, so there is hope that the universe beyond our solar system contains other signs of life, thus there is hope. Real-time monitoring of transient astronomical phenomena enhances our ability to detect and analyze events that might be missed using traditional methods. This advancement expands our understanding of dynamic cosmic processes, including planetary systems and their potential habitability. Real-time monitoring of transient astronomical phenomena enhances our ability to detect and analyze events that might be missed using traditional methods. This advancement expands our understanding of dynamic cosmic processes, including planetary systems and their potential habitability.

12.3.3 Advancing Astrophysics with AI: Monitoring, Simulation, and Prediction

Of course, AI's impact is also notable in other areas of astronomy apart from exoplanets, particularly transient astronomical phenomena, which are short-lived and sudden, including supernovae, fast radio bursts (FRBs) [19], and gamma-ray bursts (GRBs) [20]. These events are fundamental in describing some of the most violent and energetic occurrences in the world, but they are brief and hard-time responsive. We have also used AI's capacity to process data perpetually and concurrently to capture these transient occurrences. Supernovae or FRBs—which appear in the night sky and are easily detected by telescopes—can be distinguished from other signals by using machine learning models to identify unique patterns in the data that have been flagged for additional investigation. These systems are capable of watching huge areas of the sky around the clock and so allow astronomers to observe and record these short-lived events in their infancy. This capability is especially important for defining the rates at which stars evolve to become pre-supernova stars, the formation of black holes [21] as well as neutron stars [22] and other related phenomena that occur in relatively short timescales. For instance, systems that analyze data from radio telescopes have recently helped to identify fast radio bursts—short and intense high-frequency radio emissions that last for only milliseconds. These still relatively unexplained events give us a glimpse of the conditions in neutron stars or intergalactic space. Therefore, through generating real-time capabilities as they constantly observe the development of FRBs, AI is providing the solution to the realization of their intrinsic properties and origins [10].

AI has also transformed cosmological simulations, since the technology is now used to design more accurate astronomical models and systems. These numerical reconstructions of the universe and its structures—galaxies, clusters, or even dark matter—are computational and demand more time. These systems act according to complex physical rules and necessitate the assessment of high-dimensional data. Some basic concepts of deep learning are explained in detail including how they have been used in the simulation of cosmological phenomena, as illustrated in Equ. (12.2):

$$AI\ Driven\ Simulations = Data + Physics + Deep\ Learning \tag{12.2}$$

For example, in forthcoming deep space models, environments like galaxy formation or dark matter distribution as well as black hole behavior are already being explored using AI. These simulations do more than help understand the fundamental processes that govern the universe, they also contribute to improving the ideas that are currently used in the branches of space study. Additionally, AI is now used as one of the main ways to visualize astronomical data. Advanced AI analytical tools can model astronomical objects and processes, and their results are 3D models that can be manipulated in ways impossible before. In addition to serving their scientific purpose, these visualizations serve as effective means to share otherwise hard-to-grasp astronomical phenomena with the general public, thereby increasing awareness and potential participation in the discipline.

AI's actual scientific metric performance in terms of prediction of future astronomical occurrences or trends is another area where its change-making potential is apparent. Current data acquired through observations and mathematical analysis allow AI systems to predict the behavior of space objects in the future. For instance, machine learning models can predict how stars, galaxies, and other astronomical systems are likely to change in the future, which makes scientists expect what they are likely to observe later on. One of the well-used applications of the predictive AI models has been in the case of variable stars, stars that have changes in brightness over some time. These models can monitor patterns of variability from observations made in the past and then forecast the pattern of variability in the future so that astronomers can make detailed observations of these stars. In the same way, AI has been used in the forecasting of the sun's activity levels inclusive of incidents like the flares and coronal mass ejections that go a long way to determine the sphere of space weather and the interaction with satellites.

12.3.4 Quantum AI and Ethical Considerations in Astronomy

The future of AI and its applications in astronomy is getting bright as the developments in technology it identifies with—including quantum computing—proceed. Quantum computers under development today look like these computers can perform complex computations that are currently impossible, such as modeling the universe. When integrated with AI, quantum computing may help scientists create models of dark matter and dark energy, model the behavior of black holes, or recreate the Big Bang accurately. Of equal or greater importance, they may eventually take scientists to the answers to the universe's hardest questions.

AI indeed has been instrumental in the advancement of astronomical science, yet, its integration entails a few difficulties. There are issues to do with preexisting biases that are inherent in algorithms that are fed with poor or biased datasets. If these biases are not controlled, we could end up making wrong decisions or, even worse, missing great discoveries. Moreover, AI models rely on big data and often involve high-quality datasets, which are not easy to build. Another ethical issue that is of great significance in the use of artificial intelligence is the interpretability of the AI model. Lately, AI systems have appeared as "black boxes" where the systems make a decision that researchers cannot easily analyze. Given that AI solutions are progressively used in astronomical research, it is vital to create tools that would open the black boxes and make results interpretable.

12.4 Classification of Images and Its Application in Astronomy

The sheer volume and rapid growth of astronomical data from modern telescopes and observatories pose significant challenges for manual analysis. Deep learning, a subset of AI, has transformed how astronomers process and interpret these data. Using deep learning models, such as convolutional neural networks (CNNs), researchers can efficiently classify objects, detect patterns, and identify anomalies at unprecedented speeds. CNNs achieve this by sequentially extracting features through convolution and pooling layers, followed by fully connected layers that classify input data into probabilistic outputs. This approach enables the transformation of raw observational data into meaningful predictions through learned hierarchical features.

12.4.1 Deep Learning in Astronomical Image Processing

The application of deep learning has been exceptionally effective in contextualizing image material due to its capacity to allow the extraction of features of the raw data [23]. While most classical semantic algorithms used for image analysis need feature extraction by the programmer, deep learning models, specifically CNNs [24], acquire feature hierarchies automatically from images. This makes them suitable for analyzing the large and varied format of images that are received from astronomical surveys. For example, astronomical images are noisy and may contain artifacts and faint objects that would be hard to examine by the human eye. To this end, outcomes of deep learning models can eliminate noise and amplify essential patterns in the data. Moreover, these models are not just able to individually identify material objects like stars or galaxies but to also pick out characteristics of the environment like faint glowing emissions or lensing effects that would otherwise be almost impossible for the naked eye to perceive. One such example is the use of deep learning in analyzing data from radio telescopes, as a massive volume of datasets is produced from sky mappings. Deep learning architectures are capable of performing analysis on terabytes of image feeds frequently finding sources of interest like pulsars, distant galaxies, or transient radio bursts.

12.4.2 Advanced Machine Learning for Astronomical Object Detection and Discovery

Neural networks play a crucial role in astronomy, particularly in object classification, one of their most significant applications. Stars, galaxies, asteroids, and other celestial bodies exhibit diverse shapes and structures, making their classification highly complex. Deep convolutional neural networks (DCNNs) have proven to be the most effective models for handling such image classification tasks. Their ability to learn from large labeled datasets and generalize across a wide range of astronomical objects makes them invaluable for modern astronomical research.

- CNNs process astronomical data step by step, beginning with preprocessing images or light curves before feeding them into convolution layers.
- Convolution layers act as filters that detect local features such as edges, textures, and shapes.
- Pooling layers condense information, retaining only the most significant details for classification.

CNNs have been particularly effective in classifying galaxies based on their morphological characteristics—such as distinguishing between spiral, elliptical, and irregular galaxies—with high precision, thanks to large-scale surveys like the Sloan Digital Sky Survey (SDSS). Additionally, CNNs can classify stars by analyzing their spectra and detect exoplanets from light curve data. Their topology-based architecture enables them to extract both high-level and lower-level features, allowing them to refine complex patterns in images beyond what traditional methods can achieve.

Recurrent neural networks (RNNs) and long short-term memory (LSTM) networks have proven to be highly effective in analyzing time-series data in astronomy. These models process astronomical time-series data, such as light curves, by transforming raw input into temporal patterns through RNN layers [25]. LSTM layers, in particular, capture long-term dependencies in the data, while fully connected layers classify transient astronomical events. This approach has significantly improved feature extraction from large and complex datasets, revealing hidden patterns that traditional methods might overlook. This is particularly valuable in studying star formation, galaxy evolution, and gravitational wave propagation, where deep learning's ability to identify abstract features is indispensable. Key features of RNNs and LSTMs in astronomy:

- **RNNs** analyze time-series data by identifying short-term dependencies in astronomical object behavior.
- **LSTMs** extend memory capabilities, preserving long-term patterns crucial for understanding phenomena such as supernovae and gravitational waves.

RNNs and LSTMs follow a structured hierarchical flow to handle time-domain astronomical data. The process begins with signal preprocessing, such as normalization or denoising, followed by RNN layers that extract short-term dependencies. LSTM layers provide extended memory, capturing intricate temporal patterns essential for understanding long-duration events like supernovae and gravitational waves. Finally, the learned features are processed by fully connected layers, which classify astronomical objects based on their variability. This ability to track and categorize transient celestial events is fundamental to studying stellar evolution and uncovering new astronomical phenomena.

It should also be noted that deep learning's ability to learn patterns is also its primary strength in the detection of anomalies, a critical process in astronomy. Abrupt patterns in astronomical images can range from inaccurate imaging techniques to the exciting discovery of unknown forms of celestial objects. These outliers are increasingly detected using neural networks, and the accuracy of these networks helps capture rare events like gravitational lensing, gamma-ray bursts, or possibly a new class of astrophysical objects. For instance, deep autoencoders, as unsupervised deep learning models, have been used to detect anomalies in astronomical images by learning feature representations of the data. Since these models are trained on "normal" images, they can alert researchers to images that are anomalous in some way, prompting further examination. This method has been applied in the search for fast radio bursts (FRBs), where comparably short-lived signals occurring on the scale of milliseconds must be shifted from large amounts of radio data.

12.4.3 Neural Networks in Multispectral and Multimodal Imaging

Astronomical images are taken in various bands of wavelength including optical, infrared, ultraviolet, and radio wave bands. The multiple-channel data can be easily incorporated into deep learning models to offer a better and richer view of celestial objects. Imaging across multiple wavelengths yields a larger set of data from which to analyze astrophysical events.

To handle such inputs, neural networks are designed to be capable of analyzing these inputs in a mechanical manner enabling correlation of the spectral bands to enhance feature detection and object classification. That is especially beneficial for analyzing galaxies, as the combination of optical and infrared data can indicate both the visible structure of the galaxy and the hidden extras that the dust eclipses. Using such deep learning models, astronomers can discover the universe in ways that were practically impossible in the past. For instance, the asymmetric combination of radio and X-ray data from two observatories resulted in the identification of relativistic jet emissions from black holes and the projection of diverse dark matter distributions in galaxy clusters. This has an important function of assembling these multimodal datasets, thereby revealing aspects of the universe not easily discerned.

In the future, it will remain very important in astronomical discovery since the new generation of observatories like JWST and SKA will produce still larger datasets. AI models will become more advanced, distinguishing between objects and recognizing abstract patterns and physically comprehensible predictive metrics of celestial events will be offered by these models. Fourthly, richer first-stage architecture, better transfer learning, and self-supervised learning will allow deep learning models to be trained on small samples of domain-specialized datasets, which will enable astronomers to use state-of-the-art approaches in the specialties of astronomy. As we continue to advance technologies in artificial intelligence, I envisage new horizons for deep learning, especially in processing images from space and categorical tasks.

12.5 Cosmological Simulations

Cosmological simulations have been critical in the study of large-scale structure formation, where the interplay between dark matter, dark energy, and baryons plays a crucial role. These simulations probe the gravity and hydrodynamic processes that underpin phenomena such as galaxy formation, the progression of cosmic filaments, and the arrangement of large-scale structures. However, these techniques are computationally expensive, as even a small portion of the observable universe must be modeled using supercomputers. As shown in Figure 12.3, this highlights the immense computational resources required for such tasks. New approaches, like artificial intelligence (AI) and machine learning (ML), have emerged as promising tools to enhance and expedite the process of cosmological simulations.

Figure 12.3 Computational demands of cosmological simulations, highlighting the role of AI and ML in alleviating this burden. (Credit: NASA).

Recent advances in AI are transforming how researchers in astronomy and astrophysics model cosmic events. By alleviating the computational burden, AI unveils the potential to predict and analyze vast cosmic structures more effectively. This section examines the role of AI in cosmological modeling and simulations, specifically focusing on the application of ML models to forecast the large-scale structures of the universe. AI-driven cyber-physical models are now employed to simulate cosmic phenomena, reducing the computational load while enhancing the ability to predict and interpret these structures. The following discussion explores how AI is being utilized to create cosmological simulations and apply ML models to predict the universe's large-scale architecture.

12.5.1 AI-Driven Simulations of Cosmic Phenomena

Even though cosmology can be regarded as one of the most successfully addressed issues in AI application, the primary significance of AI for cosmologists is in the enhancement of simulation approaches. Cosmological simulations in which the positions and velocities of CDM, galaxies, and gas are followed over billions of years to the present day are computationally expensive. To overcome the problem of high computational cost, several AI models, including DNNs and GANs, have been designed to learn the actual physical laws from a portion of high-fidelity simulations and produce approximate responses at minimal computational cost. Smoothed-out deterministic models can also be produced by AI models as fast and approximate versions of these detailed simulations are produced. For instance, neural networks can be trained to forecast the fate of dark matter structures by conditioning them on the initial state and strapping the outputs at different stages of evolution. Thus, using learning exceptions from comparatively small-scale full-scale simulations, the models can make generalizations toward the greater large-scale unexplored regions of the universe. This greatly reduces the amount of time for carrying out cosmological simulation and thus can simulate ahead of the portions of the universe that can be tiresome to simulate. The object's governing physical laws have also been integrated recently into the learning process, which is known as physics-informed neural networks (PINNs) [26]. Equ. 12.3 emphasizes that PINNs are standard neural networks constrained by physics-based equations. When these equations are integrated within the network, the AI models guarantee that the computed output is physically plausible and is computed significantly faster than the conventional simulations.

$$\textit{PINNs} = \textit{Neural Networks} + \textit{Governing Equations} \tag{12.3}$$

For instance, PINNs have been applied to the simulation of the structural development of the cosmic web—the large-scale cosmic structure formed of dark matter filaments and voids—with great efficiency. Following the work of PINNs, the relative novelty of PINNs Former, proposed by Zhiyuan Zhao in 2024 [27], has pushed forth the development of other improvements to physics-informed networks. Handling spatiotemporal data works effectively for traditional PINNs, as it is based on differentiation equations; however, it does not efficiently distinguish between dependencies beyond the five neighboring points, which is where PINNs Former comes in and handles the job more efficiently given the transformer's computing ability.

- PINNs Former combines PINNs and transformers, effectively modeling complex astrophysical systems like galaxy formation and cosmic structure evolution by capturing dark matter filament growth and galaxy cluster formation.
- This approach improves large-scale cosmological simulations by ensuring accuracy, adherence to physical laws, and enhanced reliability in AI-based models.

Recently introduced generative models like GANs have made it possible to create realistic super high-resolution cosmic structures [28]. These models can be used to create synthetic universes given a set of real or simulated astronomical observations. For instance, GANs have been used in improving the resolution of cosmological simulations by inpainting the missing features from low-resolution simulations. This technique known as super-resolution enables AI algorithms to convert low-fidelity data into high-fidelity answers while retaining detailed characteristics of the cosmos such as galaxy mergers, gas flows, and the distribution of dark matter. It makes it possible for researchers to approximate galaxies, ionized gas clouds, and other cosmological structures at a considerably lower computational expense, and as a result, the ability to model a much larger piece of space at a good resolution. This approach leads to simulations offering important information about phenomena like the distribution of dark matter, formation of galaxies, and cosmic structure development in time. In this paper, I will present two machine learning models for identifying vast cosmological formations and analyzing high-resolution cosmic structures. These models are particularly effective in generating synthetic universes by learning from a dataset of real or simulated astronomical observations. For example, GANs have been employed to enhance the resolution of cosmological simulations by filling in missing details from low-resolution simulations. This technique, known as super-resolution, allows AI to "upscale" low-fidelity data into high-resolution outputs, preserving fine-grained cosmic features like galaxy mergers, gas flows, and dark matter distributions. This approach enables researchers to simulate galaxies, gas clouds, and other cosmic phenomena at a fraction of the computational cost, making it feasible to model larger regions of the universe without sacrificing detail. The result is faster, more efficient simulations that provide insights into phenomena such as dark matter distribution, galaxy formation, and cosmic structure evolution over time.

12.5.2 Machine Learning Models for Predicting Large-Scale Cosmic Structures

AI is helping not only to improve simulations but also in the fundamental process of predicting the formation of large-scale structures of the universe. They are walls, filaments, dense clusters such as galaxy clusters, voids, or large-scale cosmic scaffolding set up by the interplaying forces of gravity, dark matter, and dark energy. The complexity of their formation, as well as their further development over billions of years, is one of the main problems of contemporary cosmologists. CNNs [29] and RNNs have proved to be promising in predicting large-scale structures. These models are then taught with large datasets of observed or simulated cosmology data consisting of which regions of space have dense matter concentrations akin to galaxy clusters or cosmic filaments. Starting from every different cosmological parameter such as CMB anisotropies, an ML model can take the measure of the present universe mega years later on to measure the large-scale structure. For example, CNNs were applied to predict the distribution of dark matter, which will turn into galaxy clusters and cosmic filaments. These predictions are useful for tracing the development of the universe and for comparing the theory with observations with the help of available large-scale surveys for the sky such as SDSS and DES. Perhaps the most unique of all difficulties in cosmology is the question of determining the distribution density of dark matter and dark energy given sparse data. Machine learning has been created with the focus on avoiding this problem by training on observed and simulated data where traditional methods fail. For instance, ML algorithms can forecast how the distribution of dark matter looks and *via* gravity lensing or *via* large-scale distribution of galaxies and voids. They also appear to provide an accurate tool for analyzing the impact of dark energy that is behind the phenomenon of accelerated expansion of the universe. When exposed to simulated universes of varying dark energy characteristics, the ML models can estimate the effects that different cosmological parameters will have on the structure and formation of the universe at different epochs. These predictions assist scientists in adjusting their simulation of the universe, hence resulting in near-accurate cosmological theories.

12.5.3 AI Bridging Simulations and Observational Data

Another revolutionary use of AI in cosmology is the bringing together of simulation data with actual observational data. Mock data are as valuable as observational data, except for such imperfections as noise, resolution, and lost input. Machine learning programs like autoencoders, denoising neural networks, and other high-end types of tools can learn to sanitize observational data and make them more compatible for simulation. Traditional AI methods like data augmentation and inpainting are very useful when dealing with partial datasets in cosmology. In large surveys, data loss is always a problem, even more so in the areas where the signal is weak or masked by other objects. AI-based inpainting models can generate lost parts of the dataset from the ground up by training on the surrounding material and can provide a much better and continuous map of the universe. These enhanced datasets can then be injected into the cosmological simulation to get better cosmological forecasts on the large-scale structure. Furthermore, due to these models, researchers can build generative universes to fill the gaps in observations and provide meaningful substitutes for regions of the universe not easily accessible to observation. This not only improves the current dataset but also provides a way of regularly fine-tuning the theoretical models.

12.5.4 The Future: Integrated Real-Time Socio-Simulations and AI in Astrophysical Models

Studying the application of AI in numerical cosmology models suggests it will significantly advance the field in the coming decades. With the ongoing development of AI and the availability of vast datasets, real-time simulations of the universe will become feasible, allowing astronomers to create and analyze cosmic phenomena as they occur. These live simulations will enhance model predictions and deepen our understanding of the universe, particularly with the data expected from future missions like the Rubin Observatory and the James Webb Space Telescope (JWST) (Figure 12.4).

In the future, AI-driven autonomous simulations could take the lead, enabling models to explore parameter spaces independently and improving the simulation of the universe's evolution. This approach promises to uncover new physical laws and offer deeper insights into phenomena like

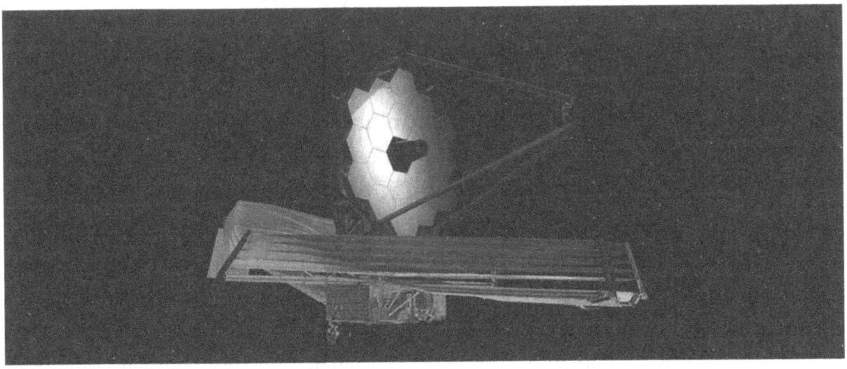

Figure 12.4 The James Webb Space Telescope (JWST), crucial for future real-time cosmic simulations and analysis (Source: NASA).

dark matter, dark energy, and the formation of large-scale structures in the universe. The integration of AI with cosmology will play a pivotal role in expanding our understanding of these fundamental cosmic forces.

12.6 The Future: AI and Quantum Computing in Astronomy

It focuses on the broad introduction and newly appeared trends in astronomical research and the computational difficulties that accompany them. Although traditional approaches to data analysis and simulation remain very useful, they are now reaching their peak capabilities because of the overwhelming amounts of data provided by new-generation observatories and the richness of the cosmological models being used. Thus, the opportunity to combine AI and QC might provide a solution to this problem, which may transform how we analyze and simulate space. Quantum computing is just around the corner of revolutionizing practically all domains of science, and astronomy is not an exception. The joint use of AI and QC could likely bring astronomers a revolution in computational capability, which marks that problems that could not have been solved with classical computing could be solved with the combined tools. In the following subsection, we will understand the principles of how QC can improve AI in astronomy and how it has the potential to bring further advancement in simulation and analysis as a field.

12.6.1 Introduction to Quantum Computing

Of particular interest is that quantum computing is a completely new paradigm compared to classical computation. This permits quantum computers to evaluate many possibilities simultaneously, giving an exponential boost to suit certain types of issues. It also employs entanglement, a situation whereby the state of a given qubit is dependent on or correlated with another either spatially or temporally. For these reasons, quantum computers are especially apt for problems dealing with high data and high dimensions, another typical category in astronomy problems. Despite the current infancy of QC and many technical challenges to overcome, progress in quantum hardware and new quantum algorithms imply that QC will be a real-world technology in the not-too-distant future. Consequently, such a shift's prospect puts into perspective how it might revolutionize AI applications in astronomy.

12.6.2 Enhancing AI's Role in Astronomy with Quantum Computing

For purposes of applying AI in astronomy, the application of quantum computing can significantly increase the pace of this activity, since it has challenges in classical computing. Deep learning networks in particular need huge computational power to be able to process and learn from astronomical data. Such models as CNN for images or RNN for time series frequently collapse when being scaled to deal with petabytes of data produced by surveys of the observable universe. Quantum computing has the potential to overcome the limitations faced by classical computing in handling vast astronomical datasets. By leveraging quantum algorithms, AI models can efficiently process and learn from data, accelerating discoveries in astronomy. For example, specialized quantum hardware, like the one shown in Figure 12.5, is designed to support these large-scale computations, enabling deep learning networks to manage the massive data from cosmic surveys without collapsing.

As shown in the image, quantum computing hardware, such as the one developed by NASA, offers the necessary computational power to handle astronomical datasets at an unprecedented scale. This hardware can significantly enhance the capabilities of AI models in processing petabytes of data from astronomical surveys.

Figure 12.5 Quantum computing hardware designed to support large-scale AI computations for astronomical data analysis (Source: NASA).

12.6.2.1 Quantum Machine Learning

The closest way that QC may directly affect AI would be seen in the creation of QML algorithms. These algorithms leverage the square of the quantum AI model to perform operations that ordinary AI models cannot perform. For example, quantum neural networks (QNNs) can learn and analyze astronomical data that are hard to learn by classical neural networks. Quantum gates applied to these networks process multidimensional data faster, and the high-dimensional scheme extracted from simulations of the cosmos is an instance. The field of quantum machine intelligence has witnessed significant growth, as evidenced by the increasing number of research papers published between 2012 and 2018. This surge reflects a growing interest in leveraging quantum computing to enhance machine learning algorithms and address complex computational challenges. QML can decrease training and inference time for AI significantly, according to experts. A QML model could for example quickly distinguish between types of galaxies or determine the density of dark matter in the universe from large-scale datasets much faster than in current classical machine learning models. This speed-up is especially crucial to analyze the huge amount of data produced by contemporary astronomical surveys such as the Vera C. Rubin Observatory or the James Webb Space Telescope.

12.6.2.2 Optimization and Efficiency

AI algorithms such as cosmological simulation suffer optimization problems when dealing with huge parameter spaces. Such simulations involve the manipulation of many parameters: one of these includes the density of dark matter, and other parameters including the state at which the universe began, and the part played by dark energy. In traditional simulations, the employment and the alteration of these parameters are computationally expensive and a time-consuming process. Through its NP-hard optimization abilities, quantum computing speaks to the ability to process many relevant potential solutions that will improve the optimization tasks in AI. Quantum annealing is a quantum algorithm aimed at optimization problems. This would mean that cosmological simulations that are currently AI-driven could be optimized for parameters by use of quantum annealing and would take less time to compute and would be accurate. This would allow the computation of large maps and observing realistic simulations of large-scale cosmic events such as galaxy formation and black hole coalescence at a level of detail not possible hitherto.

12.6.3 The Potential for More Complex Simulations

Perhaps one of the most exciting opportunities involved with quantum computing originates from its improved ability to simulate the universe. Conventional approaches to modeling cosmic events like galaxy formation, star clusters or the formation of the cosmic web involve the solution of a large number of differential equations over billions of years of cosmic time. Many of these simulations take weeks or in some cases even months to be processed in the most powerful supercomputers around the globe. However, with the help of quantum computing, the kind of simulations that can be done can be extended. It turns out that quantum algorithms natively apply to solvers for systems of partial differential equations (PDEs) that are critical for simulating the underlying physics of the universe. This speed-up is especially crucial to analyze the huge amount of data produced by contemporary astronomical surveys such as the Vera C. Rubin Observatory or the James Webb Space Telescope.

12.7 Challenges and the Path Forward

While the potential benefits of quantum computing in astronomy are immense, it is important to acknowledge that the field is still in its infancy. Current quantum computers, though promising, are not yet powerful enough to outperform classical computers for most real-world applications. Additionally, integrating quantum algorithms with AI models in a way that consistently outperforms classical approaches remains a significant challenge due to noise and error rates in existing quantum systems. Nevertheless, the pace of progress in quantum hardware and quantum algorithm development suggests that we are on the cusp of a new era. With continued advancements, quantum computers could soon be powerful enough to tackle the complex challenges posed by astronomical data and simulations. Quantum-AI hybrid models, combining the strengths of both paradigms, will likely play a crucial role in the future of cosmology and astrophysics. As quantum computing matures, its fusion with AI will lead to unprecedented breakthroughs in our understanding of the universe. The ability to process and analyze vast datasets at quantum speeds, coupled with the power of AI to learn and infer from these data, will open new frontiers in astronomy. Whether it is uncovering the mysteries of dark matter and dark energy, simulating the birth of galaxies, or detecting the faintest signals from the distant universe, the combination of AI and quantum computing holds the key to unlocking the next chapter of cosmic discovery.

12.7.1 Algorithmic Bias in AI Models

One of the most pressing concerns surrounding the use of AI in astronomy is algorithmic bias. Biases can infiltrate AI systems due to the quality of training data or the design of algorithms, potentially leading to skewed or incorrect outcomes. In astronomy, this could have profound consequences, as biased models might yield false conclusions about fundamental aspects of the universe, such as the distribution of matter or the classification of celestial bodies.

12.7.2 Data Bias and Incomplete Datasets

Astronomical data are inherently biased due to how they are collected. Observational instruments, telescope locations, and specific research interests shape the data that are recorded. For instance, deep-sky surveys often target specific regions of the sky and detect objects within certain luminosity limits. These constraints mean that the resulting datasets may not represent the full variety of objects in the universe, leading to data bias in AI models trained on these datasets. A machine learning model trained on data from the SDSS, which focuses on galaxies in the visible spectrum, may underperform when tasked with classifying galaxies observed in other wavelengths, such as infrared or ultraviolet. This data bias can result in models that favor the detection and classification of easily observable bright galaxies while overlooking fainter distant galaxies, potentially skewing our understanding of cosmic evolution.

- **Labeling Bias in Supervised Learning:** In supervised learning, biases in human labeling can affect AI predictions, especially in complex tasks like classifying galaxy morphologies. Inconsistent labeling, such as in the Galaxy Zoo project [30], can lead to biased models that misclassify galaxies based on inaccurate or subjective data.
- **Bias Amplification:** AI models can amplify existing biases in datasets, such as overrepresenting certain astronomical objects like quasars. This can skew predictions and distort our understanding of phenomena, leading to inaccurate conclusions about the distribution of objects like quasars in the universe.

12.7.3 Data Requirements and Challenges

AI models thrive on data, and in the realm of astronomy, the availability of vast datasets is both a blessing and a challenge. While modern telescopes and space observatories generate petabytes of data, the nature of these datasets can present significant obstacles for AI. Issues such as the high dimensionality of data, noise, incomplete information, and data labeling all play critical roles in the effectiveness of AI models in astronomy.

- **The Curse of Dimensionality:** Astronomical datasets are often high-dimensional, meaning they contain numerous features or variables, such as object positions, brightness across multiple wavelengths, redshifts, and spectral properties. As the dimensionality of the data increases, it becomes more challenging for AI models to efficiently process and learn from the data. The curse of dimensionality refers to the exponential increase in computational complexity as the number of dimensions in the data grows. AI models, particularly deep learning networks, may struggle to discern meaningful patterns from noisy high-dimensional data without overfitting to the training set. In cosmological simulations, data can include galaxy positions, velocities, dark matter distributions, cosmic microwave background radiation, and more. These complex multidimensional datasets can overwhelm even sophisticated AI models, making it difficult to extract meaningful insights without advanced techniques like dimensionality reduction or feature selection.
- **Data Quality and Availability:** Although large-scale surveys such as the Large Synoptic Survey Telescope (LSST) and the SDSS have provided unprecedented amounts of data, challenges remain regarding data quality and completeness. AI models rely on high-quality, well-curated data to function effectively, yet many astronomical datasets contain missing values, high levels of noise, or incomplete observations. This can degrade the performance of AI systems, leading to erroneous predictions or unreliable inferences. In the detection of gravitational waves using data from instruments like LIGO and Virgo, the signals are extremely faint and often buried in noisy data. AI models must carefully handle the noisy environment in which these waves are detected,

as even small errors in data processing could lead to false positive detections or missed events, undermining efforts to study these cosmological phenomena.

- **Data Labeling and Rare Events:** For supervised learning, where AI models require labeled datasets to learn effectively, the challenge of labeling vast astronomical datasets can be immense. Additionally, in many areas of astronomy, the objects of interest—such as supernovae, FRBs, or exoplanets—are rare events. Obtaining enough labeled examples of these phenomena for AI models to train on can be difficult, making it challenging to build accurate models. In time-domain astronomy, where transient phenomena like supernovae are studied, AI models may struggle to generalize well from limited labeled datasets. Since supernovae occur relatively infrequently, astronomers have limited examples with which to train models, complicating the development of AI systems capable of reliably identifying these events in new data.

Addressing these challenges requires advanced AI techniques, improved data preprocessing, and collaborative efforts to curate high-quality astronomical datasets."

12.7.4 Ethical Considerations in AI-Driven Astronomy

While the technological advancements driven by AI offer promising avenues for astronomical research, the rise of AI also brings with it important ethical considerations. As with any tool of significant power, the potential for misuse or unintended consequences must be carefully considered, particularly in an area as data-intensive as astronomy.

A common criticism of AI models, especially deep learning networks, is their black-box nature—the tendency for these models to make predictions without offering a clear explanation for how they arrived at those conclusions. In astronomy, where understanding the underlying physical processes behind observed phenomena is a fundamental goal, the lack of interpretability in AI systems can create problems. Researchers may become hesitant to rely on AI-generated results if they cannot trust or explain the models' decision-making processes. An AI model might be used to predict the mass distribution of dark matter in the universe based on galaxy surveys. While the model may generate accurate predictions, if the reasoning behind these predictions is opaque, astronomers may struggle to extract physical insights or validate the results. This lack of transparency could hinder progress in understanding the nature of dark matter and its role in cosmic structure formation.

As AI takes on a greater role in the discovery and analysis of astronomical phenomena, the issue of accountability becomes more pressing. When an AI model makes an incorrect prediction or fails to detect a significant event, it is often unclear who should be held responsible—the developers of the AI, the researchers using it, or those who provided the training data. This issue is particularly relevant in high-stakes areas such as exoplanet detection of gravitational wave analysis, where false positives or missed detections can have costly consequences. In the context of exoplanet discovery, models used to identify planet candidates from transit data have led to significant discoveries. However, these models also occasionally yield false positives, misidentifying stellar activity or other phenomena as potential planets. In cases where a false positive leads to costly follow-up observations, questions about accountability—whether to blame the model, the data, or the scientists—arise.

While most astronomical data are publicly accessible, certain datasets, such as proprietary early-access observations from space missions or data generated by private collaborations, raise issues of data privacy and ownership. As AI becomes more prevalent, the question of who has the right to access, use, and derive value from the results of AI analysis becomes crucial. Furthermore, the intellectual property generated by AI models based on these datasets introduces additional complexity regarding data sharing and usage rights. The data collected by the James Webb Space Telescope (JWST), for instance, may initially be reserved for the astronomers who proposed the observations. However, if an AI system trained on proprietary JWST data generates significant new scientific findings, questions of intellectual property rights arise—Should the data's original collectors receive credit or should the findings be publicly shared for the broader benefit of the scientific community?

12.8 Strategies for Mitigating Challenges in AI-Driven Astronomy

As AI continues to revolutionize the field of astronomy, addressing the associated challenges is crucial for ensuring the integrity and reliability of scientific discoveries. The difficulties outlined in the previous sections—such as algorithmic bias, data requirements, and ethical concerns—necessitate well-structured strategies for mitigation. In this section, we will explore the most effective approaches to overcoming these challenges, with a focus on creating robust, transparent, and ethical AI models for use in astronomical research.

12.8.1 Mitigating Algorithmic Bias

Algorithmic bias, one of the most critical issues in AI, can skew the results of AI models, leading to inaccurate or unreliable findings in astronomy. Since these biases often stem from the data used to train models, or from the algorithms themselves, a multifaceted strategy is required to reduce bias in AI-driven astronomy.

12.8.1.1 Diverse and Representative Datasets

One of the most effective ways to combat algorithmic bias is to ensure that the datasets used to train AI models are as diverse and representative of the universe as possible. This means gathering data from different regions of the sky, observing at multiple wavelengths, and incorporating information from various sources and instruments. By training models on broader datasets that capture the full range of astronomical phenomena, AI systems are less likely to develop biases that favor specific types of celestial objects or environments. To mitigate bias in galaxy classification models, data from both visible light and infrared surveys should be used. Infrared observations can detect galaxies that are obscured by dust in visible wavelengths, leading to a more comprehensive dataset that reduces the likelihood of the model favoring bright dust-free galaxies.

Another strategy to reduce bias involves creating balanced training sets that contain an equal representation of different classes or types of objects. For example, if a model is being trained to classify galaxies, care should be taken to include a balanced number of spiral, elliptical, and irregular galaxies. This ensures that the model does not become biased toward a specific type of galaxy simply because one class was overrepresented in the training data. In the case of supernova classification, AI models should be trained using data that equally represent the various types of supernovae (Type Ia, Type II, etc.). If Type Ia supernovae dominate the training data, the model may struggle to correctly identify fewer common types of supernovae in new observations.

12.8.1.2 *Bias Audits and Fairness Metrics*

Conducting bias audits and implementing fairness metrics are crucial steps in identifying and mitigating bias in AI models. These audits involve testing the model across different subgroups or categories of data to determine whether it performs consistently across all cases. Fairness metrics can quantify how evenly the model treats different types of objects or observations, providing valuable insight into where biases may exist. Before deploying an AI model to classify stars based on their spectra, a bias audit could examine how well the model performs on stars of different magnitudes, temperatures, and metallicities. If the model consistently underperforms for dimmer stars, further training may be required to address this imbalance.

To further reduce bias, human-in-the-loop (HITL) [31] approaches can be used. These systems involve human experts at key stages of the AI pipeline to correct or adjust any biases that may have been introduced during the training process. By involving astronomers in reviewing AI predictions, errors and biases can be caught before the model is fully deployed, ensuring that the AI's conclusions are scientifically sound. In galaxy morphology classification, astronomers could review a subset of AI-generated predictions to identify whether the model is consistently misclassifying certain edge cases (e.g., borderline spirals or irregulars). Based on this feedback, the model can be retrained with additional or more balanced data. A responsible AI workflow involves human intervention to ensure the accuracy and ethical alignment of AI systems. The cycle begins with data validation by humans, followed by dataset optimization for training. An AI model is then developed, and its output is reviewed and corrected by humans as needed. This process includes validating both data and model output, forming a continuous loop where humans enhance AI decisions and outcomes, fostering responsible and trustworthy AI deployment.

12.8.2 Addressing Data Requirements and Challenges

The success of AI in astronomy depends heavily on the quality and availability of data. However, given the complexity of astronomical data—characterized by high dimensionality, noise, and incomplete information—strategies must be developed to manage and optimize these datasets for AI applications.

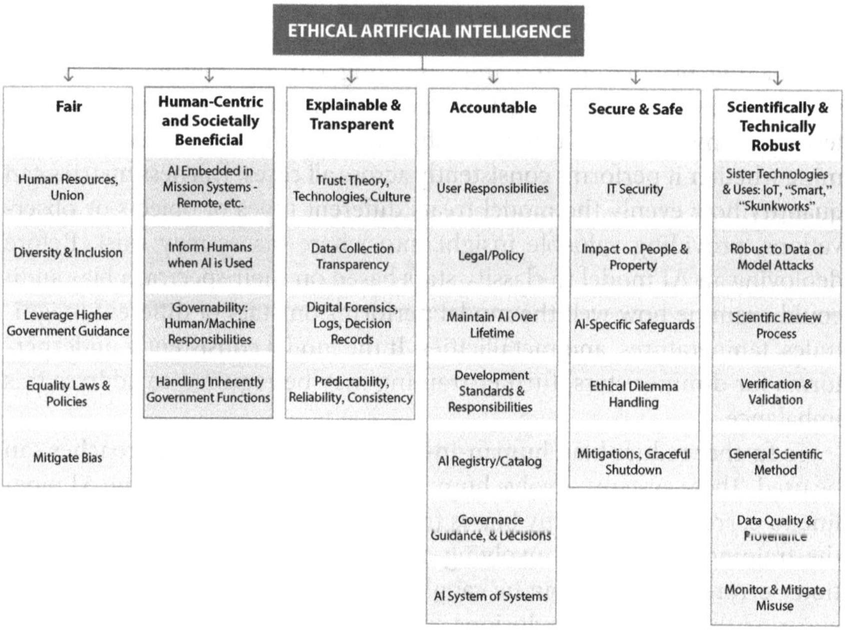

Figure 12.6 Ethical AI principles guiding data optimization in astronomy (Credit: NASA).

One approach to addressing these challenges is the integration of ethical AI principles to ensure transparency and reliability in data-driven discoveries (Figure 12.6). By refining data preprocessing techniques, improving noise reduction methods, and leveraging advanced algorithms, astronomers can enhance the accuracy and robustness of AI models in analyzing vast astronomical datasets.

12.8.2.1 Noise and Dimensionality Reduction

Effective data preprocessing is essential for dealing with the noisy, incomplete, and high-dimensional data commonly found in astronomical datasets. Preprocessing techniques such as noise reduction, outlier removal, and data normalization can significantly improve the performance of AI models. These processes ensure that AI systems learn from the most accurate and relevant information, minimizing the risk of erroneous predictions. In gravitational wave detection, preprocessing the data to filter out noise from seismic activity, atmospheric disturbances, and instrumental errors is crucial. AI models trained on cleaned data are more likely to correctly identify gravitational wave events while minimizing false positives.

High-dimensional data can overwhelm AI models, leading to slower processing times and overfitting. To mitigate this, dimensionality reduction techniques, such as principal component analysis (PCA) [32] or t-distributed Stochastic Neighbor Embedding (t-SNE), can be employed to reduce the number of features while preserving the most important information. This enables AI models to focus on the key patterns in the data without becoming distracted by irrelevant or redundant features. In cosmological simulations, where data on galaxy positions, velocities, dark matter distributions, and cosmic microwave background (CMB) radiation may contain thousands of variables, PCA can be used to reduce the dataset's dimensionality. This allows the AI model to efficiently predict large-scale structure formation while avoiding overfitting to extraneous details.

12.8.2.2 Semi-Supervised and Unsupervised Learning

Given the difficulty of labeling vast astronomical datasets, particularly for rare phenomena, semi-supervised and unsupervised learning techniques [33] can be employed to mitigate the need for labeled data. Semi-supervised learning allows models to leverage a small amount of labeled data alongside a larger pool of unlabeled data, while unsupervised learning algorithms can detect patterns and relationships without requiring labeled datasets. In the study of FRBs, which are rare and transient events, semi-supervised learning could allow an AI model to identify potential FRBs in new data by training on a limited set of labeled FRBs. Unsupervised learning, on the other hand, could help discover unknown types of transient events without requiring prior classifications.

12.8.2.3 Collaborative Data Sharing and Standardization

To ensure that AI models are trained on high-quality and diverse datasets, collaborative efforts between observatories, institutions, and researchers are essential. Data sharing agreements and standardization of data formats allow AI models to be trained on a broader range of datasets, ensuring that they are more representative of the universe and less prone to bias or overfitting. Initiatives like the Virtual Observatory (VO), which aims to standardize and integrate astronomical datasets from around the world, provide a rich resource for training AI models. By accessing standardized data from a variety of telescopes and instruments, AI models can be trained more effectively and can perform more reliably across different types of observations.

12.8.3 Ethical Considerations and AI Governance

As AI's role in astronomy grows, so too does the importance of addressing the ethical implications of its use. Ensuring that AI systems operate transparently and have accountability, while respecting issues of data privacy and intellectual property, is crucial for maintaining trust in AI-driven discoveries.

12.8.3.1 Explainability and Transparency

One of the key strategies for addressing ethical concerns is the development of AI models that are explainable and transparent. Explainable AI (XAI) [34] techniques seek to make the decision-making processes of AI systems more understandable to human users, which is particularly important in scientific fields like astronomy, where insights into physical processes are as important as the predictions themselves. By designing models that provide clear explanations for their conclusions, researchers can ensure that AI discoveries are scientifically valid and interpretable. In the context of AI-driven exoplanet discovery, XAI could help astronomers understand why an AI model classified a specific light curve as indicative of a planet transit. By providing explanations in terms of the features it identified (e.g., the depth and duration of the transit), XAI can make AI predictions more trustworthy and easier to validate with follow-up observations.

12.8.3.2 Ethical AI Development Frameworks

Ethical concerns can be addressed by adhering to ethical AI development frameworks, which provide guidelines for the responsible creation and use of AI systems. These frameworks emphasize principles such as fairness, accountability, and transparency, and encourage the development of AI models that respect privacy, avoid discrimination, and promote the broader good. For the astronomy community, adopting such frameworks can help ensure that AI systems are used responsibly and ethically. Research collaborations like the SKA could adopt ethical AI frameworks to guide their use of AI in the processing of radio astronomy data. By ensuring that their AI systems adhere to these principles, they can mitigate risks such as bias or unfair treatment of certain types of data, while fostering trust in AI-driven results.

12.8.3.3 Collaborative Governance and Accountability

Ensuring that AI systems are used responsibly in astronomy requires the creation of collaborative governance structures that involve stakeholders from multiple disciplines, including AI researchers, astronomers, ethicists, and policymakers. These governance structures can establish clear protocols for how AI systems are deployed, monitored, and evaluated, ensuring that AI-driven discoveries are held to rigorous scientific and ethical standards. For AI models used in time-domain astronomy to detect transient events like supernovae or gamma-ray bursts, a governance structure could be established to review the results, assess the performance of the AI, and make decisions on follow-up observations. This would ensure that AI-driven discoveries are subject to oversight and are treated with the same rigor as traditional observations, helping maintain the scientific integrity of AI contributions.

12.8.3.4 AI Ethics Committees

To ensure ongoing ethical oversight of AI use in astronomy, the formation of AI ethics committees within major research institutions and collaborations can be instrumental. These committees can evaluate the implications of AI deployments, especially when handling sensitive data or making potentially groundbreaking discoveries. Such bodies could also establish guidelines for data privacy, ownership, and consent, ensuring that data are handled responsibly and that the astronomical community remains accountable for how AI-driven research is conducted. An AI ethics committee within a project like the Large Synoptic Survey Telescope (LSST) [35] could monitor the use of AI to detect rare astronomical events. This body would not only oversee the AI models themselves but would also ensure that the data collected are used in a manner consistent with ethical guidelines, including respecting intellectual property rights and data sharing agreements among international research teams.

12.8.4 Technical Solutions for Scaling Data Requirements

With the increasing volume of astronomical data generated by next-generation telescopes and observatories, strategies are needed to address the scale and complexity of the data that AI models must process. Managing these large datasets effectively will be critical to AI's continued success in astronomy.

12.8.4.1 *Distributed Computing and Cloud Infrastructure*

To process the vast amounts of data from observatories such as the SKA or the JWST, astronomers can leverage distributed computing and cloud infrastructure [36]. These technologies allow researchers to harness the computing power of large-scale networks and data centers, enabling AI models to handle increasingly complex datasets efficiently. The SKA will generate exabytes of data annually, far exceeding the capacity of most traditional computing systems. By using cloud infrastructure, AI models can access and process these data in real time, facilitating rapid analysis and detection of cosmic phenomena like pulsars or fast radio bursts.

12.8.4.2 *Edge Computing and AI*

In tandem with distributed computing, edge computing [37]—the practice of processing data closer to where it is generated—can enhance AI's ability to manage large datasets. By conducting initial data processing and analysis on-site at observatories, edge computing can reduce the need to transmit enormous datasets across long distances, allowing AI models to perform faster and more efficiently. Edge computing could be used at the Vera C. Rubin Observatory, where initial data processing is conducted on-site to filter out noise and identify promising transient events before transmitting relevant data for further analysis. AI models could then prioritize data based on its scientific potential, reducing the strain on centralized data centres.

12.9 Conclusion: Embracing the Future of Astronomical Discovery

As we stand at the crossroads of astronomy and artificial intelligence, it becomes clear that we are entering a new era of discovery. The integration of AI, ML, DL, and soon, quantum computing, is not only accelerating our understanding of the universe but also enabling us to anticipate and explore realms previously inaccessible through traditional methods. These tools are reshaping the boundaries of what we thought possible, offering unparalleled precision in observing cosmic phenomena, modeling the evolution of galaxies, and predicting large-scale structures in the cosmos. The applications we have explored—cosmological simulations, real-time data analysis, and the transformative potential of quantum-enhanced AI—demonstrate a shift in how research in astronomy is conducted. AI's role in improving both the speed and accuracy of simulations allows us to study the universe with an efficiency that was once unimaginable. This synergy between machine learning algorithms and observational data is already leading to significant breakthroughs, from detecting gravitational waves to identifying new exoplanets. However, with these advancements come challenges that must be addressed. The risks of algorithmic bias, the increasing demand for vast datasets, and the ethical considerations surrounding AI's use in research are significant hurdles that will shape how these technologies are applied in the future. Ensuring the responsible use of AI in astronomical research will be critical to ensuring that its benefits are equitably distributed and that the data it processes are handled ethically. As we look ahead, the potential of quantum computing looms large. By bringing together the computational power of quantum systems with AI, we may soon be able to simulate the cosmos in ways that were previously thought impossible. These simulations will not only be faster but also allow us to tackle more complex problems, unlocking insights into the fundamental nature of the universe. In doing so, we stand on the brink of discoveries that could redefine our understanding of space, time, and the very building blocks of existence. In conclusion, the marriage of AI, ML, and quantum computing in astronomy promises to redefine our pursuit of knowledge. As these technologies evolve, so too will our capacity to explore the cosmos, unveiling its deepest mysteries. What lies ahead is a future where AI helps us not just observe the universe but anticipate the unseen, transforming both our scientific approach and our understanding of the cosmos itself.

Data Availability

The data referenced and discussed in this chapter are derived from publicly available astronomical datasets and survey archives, including but not limited to those from NASA, ESA, and other collaborative research observatories. Specific datasets mentioned, such as those from the Kepler mission, LSST, and SDSS, can be accessed through their respective repositories. For further inquiries or specific requests, please contact the corresponding author.

Acknowledgement

The authors would like to extend their gratitude to the Department of Physics at Soban Singh Jeena University, Almora, India, for providing the academic environment and resources necessary for the completion of this work.

References

1. Patil, B., AI in Astronomy. *Int. J. Adv. Res. Sci. Commun. Technol.*, 3, 1–10, 2023.
2. Zhang, H., Wang, J., Zhang, Y., Du, X., Wu, H., Zhang, T., Review of artificial intelligence applications in astronomical data processing. *Astron. Tech. Instrum.*, *1*, 1, 1–15, 2023.
3. Russell, S.J. and Norvig, P., *Artificial intelligence: a modern approach*, Pearson, US, 2016.
4. Djorgovski, S.G., Mahabal, A.A., Graham, M.J., Polsterer, K., Krone-Martins, A., *Applications of AI in Astronomy*, World Scientific, Singapore, 2022, *arXiv preprint arXiv:2212.01493.*
5. Huang, K., Hu, T., Cai, J., Pan, X., Hou, Y., Xu, L., Cui, X., Artificial Intelligence in Astronomical Optical Telescopes: Present Status and Future Perspectives. *Universe, 10*, 5, 210, 2024.
6. Borgman, C.L. and Wofford, M.F., *From data processes to data products: Knowledge infrastructures in astronomy*, Harvard Data Science Review, USA, 2021, *arXiv preprint arXiv:2109.01707.*
7. Weddell, S. and Webb, R., Data preprocessing on sequential data for improved astronomical imaging, in: *Proceedings of Image and Vision Computing*, pp. 1–8, 2005.
8. Malik, A., Moster, B.P., Obermeier, C., Exoplanet detection using machine learning. *MNRAS*, *513*, 4, 5505–5516, 2022.
9. La Plante, P., Williams, P.K., Kolopanis, M., Dillon, J.S., Beardsley, A.P., Kern, N.S., Zheng, H.A., Real Time Processing system for big data in astronomy: Applications to HERA. *Astron. Comput.*, *36*, 100489, 2021.
10. Groenewald, E.S., Kumar, N., Groenewald, C.A., AI-Assisted Astronomical Data Analysis Unveiling Patterns and Phenomena in the Universe. *Naturalista Campano*, *28*, 1, 1916–1925, 2024.
11. Cuéllar, S., Granados, P., Fabregas, E., Curé, M., Vargas, H., Dormido-Canto, S., Farias, G., Deep learning exoplanets detection by combining real and synthetic data. *PLoS One*, *17*, 5, e0268199, 2022.
12. Kaltenbaek, R., Acin, A., Bacsardi, L., Bianco, P., Bouyer, P., Diamanti, E., Bassi, A., Quantum technologies in space. *Exp. Astron.*, *51*, 3, 1677–1694, 2021.

13. Feigelson, E.D. and Babu, G.J., *Statistical challenges in modern astronomy*, Springer New York, USA, 2004, *arXiv preprint astro-ph/0401404*.
14. Zhang, Y. and Zhao, Y., Astronomy in the big data era. *Data Sci. J.*, *14*, 11–11, 2015.
15. Babu, G.J. and Djorgovski, S.G., Some statistical and computational challenges, and opportunities in astronomy. *Stat. Sci.*, 19, 322–332, 2004.
16. Vogelsberger, M., Marinacci, F., Torrey, P., Puchwein, E., Cosmological simulations of galaxy formation. *Nat. Rev. Phys.*, *2*, 1, 42–66, 2020.
17. Garofalo, M., Botta, A., Ventre, G., Astrophysics and big data: Challenges, methods, and tools. *Proc. Int. Astron. Union*, *12*, S325, 345–348, 2016.
18. Sarker, I.H., Deep learning: a comprehensive overview on techniques, taxonomy, applications and research directions. *SN Comput. Sci.*, *2*, 6, 420, 2021.
19. Conroy, G., This fast radio burst took 8 billion years to reach Earth. *Nature*, 2023.
20. Troja, E., Fryer, C.L., O'Connor, B., Ryan, G., Dichiara, S., Kumar, A., Castro-Tirado, A.J., A nearby long gamma-ray burst from a merger of compact objects. *Nature*, *612*, 7939, 228–231, 2022.
21. Volonteri, M., The formation and evolution of massive black holes. *Science*, *337*, 6094, 544–547, 2012.
22. Lattimer, J.M. and Prakash, M., The physics of neutron stars. *Science*, *304*, 5670, 536–542, 2004.
23. Meher, S.K. and Panda, G., Deep learning in astronomy: a tutorial perspective. *Eur. Phys. J. Spec. Top.*, *230*, 10, 2285–2317, 2021.
24. Alzubaidi, L., Zhang, J., Humaidi, A.J., Al-Dujaili, A., Duan, Y., Al-Shamma, O., Farhan, L., Review of deep learning: concepts, CNN architectures, challenges, applications, future directions. *J. Big Data*, *8*, 1–74, 2021.
25. Ma, S., Xiao, B., Hong, R., Addissie, B., Drikas, Z., Antonsen, T., Anlage, S., *Classification and prediction of wave chaotic systems with machine learning techniques*, Cornell University Library, USA, 2019, *arXiv preprint arXiv: 1908.04716*.
26. Raissi, M., Perdikaris, P., Karniadakis, G.E., Physics-informed neural networks: A deep learning framework for solving forward and inverse problems involving nonlinear partial differential equations. *J. Comput. Phys.*, *378*, 686–707, 2019.
27. Ullmo, M., Decelle, A., Aghanim, N., Encoding large-scale cosmological structure with generative adversarial networks. *Astron. Astrophys.*, *651*, A46, 2021.
28. Zhao, Z., Ding, X., Prakash, B.A., *Pinnsformer: A transformer-based framework for physics-informed neural networks*, Conference paper at ICLR, 2023, *arXiv preprint arXiv:2307.11833*.
29. Schaefer, C., Geiger, M., Kuntzer, T., Kneib, J.P., Deep convolutional neural networks as strong gravitational lens detectors. *Astron. Astrophys.*, *611*, A2, 2018.

30. Lintott, C.J., Schawinski, K., Slosar, A., Land, K., Bamford, S., Thomas, D., Vandenberg, J., Galaxy Zoo: morphologies derived from visual inspection of galaxies from the Sloan Digital Sky Survey. *MNRAS*, *389*, 3, 1179–1189, 2008.
31. Mosqueira-Rey, E., Hernández-Pereira, E., Alonso-Ríos, D., Bobes-Bascarán, J., Fernández-Leal, A., Human-in-the-loop machine learning: a state of the art. *Artif. Intell. Rev.*, *56*, 4, 3005–3054, 2023.
32. Jolliffe, I.T. and Cadima, J., Principal component analysis: a review and recent developments. *Philos. Trans. R. Soc. A Mathematical Phys. Eng. Sci.*, *374*, 2065, 20150202, 2016.
33. Naeem, S., Ali, A., Anam, S., Ahmed, M.M., An unsupervised machine learning algorithms: Comprehensive review. *Int. J. Comput. Digital Syst.*, 13, 911–921, 2023.
34. Gohel, P., Singh, P., Mohanty, M., *Explainable AI: current status and future directions*, IEEE Access, 2021, *arXiv preprint arXiv:2107.07045*.
35. Zhan, H. and Tyson, J.A., Cosmology with the large synoptic survey telescope: an overview. *Rep. Prog. Phys.*, *81*, 6, 066901, 2018.
36. Berriman, G.B., Juve, G., Deelman, E., Regelson, M., Plavchan, P., The application of cloud computing to astronomy: A study of cost and performance, in: *2010 Sixth IEEE International Conference on e-Science Workshops*, 2010.
37. Paripati, L.K. and Hajari, V.R., Edge Computing and its Role in Enhancing Artificial Intelligence and Machine Learning Applications in the Cloud. *J. Electr. Syst.*, 20, 2958–2969, 2024.

Index

Printed in the USA/Agawam, MA
September 16, 2026

910661.161